New Perspectives in Personal Construct Theory

New Perspectives in Personal Construct Theory

Edited by D. Bannister

High Royds Hospital, Menston,
Ilkley, West Yorkshire, England

1977

Academic Press

LONDON · NEW YORK · SAN FRANCISCO

A Subsidiary of Harcourt Brace Jovanovich, Publishers

ACADEMIC PRESS INC. (LONDON) LTD.
24/28 Oval Road
London NW1

United States Edition published by
ACADEMIC PRESS INC.
111 Fifth Avenue
New York, New York 10003

Library of Congress Catalog Card Number: 776483
ISBN: 0-12-077940-4

Text set in 10/12 pt. Monotype Times, printed by letterpress, and bound in Great Britain at The Pitman Press, Bath

List of Contributors

D. BANNISTER, *Psychology Department, High Royds Hospital, Menston, Near Ilkley, West Yorkshire, LS29 6AQ, England.*

FAY FRANSELLA, *Academic Department of Psychiatry, Royal Free Hospital, London, NW3, England.*

JOHN W. GROUTT, *Upward Bound Program, University of Maryland, Eastern Shore, Maryland, 21853, U.S.A.*

THOMAS O. KARST, *Department of Psychiatry, Medical College of Ohio at Toledo, Toledo, 43614, U.S.A.*

† GEORGE A. KELLY.

MILDRED M. MCCOY, *Department of Psychology, University of Hong Kong, Hong Kong.*

J. M. M. MAIR, *Department of Psychological Research, Crichton Royal Hospital, Dumfries, Scotland.*

J. BRENDA MORRIS, *Department of Psychology, Stone House Hospital, Dartford, Kent, England.*

MARGARET NORRIS, *Department of Sociology, University of Surrey, Guildford, Surrey, GU2 5XH, England.*

JOY O'REILLY, *Educational Research Centre, St. Patrick's College, Dublin, 9, Ireland.*

ALAN RADLEY, *Department of Social Sciences, Loughborough University, Loughborough, Leicestershire, LE11 3TU, England.*

A. T. RAVENETTE, *Child Guidance Clinic, West Ham Lane, London, E15, England.*

CHARLES STEFAN, *The Binghamton Psychiatric Center, Binghamton, New York State, U.S.A.*

PETER STRINGER, *Department of Psychology, University of Surrey, Guildford, Surrey, GU2 5XH, England.*

FINN TSCHUDI, *Institute of Psychology, University of Oslo, Blindern, Oslo, 3, Norway.*

†Deceased.

List of Contributors

Preface

The original collection of essays under the title "Perspectives in Personal Construct Theory" was published in 1970. The contributors were invited to choose a theme and write about it in the light of personal construct theory. They worked entirely independently. The aim was freely to illustrate the ways in which personal construct theory had inspired, confused, aroused, informed, a varied range of writers on a varied range of themes. This second volume of essays has been gathered together in exactly the same way.

One difference between this volume and the first is that this one represents a "second generation" of personal construct theorists. Of the contributors to the original volume only three (Bannister, Fransella and Mair) have been retained to represent the "Old Guard". Ten of the contributors to this volume did not appear in the first "Perspectives" and, with some exceptions, have not published extensively in the field of construct theory before.

Comparisons between first and second volumes are the readers' business but a point of interest editorially is that this set of essays is more difficult to classify in terms of traditionally labelled psychological topics. It may be that, increasingly, construct theory is inspiring psychologists to ask new questions as well as to find new answers to old questions. This trend would certainly have delighted George Kelly.

The book contains one hitherto unpublished essay by the late George Kelly called "The Psychology of the Unknown" and a quotation seems appropriate for this preface.

> "There is a psychology for getting along with the unknown. It is a psychology that says in effect, 'why not go ahead and construe it to be organised—or disorganised if you prefer—and do something about it. In the world of unknowns seek experience, and seek it full cycle. That is to say, if you go ahead and involve yourself rather than remaining alienated from the human struggle, if you strike out and implement your anticipations, if you dare to commit yourself, if you prepare to assess the outcomes as systematically as you can, and if you master the courage to abandon your favourite psychologisms and intellectualisms and reconstrue life altogether; well, you may not find that you guessed right, but you will stand a chance of transcending more freely those 'obvious' facts that now appear to determine your affairs, you may just get a little closer to the truth that lies somewhere over the horizon."

The essays in this book seem to me to be seeking, to some purpose, for such a psychology, a psychology for "getting along with the unknown".

February, 1977 D. BANNISTER

Acknowledgements

The Editor is grateful to Mrs. Gladys Kelly for agreeing to the publication of the essay "The Psychology of the Unknown" by the late George Kelly.

The Editor and the publisher also wish to thank the following for their kind permission to reproduce material quoted from other volumes:

W. W. Norton & Co., Inc. for quotations from "The Psychology of Personal Constructs" by George A. Kelly; Houghton Mifflin Company for a quotation from "The Outsider" by Colin Wilson; Cambridge University Press for a quotation from "Remembering" by F. C. Bartlett; Routledge and Kegan Paul Ltd. for a quotation from "Social Development of Young Children" by S. Isaacs; Methuen and Co. Ltd. for a quotation from "Approaches to Personality Theory" by D. Peck and D. D. Whitlow, a quotation from "Sketch for a Theory of the Emotions" by J. P. Satre and a quotation from "Frames of Mind" by Liam Hudson; Constable and Co. Ltd. for a quotation from "The Principles of Psychology" by W. James; University of Chicago Press for a quotation from "Mind, Self and Society" by G. H. Mead; John Wiley and Sons Ltd. for a quotation from "Clinical Psychology and Personality: Selected Papers of George Kelly" edited by B. A. Maher.

Contents

The Psychology of the Unknown

George A. Kelly

The psychologist who attempts to understand human behaviour often finds it convenient to envision man's actions in terms of what the man knows—or thinks he knows. When this device works it is said that the behaviour is *rational*. Strictly speaking, however, it is not the behaviour that is rational; it is the explanation that is rational. This is an important point, and I think that it, or something like it, will keep cropping up as a recurrent theme in what I have undertaken to say.

When, on the other hand, the device fails and the psychologist tries in vain to see a logical relationship between what his friend knows and does, he is inclined to give up and say simply that the behaviour is *irrational*. But here again the psychologist's failure to apply the rationality paradigm successfully can just as well be regarded as a failure of the paradigm to meet the situation as it can be attributed to some essential property of the behaviour.

In order not to be left high and dry when his rationality paradigm fails him, the psychologist is likely to seek other modes of explanation when he is confronted with what he has called "irrational behaviour" in a friend. Most often he attributes such behaviour to some special force or motive which operates semi-autonomously within his friend's personality and distorts the normal process of making inferences. The poor man, it seems to the psychologist, is not altogether free to cope with situations as he sees them; he must, instead, continually compromise with the intruding psychodynamics of his own personality.

As a by-product of this kind of psychologising we have in our society a growing belief that, while a person may be held responsible for his "rational behaviour", it is unfair to make him account for his "irrational behaviour". In the subcultures where this belief has taken root it becomes quite clear how any of us who finds personal responsibility bothersome may conduct himself so as to be rid of it. All he has to do is to act as if he were the unwitting victim of his psychodynamics. No one can hold him responsible for that!

Cognitive *versus* Dynamic Theories of Personality

When a personality theory is constructed in a rational form it becomes known, quite naturally, as "a rational theory". And, because we find it so easy to confound the rationality of the constructs we erect with the presumed intrinsic rationality of what they seek to explain, we are likely to take it for granted that such a theory is useful only in understanding those behaviours we have long regarded as essentially "rational". Moreover, because "rational behaviour" is supposed to be based on what the behaving person knows, such a theory is likely to be classified further as "a cognitive theory", and its applications are presumed to be limited to the knowing process in man, that is to say, to *epistemology*.

But a theory may have a rational structure without the theorist having to claim that his own rationale is identical with that of the person he seeks to understand. Certainly the purpose of a psychological theory is not simply to reproduce symbolically the human processes through which man's behaviour emerges, but, more appropriately, to enable the psychologist to gain some overview of them, and thus to achieve a level of understanding that will enable him to transcend them. Psychology is not the simulation of human behaviours, hither, thither, and yon, but the comprehensive overview of man—man whose full range of behaviours has yet to unfold itself. It is therefore quite appropriate to erect a theoretical structure on rational grounds and apply it to any man, no matter how incompetent he is, or to any act of man, no matter how impulsive it is.

Much the same can be said about theories that have been designed to explain "irrational behaviour". Since we have been inclined to think of theories as simulations of reality, rather than approaches to it, we are likely to suppose that such experiences as love, hostility, and anxiety cannot be regarded as the calculated outcomes of one's intentions, but as the involuntary products of such corresponding agencies as "motives", "drives", and "psychodynamics". For example, a person in love is not likely to claim that he reached this state of mind by adding up facts. He is more likely to say that love was something that came over him, perhaps "overwhelmed" him. If, then, the theorist is to explain this man's experience it may seem as if he, too, must explain it in terms of "something that came over him". Again, this is a case of the subject matter being confounded with the constructs one has devised in order to understand it. This confounding of objects and constructs is what characterises, it seems to me, the so-called dynamic theories in psychology.

Representing Events *versus* Construing Them

The man in love may see nothing rational in his experience, and he may go so far as to regard himself as the unwitting victim of psychodynamics or love potions. But that does not mean that we must limit ourselves to the same terms he uses. Our job is to understand his experience in general, not merely to simulate it in particular. To do this with any perspicacity we must devise our own constructions. Our constructs must enable us to subsume his constructs, not merely simulate them. If he thinks in terms of psychodynamics, that is something we ought to understand and appreciate. But it is not necessary for us to resort to psychodynamic explanations ourselves in order to understand his construct of "psychodynamics".

Furthermore, it seems clear that if we are to talk psychologically about any subject matter at all we cannot allow ourselves to rely upon *simulative* terms alone. We must use *constructive* terms, and these are terms for which we must hold ourselves—not their objects—firmly responsible. Thus, in the natural sense, even the words, "love", "hostility", and "anxiety" refer to notions we have ourselves erected, not to events so obvious as to be invulnerable to human interpretation.

So it remains important for us to keep in mind that, even though our only access to events is through the psychological devices we create for looking at them, it is always proper to raise pointed questions about which is which, and about the appropriateness of our personal constructs to their intended objects. "Love", for example, may hold an ever so important position in one's construct system—as I, for one, believe it should. But it may be fitting, nonetheless, for you to ask if what I regard as "love" might not better be construed as "dependency", or even as "hostility". Similarly, one may, in some cases, ask if what a reluctant mother dreads as "self-denial" she might not better experience as "love".

But is love *actually* rational? The answer to this question must, I believe, remain unknown in our generation. Yet our psychology must enable us to cope with such unknowns. If we are careful to distinguish between love as an experience, on the one hand, and our rational construction of it, on the other, the question need not arise. Whatever it will turn out to be in the end, we should not be inhibited in examining it through the spectacles of rationality. As long as we remember what spectacles we are wearing, love, itself, need not be distorted by such inspection.

But a second question: Can love be completely understood if it is regarded in rational terms only? Probably not; we have not yet developed a completely rational understanding of much of anything, even things that seem much less complex than love. Does this not mean, therefore, that we ought to resort to irrational psychodynamic terms? Not necessarily that either; they may cause us to miss even more of what there is to be seen.

And now a third question: Is not the understanding of love implicit in the experience of love? If you can fall in love, does that not mean that you understand it? The experience does indeed provide a kind of understanding, but being in love with someone does not always carry with it a certain understanding of the love that another feels. And that is the root of many a tragedy.

It is sometimes said that man is the only animal who can represent his environment. The implication is that this ability to simulate reality is the principal psychological feature of man, and, hence, something we all ought to cultivate. To be sure, the reproduction of reality is one of the *post hoc* tests of understanding—prediction, for example, is another. But the mere reproduction of reality adds precious little to one's understanding of it. Man can and does do better than that! This, of course, is something every artist knows.

As for the artist, so for all of us; to construe the surrounding world is to visualise it in more than one dimension. And this is no less true for the psychological theorist. He, like the artist—and man—must transcend the obvious. A simulative theory is not enough; in fact it is no theory at all, only a technician's blueprint for the reproduction of something, say, for example, the reproduction of some well known bit of behaviour through the presentation of a "stimulus". To serve as a theory one's psychological formulation must not only give him mechanistic prediction and control over the familiar acts of man, but provide access to human potentialities not yet realised. This means that a theory must be *constructive*, not merely *representative*.

To transcend the obvious—this is the basic psychological problem of man. Inevitably it is a problem we must all seek to solve, whether we fancy ourselves as psychologists or not. What has already happened in our experience may seem obvious enough, now that we have been through it. But literally it is something that will never happen again. It can't, for time refuses to run around in circles. If then, as we live our lives, we do no more than erect a row of historical markers on the spots where we have had our experiences, we shall soon find ourselves surrounded by a cemetery of monuments, and overburdened with biographical mementos.

But to represent an event by means of a construct is to go beyond what is known. It is to see that event in a way that could possibly happen again. Thus, being human and capable of construing, we can do more than point realistically to what has happened in the past; we can actually set the stage for what may happen in the future—something, perhaps in some respects, very different. Thus we transcend the obvious! By construing we reach beyond anything that man has heretofore known—often reach in vain, to be sure, but sometimes with remarkable prescience. This paper has to do with how man, from his position of relative ignorance, can hope to reach out for knowledge that no one has yet attained. This, as I see it, is a primary problem for the psychologist, though I doubt that most psychologists see it that way.

Constructive Alternativism

It is precisely because we may venture to look ahead only by construing never-to-be-repeated events, rather than merely recording or duplicating them, that we must continually and adventurously hold all matters open to the possibility of fresh reconstruction. No one knows yet what all the alternative constructions are, except that, if the history of human thought offers us any clue, there must still be an awful lot of them.

Even the constructions we daily take for granted are probably open to an incalculable number of radical improvements. But, our imaginations being limited as they are, it may be a long time before we get around to looking at familiar things in new ways. What we tend to do is to accept familiar constructs as downright objective observations of what is really there, and to view with great suspicion anything whose subjective origin is recent enough to recognise. The fact that familiar constructs have equally subjective—though possibly more remote—origins usually escapes us. We continue to refer to them as objective observations, as the "givens" in the theorems of daily existence. Yet it is doubtful that any of the "givens" we accept so "realistically" has yet been cast in its final form.

It may, on first thought, be disturbing to visualise ourselves trying to make progress in a world where there are no firm points of departure immediately accessible to us, no "givens", nothing that we start out by saying we know for sure. There are, of course, those who stoutly maintain that this is not indeed the case, that we do have such unimpeachable sources of evidence, that they know what those sources are, and that we too had better believe in them. One such person—the idealist, in the philosophical sense—may say that his anchor point is a god who maintains a certain point of view, or he may hold to a principle or dogma from which he attempts to deduce everything else he needs to know. On the other hand, the materialist, always sceptical of abstractions, anchors his ideas in his senses, or, in the case of the dialectical materialist, in a line of inference which seeks to synthesise physical and social facts. Each, however—idealist and materialist—thinks he must initially have an anchor point, a known reality from which he may safely launch out, and without which it would be futile to proceed.

Yet the "known realities" keep slipping out from under us. Our senses play all kinds of tricks and prove themselves to be the most unreliable informants. And our theologies, far-seeing as they appear to be, do, in time, lead to such indecent practices that sensitive men refuse any longer to take them literally. Thus we find ourselves repeatedly cut off from what once we thought we knew for sure, and we must reluctantly abandon the very faiths from which we originally launched our most fruitful enterprises.

The upshot of all this is that we can no longer rest assured that human progress may proceed step by step in an orderly fashion from the known to the unknown. Neither our senses nor our doctrines provide us with the immediate knowledge required for such a philosophy of science. What we think we know is anchored only in our own assumptions, not in the bed rock of truth itself, and that world we seek to understand remains always on the horizons of our thoughts. To grasp this principle fully is to concede that everything we believe to exist appears to us the way it does because of our present constructions of it. Thus even the most obvious things in this world are wide open to reconstruction in the future. This is what we mean by the expression *constructive alternativism*, the term with which we identify our philosophical position.

But let us assume also that there is indeed a real world out there, one that is largely independent of our assumptions. Moreover, it is a world that moves along and we are, willy nilly, caught up in its stream of time. To understand this objective world is to be able to do more than trace its permanent features; we must anticipate the flow of its events. Now it is precisely at this point that our line of reasoning veers away from classical phenomenology. While we do hold that perceptions are anchored in constructs, we hold also that some constructions serve us better than others in our efforts to anticipate comprehensively what is actually going on. The big question is, of course, which ones, and how do we know.

In Which Direction Lies the Truth?

Since our experiences lie mostly behind us, we are likely to think of the past as defining as well as encompassing all the things we know for sure. And, since, at any given moment, we cannot tell what the future holds in store, we are likely to regard everything that lies up front as somewhat chaotic. If, then, we are asked to say something that is "true" we are likely to mention something we have experienced—something from our past—rather than something we are looking for.

Moreover, when we sit down to try to figure out what will happen in the future, it usually seems as if the thing to do is to start with what we already know. This progression from the known to the unknown is characteristic of logical thought, and it probably accounts for the fact that logical thinking has so often proved itself to be an obstacle to intellectual progress. It is a device for perpetuating the assumptions of the past.

Perhaps at the root of this kind of thinking is the conviction that ultimate truth—at least some solid bits of it—is something embedded in our personal experience. While this is not the view I want to endorse, neither would I care to spend much time quarrelling with it. It does occur to me, however, that one of the reasons for thinking this way is our common preference for certainty over meaning; we would rather know some things for sure, even though they don't shed much light on what is going on. Knowing a little something for sure, something gleaned out of one's experience, is often a way of knowing one's self for sure, and thus of holding on to an identity, even an unhappy identity. And this, in turn, is a way of saying that our identities often stand on trivial grounds. If I can't be a man I can, at least, be an expert.

There is another reason for the wide-spread conviction that we must look backwards if we want to see truth. If we regard the attainment of knowledge as a matter of the accumulation of information, then we will see truth as something put together bit by bit, with each individual fact verified and then nailed down to stay. If you want to point to truth, then, you have only to look back at what has been nailed down. A scientist is supposed to start his inquiries with what is already known and published. Only from such a point of departure is it believed that he can best explore the realm of uncertainty.

Having accepted this viewpoint, some scientists in the last century speculated as to what proportion of all possible knowledge had already been accumulated. It was a large proportion, if I remember correctly, somewhat over half. And they guessed that within a century or so everything would be pretty well settled. Now, however—a century later—the day when we will know everything of importance seems a good deal further off. Moreover, a lot of those matters we thought we had settled are having to be reconsidered.

Experience, Knowledge and Truth

I suppose that science can be regarded as moving ahead step by step—whatever that means. But with each step that brings into focus some new facet of the universe something, which before we thought was all settled, begins to look questionable. It is not that each new fact displaces an old one, but that gradually, almost imperceptibly as our ventures progress, a darkening shadow of doubt begins to spread over the coastline behind us.

A student, for example, who starts his career by making certain conventional assumptions about scientific determinism thereafter comes across many things in his lifetime, and may end up with serious doubts about scientific determinism itself. This is not to say that his assumptions trapped him in blind alleys. Indeed, only by pursuing their implications as avidly as he did was he able to arrive where he did. Still, from the very vantage point to which they eventually led him, those assumptions appeared much more questionable than they originally did, and he would now hesitate to use them again.

Now the question is, did the truth lie ahead of him or behind him? Did he, using his partial knowledge as a base, launch an invasion of the remaining world of things he did not know? Or did he, starting from a position of admissible ignorance and using only his faulty vision of an ultimate truth, proceed toward what he—ignorant man that he was—dared hope to understand?

Perhaps we had better make a distinction between knowing something and believing it to be the truth. Since knowing something can also be regarded as a subjective experience—that is to say an *awareness*, and need not imply that what is thus known is significant of anything except the experience itself, it would seem reasonable to say that the student had indeed proceeded from the "known" into the realm of what had once been for him the "unknown". What he knew was how he felt, and what he did not know was how he would later feel. Yet his point of departure seemed less valid when he looked back on it. And if he had been forewarned of the extent to which the outlooks of scholars are likely to change during a lifetime, he would not have felt, when he started out, that he really knew very much at all.

Now we can say that truth becomes known only *through experience*. This is not the same as the more trivial proposition, mentioned a moment ago, that bits of truth are embedded in one's experience. One can make a stronger case for the *through experience* proposition, and I shall have something to say presently about the importance of it. But, before I do, let me remark that it is tempting to infer from this proposition that there is no point in venturing beyond one's own experience in searching for truth.

I suppose it does seem a little silly for man to attempt to transcend himself? That is a prerogative which, since the Greeks, we have left to the gods, is it not —or to the priests who, after all, have proved themselves much more articulate than gods ever were, or, nowadays, to the medics who have turned out to be more enterprising in telling us just what to do with ourselves than either gods or priests? If, then, it is personal experience alone that counts, I suppose it would seem logical for us mortals to focus the attention of all psychology upon the self, the self which is both the agent and the beneficiary of all experiential outcomes.

There is indeed a point here, and if man were no more than a bystander to that procession we call the universe, or if the universe were itself no more than a spatially distributed display of interesting objects, then we might reasonably regard experience and truth as facsimiles of each other. But what man thinks he sees leads him to conjure with what he has not seen, and what he has experienced makes him wonder what he has missed. So imagination, once stirred, often leads to initiative, and initiative to action, and action produces something unexpected for men to contemplate and experience, and, finally, the newest experience throws the recollections of prior experiences into fresh perspective, thus reducing them to the level of mere chronicler's facts, facts whose historical meaning takes its shape from present rather than past interpretations.

What, for example, is experienced on the battlefield as fear and desperation is recast as heroism and noble purpose; what once was experienced as tenderness in courtship is sometimes re-experienced in the life of a married man as possessiveness; adventure becomes boredom, and boredom in a classroom is translated by some magic of retrospect and academic acclaim into scholarly enterprise. Certainly shifting perspectives can do strange things to experience.

Where, then, lies the truth? In experience? And if in experience, then in what experience—the experience of the past, or in the experience of the future? And can it be said that experience is the key to what is known? If so, then what is it that is known? Is it only experience itself that is known—one's past experience, perhaps, or the experience of the fleeting instant—as psychologists of the actual genetic school would say—or of the phenomenal present—as the true phenomenologist might say? If so, then what about the desperate soldier on the battlefield, alternately performing acts of tender mercy and deeds of enormous cruelty—for one or the other of which he may later be decorated by Act of Congress, or perhaps even elected to public office—does he know only his immediate experience, and will he never know anything else? Is killing and rescuing and being killed all "just for kicks"? Is experience both end and means, both truth and evidence, premise and conclusion? In short, is the universe of man a vast redundancy?

Or would it not be better to say that one can never know his immediate experience until he has looked back on it again and again—long after it has ceased to be immediate—and even then will he know it only partially until he looks back on it still again? And is it not better to say that a man of experience is one who has done more than collide with a series of events, that he is, instead, a person who, by anticipating and taking account of those events, has been moved to reconstrue them in many dimensions to gain fresh perspectives on everything around him? Yes, I rather think such a definition of human experience has more to offer than the notion of immediate *self-involvement*—important as I believe *being-in-the-world* is in the full realisation of existence.

Even *commitment*, which is *self-involvement* plus *affirmative anticipation*, does not altogether unlock the potentialities of human experience, for it often leaves man a hostage to his past involvements, rather than the beneficiary of them, and it indentures him to the immature expectations of his youth. Only by adding *reconstruing* to the sequence of psychological processes can the full cycle of human experience be completed and man freed from his Sisyphean labours. The cycle of human experience remains incomplete unless it terminates in fresh hopes never before envisioned. This, as I see it, is no less true for the puzzled scientist than for the distraught person who seeks psychotherapeutic escape from the psychological redundancies that he has allowed to encompass him, or, for that matter, for the experienced sinner who finds in repentance, not reincapsulation within a dogmatic system, but the full restoration of his initiative.

The Role of Experience in a World Where Nothing Is Yet Known for Sure

It is true that most psychologists concede that the full possession of the ultimate truth is an accomplishment which still lies far ahead of us. But the structure of the scientific establishment they usually work to build is founded on the notion of an ultimate truth that is pieced together bit by bit, and fragment by fragment. This is to say that if one knows A and B to be true, he can safely proceed to the discovery of C, and be that much further along toward having the whole jig-saw puzzle of the universe properly assembled. But if he is mistaken about A or B he can scarcely hope to arrive at C. This is the *fragmental* notion of truth, and science, according to this notion, progresses by a series of technical "break-throughs". So the business of the scientist is to get things nailed down, one at a time, and, using each established fact or law as a point of departure, to make further inferences and test them by experimentation. If the data confirm the hypotheses, then one will have something more to nail down, and the total job of simulating the universe will be that much nearer completion. I suppose when the task is complete they hope to declare a scientists' "sabbath" and rest from their labours.

But what I am now led to say differs fundamentally from this notion of *fragmental* truth, though I think it has been implicit in scientific enterprise for some time. Moreover, I see in it profound implications for the science of the future, and especially for psychology. Let us say that the whole of truth lies ahead of us, rather than that some parts of it ahead and some behind. What we possess, or what we have achieved so far, are *approximations* of the truth, not *fragments* of it. Hopefully we are getting closer, in some sort of asymptotic progression, and, at some infinite point in time, science and reality may indeed converge.

What we do then is not so much to go about sifting out nuggets of truth to add to our treasure chest, as it is to invent new constructions to place upon the events of the universe in which we have engaged ourselves. The constructions are admittedly inadequate—the history of scientific thought should, if nothing else, lead us to regard them so. But, even though they are inadequate expressions of what we will eventually conclude, they do provide practical grounds for more intimate involvement, for more alert anticipation, for more courageous commitment, and, eventually, for the very reconstructions which will supplant them. None of us knows how many alternative constructions are tenable, but I think the variety must always—this side of eternity—be considerable.

So we have called this outlook on the scientific undertaking *constructive alternativism*. It is itself an alternative to the notion of *accumulative fragmentalism*, upon which it now appears we have, in fact, been building our scientific edifice. And I should hasten to add that *constructive alternativism* is a philosophical outlook that has as much significance for the distressed individual who seeks psychotherapy—or even for the artist—as it does for the scientist. But, for the moment, let us speak only to the issues of science, or to that particular programme of human effort we are momentarily inclined to call "science".

Our venture as scientists, then, is not to press down with one hand on what is presumed to be known for sure and reach out with the other into the unknown for more bits of the puzzle, but rather to proceed from propositions which are admittedly faulty, in the hope that we can complete fully the experiential cycles which will enable us to formulate new propositions that are perhaps less faulty. It is not that we, as scientists, know just so much, and are out to add to the store, but rather that our experiences, pressed full cycle, will lead us to question more freely, to be less taken in by the "obvious", to see fresh possibilities of relationships, to put facts together into more productive combinations, and to entertain sweeping alternative constructions of events where once we could only "feel" and "perceive" and "learn" what was already "known"—in the manner that "psychologists" write so much about.

Transcending the Obvious

Now what is there about this human experience, pursued in full cycle, that enables man to rise above what he thinks he knows and so often to do better than he knows how? How does he transcend the obvious? If a man, say a psychologist, remains aloof from the human enterprise he sees only what is visible from the outside. But if he engages himself he will be caught up in the realities of human existence in ways that would never have occurred to him. He will breast the onrush of events. He will see, he will feel, he will be frightened, he will be exhilarated, and he will find himself feared, hated, and loved. Every resource at his disposal, not merely his cognitive and professional talents, will be challenged. So involved will he be that, in order to survive, he will have to cope with his circumstances inarticulately as well as verbally, primitively as well as intelligently, and he will have to pull himself together physically, socially, biologically, and spiritually.

Now add to this *involvement* the further ingredient of *commitment*. Not only does he fling himself into the path of events but he also undertakes to anticipate them and to deflect their course. He becomes a significant event himself, and hence a factor in what happens. So not only must he cope with what has engulfed him, but he must bear some responsibility for the turn of events. So cast, he becomes, in some measure, the architect of his own hazards and catastrophes, and thus what was once so obvious to him as an external observer is emptied of its previous meaning.

It is true that a person so caught up in the tide of circumstances, or so committed to the control of them, can scarcely be accredited as an unbiased observer. But, from the standpoint of constructive alternativism, the issue is not bias versus unbias, but the question of what the bias is and how long it takes to see things in a new light. There are indeed times in one's life when the stream of events flows so rapidly that he becomes preoccupied with his own survival and he fails to see the world in any but a momentary perspective. But give him time to look back on an experience, not as a bystander to it, but as one who has survived it. Or let him involve himself in more massive affairs. These are the ways of gaining perspective, and not disengagement from the human enterprise in an effort to achieve the fiction of being an "unbiased" observer.

It is not an uncommon occurrence that men who have faced death or tragedy come to see their worlds in quite different ways. Is it that misfortune is merely a rude reminder of some earlier conviction? I think not. It is the experience, with its self-involvement and, in some cases, its commitment, that mobilises their reconstruing resources. Before the crisis such a man may have lived in a day-to-day world of obvious meanings. It was "obvious" that he must drive his car at high speed—else he would miss his appointment. It was "obvious" that he had to burn the candle at both ends—else he would not get his job done, or what he thought was his job. It was "obvious" that starvation was the result of perversity. But, through an experience in which he himself was involved and committed, he quickly found a way to transcend such "obvious" things.

A man who is not engaged with life has to explain his behaviour in terms of "motivations"—how else can he? But a man who is fully engaged and committed never thinks of such things. He has purposes and expectations, but not motives. A man who experiments as a way of coping with his life may devise something new, but the fellow who is merely an experimentalist can scarcely hope to see anything that he does not already presume is there. In the one case an experiment is a recourse to experience, in the other it is a formalistic exercise intended to take the place of such a grubby occupation.

Using the Full Cycle of Experience

I hope it is clear at this point that I see personal experience as a key to human progress, both individual and scientific. This is the reason I have identified myself with clinical psychology, even though I have no particular interest in applied psychology. In fact, it is precisely because I do not wish to be an applied psychologist that I have taken up the more intimate approach to man. Except as I involve myself deeply with a person whose life is at the turning point, unless I seek to anticipate the outcomes of his decisions at this point, unless I myself make some commitment to joining him in a common undertaking, and unless in this situation I am ready to *reconstrue* psychology rather than *apply* it merely, then I think that, as a psychologist, I shall accomplish little more than to accumulate a bibliography to attach to my next application for a job.

But, while *constructive alternativism* turns me to *experience* as a way of transcending the obvious, it is, as I have said before, the full cycle of experience —that is to say, my involvement in the life of man, unequivocal prediction, the recognition of confirmation and disconfirmation, and ingenious re-approximation—rather than one phase of the cycle only that must concern me. So, along with the emphasis I place on experience, I hope it is equally clear than I am not proposing phenomenology as the key to scientific advancement —nor existentialism—nor cognitive theory. I realise that *personal construct theory*, with which my name has been identified, has been placed by various writers in each of these categories; but I think it belongs in none of them. I even once found it categorised as a "learning theory", a discovery which has given me considerable amusement, since, in proposing the theory, I went to some lengths to urge psychologists to consider the advantages of abandoning the concept of "learning" altogether.

May I take a moment to point out what *constructive alternativism*, and its implementation through personal experience, means to the graduate student in psychology. A student interested in the topic of creativity, for example, would first try to see it in the flesh. What he thought he "knew" about it would be assumed to be an approximation to truth only, and completely open to reconstruction, rather than as so much known for sure with a little more about to come.

He might, of course, round up some creative people and measure them. This is what some psychologists who are interested in creativity are doing, and they are no longer graduate students. But there is not very much self-involvement in that. There would be a little more involvement if the student observed these "organisms" in the act of being creative, and still more if he found some way to participate. Then he could have a go at being creative himself, unless his dissertation adviser were too upset by that sort of thing.

But, more than this, the full use of experience implies that the student would seek to understand the difference between the creative moments and the uncreative moments in one's life, and the possible complementary relationship between them. This is not the same as trying to find dimensions in which the creative person differs from the less creative person; it has, rather, to do with the fluctuations in the creative process itself and their particular dimensions, and it provides better grounds for participating in the experience.

But, thus far, the student has undertaken only the personal involvement phase of the experiential cycle. It is the sort of thing any phenomenologist might suggest. Now we can move from involvement to commitment, as

Camus did when breaking with post-war French existentialism. And we do this by adding anticipation, that is to say, the setting of goals, trying to make something happen with a vision of what it might be, and the prediction of outcomes. Can the graduate student do something to make himself more creative at a given time, or make his associates more creative, or help them find their creative moments, and can he say explicitly what he has done, or can they? If they undertake the task together will their commitment to the task and to each other throw any light on the matter?

Next, there is the matter of assessing the outcomes of their undertakings, and, again, it is just as important, I think, for the so-called subjects in this phase of the inquiry to assess the outcomes as it is for the person who initiated the experiment. This is to say that it is proper to exploit all the available experience of all the persons who have it.

Finally, there is the reconstruing phase of the cycle. Can conventional psychological theory adequately cover both the experience and its apparent outcomes? Does the student dare tamper with theory enough to adapt it to the facts?

After all this—*after*—there is the formal experiment. That may invite some experience too, but not necessarily very much. Experiments are usually designed to test out some element of a theory, or to add another brick to the scientific establishment, but rarely are they designed to put the experimenter into an advantageous position for seeing things in a new light or for re-approximating truth.

But, nevertheless, experiments must be done if degrees are to be awarded, and, in general, I am in favour of them—both experiments and degrees. A formal experiment is a test to see if one has a reasonably good rule for predicting what will happen, or for making it happen. At this point it may be safe for the student to see his experimental design committee, though that often turns out to be a deadly thing to do.

What happens next is well known and psychologists reverently allude to it as "research"—sometimes as "re-search", which is, perhaps, a better term for what they do. To be a "scientist" one must reduce his experience to a set of rules and techniques, so that anyone can accomplish the same result without having to know what he is doing. From here on the job is legalistic and technical, and our profession abounds in the kind of people who are good at this sort of thing; some graduate departments will hardly employ anyone else.

But this is all right. The student who has made sure that he is clinically familiar with "creativity", or whatever else he has chosen to explore, who has produced the phenomenon under face-to-face operating conditions, who has experienced its personal components, and who is patient enough to listen respectfully to the research technicians without losing track of what he is doing, has little to worry about. In fact, at this point—if, indeed he has actually reached it—he probably doesn't even need a clinical psychologist as a dissertation adviser.

Faith and the Unknown

While I have proposed this before, it may not hurt to express, in this context, the idea that one has to start with his own construction of a situation—not because he believes it to be true, or out of any conviction that he knows anything for sure, or even after assuring himself that it is the best of all available alternatives. On the contrary, he must, if he has any sense of historical perspective, realise that his psychology must, in the light of coming centuries, still be considered a psychology of the unknown—this, rather than a psychology that has been partly substantiated—and that knowledge, certainly, and truth all lie a considerable distance ahead of him.

This does not, by any means, make psychology a footless undertaking. What man starts with is not a certainty of the way things are, but a *faith*, a faith that, by systematic effort, he can get a little closer to the way they are. One need not assume he has possession of any shining bits of "revealed truth", picked up either on Mt. Sinai or in a psychological laboratory. But it is important to appreciate the fact that there have been some mighty ingenious approximations to truth, and some of them can be shown to be a lot better than others. Still, ingenious as these approximations are, one may live in the *faith* that better ones can be contrived. It is this *faith* that distinguishes the psychology of the unknown from simple psychological agnosticism. And it is *experience*, sought in full cycle, that is the implementation of the *faith*.

So one's construction of a situation, for which he must always take full personal responsibility—whether he can put it into words or not—provides the initial grounds for seeking experience with events. This is to say that one's personal constructs—not physical accidents—are the springboard to self-involvement. I become aware of a situation by construing it in my own terms, and on these terms I seek to come to grips with it. Some psychologists call this "opening the self to experience". But there is more to experience than mere collision with events—a lot more. I dare anticipate what will happen and I wager my life that what happens will be different because I have intruded myself. This, as I see it, is *commitment*—what I have called "self-involvement plus anticipation".

But, as I have suggested, flinging oneself into the path of onrushing events does not make the experienced man. People do that sort of thing all the time, and over and over in the same way. Nor is experience ever complete at the instant the last impact with an event has taken place, unless, of course, the man fails to survive. No sooner has the impact occurred than the experience-prone man starts reconstruing its significance. With a little ingenuity he should, by sundown, have come up with several versions of what happened.

With his reconstructions he should be prepared to face another day, to involve himself with it in a somewhat different way, to anticipate the day's being different because of his designs upon it, and to nurse still another bloody nose when evening comes. Now that, I submit, is experience! His point of departure is characterised by what the fellow did not know. But a lifetime of such diurnal cycles should bring him a lot closer to knowing something about the way the world is run. It is not the bumps that teach him, nor his bull-headed commitment, nor what he thinks about when he is putting cold compresses on his nose, but the full cycle of experience. What he thought when he started it all will probably seem a lot different a few years later, yet it must be remembered that it was the initial basis for launching himself into the experience. And let us keep in mind that he must all along have had some faith that at least he could begin to get the hang of it.

Where Does the Responsibility for Explanation Lie?

Now I have come to the last paragraphs of this paper, and I must confess that I have not yet even mentioned the most important thing I wanted to say. But no matter; if what I have said so far is clear enough, there should be no difficulty in making my point.

You will recall that I started out by saying that psychology does not have to be irrational or to resort to motivational explanations, even in accounting for the kinds of behaviour that do not appear to make sense. The difficulties we have are intrinsic to the theories we employ and to the persons who use them, and need not be attributed to the behaviour itself. And as for psychological constructs, it is the man who applies them who must take responsibility for them. The objects of his constructions should not be blamed for his failures to predict accurately.

In going on to speak of the psychology of the unknown, I suggested that there are many alternative constructions one can place upon events—such as the acts of a friend—and the chances are that none of those that occur to us in our lifetimes will turn out to be precisely right. What we do is embark upon the psychological enterprise with a faith that, given a new vantage point through experience, we can formulate better versions of what is going on in the mind of man. It is not that we may ever start with the certainty that we know just so much and are about to uncover a little more.

So how do we make the plunge? By undertaking experience! And experience comes about in a sequence of cycles, not just in one cycle, or in part of one cycle. It comprises, first of all, self-involvement—our jumping into something with both feet. In psychology I think that means undertaking intimate relationships with man—as, for example, in clinical psychology, and, more particularly, in psychotherapy. To this is added anticipation, and thus experience includes commitment. But even that is not enough, for then reconstruction is required, else no progress is accomplished.

Experience need not be an irrational undertaking, as so many psychologists claim, though it may be very hard to account for what happens in experience on available rational grounds. That difficulty is another matter. The point here is that experience, because it mobilises all of man's resources—intellectual, inarticulate, implicit, physical, social, and even his will to change matters—enables him to transcend the obvious. And that is something that is very hard to do when you stick to the psychology of the known, or when you keep thinking about science in terms of *accumulative fragmentalism.*

Constructive alternativism, however, provides, in contrast to *accumulative fragmentalism,* an approach in which experience has a systematic role. And, in addition to putting such matters as the role of experience into a clearer psychological perspective, it suggests some startlingly fresh answers to some very old questions. Take, for example, that old favourite, *free will versus determinism.* Not only is this a basic issue in science, but it is of peculiar importance to the psychologist whose primary concern is the alleged agent of free will himself, as well as its often pitied victim.

Determinism and Freedom

I have suggested elsewhere that freedom and determinism are two sides of the same coin. That is to say that each implies the existence of the other, and, indeed, would make no sense without the presence of the other. The argument is very simple—at least it seems so to me. Free will would not make much sense unless, by exercising it, I could make something happen. That means exercising a deterministic influence on subsequent events.

If, as a free man with all the supposed rights of citizenship, I decide to tie my shoe, I make a little bargain with myself and hold myself to it until the shoe is properly tied. That means that I make a decision and subsequently bind myself by its deterministic influence. But if my decision imposes no obligation on my subsequent actions, if I am as fancy free as ever to change my mind before I so much as stoop to reach my shoe laces, the freedom with which I make the original decision is futile. Freedom is freedom only if it is the freedom to determine something. And I reap the rewards of freedom only by obligating myself. Of course, if I am running down stairs in a five-o'clock subway crowd, I may find the exercise of my constitutional freedoms somewhat hazardous.

Determinism, conversely, implies an option exercised at some prior point. I do not see how a given event can be regarded as the outcome of a deterministic linkage if there was never any possibility of its being anything other than what it is. To say that it was determined is to say that it is as is because, but for certain circumstances, it might have turned out differently. Somewhere along the line the die was cast. Up until the moment it was cast there was still the possibility of the event later turning out in some other way. So determinism implies a prior moment of decision, and, while "free will" may seem like too anthropomorphic a term, we do imply that there was an instant when matters were still free to head one way or another.

What we do, in fact, is to apply the constructs of freedom and determinism alternatively. In the sequence; A is followed by B, which, in turn, is followed by C; the term C can be regarded as "determined" by B—the point of choice. But, if we wish, we can go back to A and call it the moment of choice—the instant of "freedom". And why not?

Or if you prefer to turn the illustration around, you can start with A and define it as a point of choice to start with. This is a way of saying that B and C were determined by it. But then, if you don't like to admit that it was Grandfather A's choice that started it all, you can start, instead, with Father B—only one generation up the line—and blame what happened to C on him. Or, if you are a psychiatrist with a lot of counter transference on C, you may rather say that it was Mrs. B—the mother—who was at fault.

Now what I have been doing is applying the principle of *constructive alternativism* to a problem that is both old and urgent. Some day we may know who to blame for a child's troubles, or we may give up the construct of "blame" altogether. But as of now, all we have are some approximations to the understanding of psychological problems—not the perplexing nuggets of truth the accumulative fragmentalist thinks he has collected. Let us not say that "blame" is intrinsic to the man to whom it is applied. It is our construct, and we who use it are the ones who must take primary responsibility for its appropriateness. We are the ones who must take the blame for "blame".

Free will, with its implicit moral responsibility, and *determinism*, with its hope that everything can be accounted for in some grand plan, are alternative constructions that we currently employ in our quest for better approximations to the complete understanding of man. And they are not altogether antithetical as principles, unless one makes the mistake of regarding them as intrinsic to the events which we attempt to construe by means of them. Of course it is difficult to regard an event as *intrinsically* both a consequent and as a determinant in the same progression. But the contradiction vanishes when we realise that whether an event is regarded as a determinant or a consequent is a matter of construction, and it should not, this side of eternity, be identified as being of itself one or the other. Moreover, the constructions of determinism and free will are matters for which we must ever hold ourselves and our theories responsible, not the events themselves. But this I have said before.

There is a Psychology for Getting Along with the Unknown

But where does all this about encouraging rational theories of "irrational' behaviour, about the antics of organisms not really being explained by psychodynamics, about experience transcending "the obvious" and taking the place of inferences from known facts, about the possible abandonment of those fragments of truth that students memorise for their qualifying examinations, and the dialectic about freedom and determinism each being propped up by logic of the other—now where does all this lead us? Do I not believe the universe is organised? My answer to that is that I would not claim to know that it is. Whether it is organised or not is still one of those things that are unknown. I don't even know whether it is a good question or not.

But while I don't know the answer to the question, I need not be immobilised. There is a psychology for getting along with the unknown. It is a psychology that says in effect, "Why not go ahead and construe it to be organised—or disorganised, if you prefer—and do something about it. In the world of unknowns seek experience, and seek it full cycle. That is to say, if you go ahead and involve yourself, rather than remaining alienated from the human struggle, if you strike out and implement your anticipations, if you dare to commit yourself, if you prepare to assess the outcomes as systematically as you can, and if you master the courage to abandon your favourite psychologisms and intellectualisms and reconstrue life altogether; well, you may not find that you guessed right, but you will stand a chance of transcending more freely those "obvious" facts that now appear to determine your affairs, and you just may get a little closer to the truth that lies somewhere over the horizon.

The Logic of Passion

D. Bannister

Psychologists strive for novelty while repeating the patterns of their culture. Thus, they have, in large measure, followed the lay tradition that man is to be viewed psychologically as a collection of poorly related parts. Psychology has been structured around concepts such as learning, motivation, memory, perception, sensation, personality and so forth, all of which clearly derive from common sense language and each has been given autonomy as an *area of study*. Psychologists have invented little and contented themselves largely with refining notions which have a long and tangled intellectual history.

Perhaps the most unbreakable grip exercised by traditional thought over the formal discipline of psychology is manifest in the historic division of man into *thought* and *feeling*. The effect of this dichotomy has been to deny psychology any unity and produce what are essentially two psychologies, cognitive psychology on the one hand with sub-psychologies such as memory, perception, thinking and reasoning. On the other hand it has produced a psychology of emotion, ranging around concepts such as drive, motivation, libido and so forth.

So deeply ingrained in our culture is this division of man into his thinking and feeling aspects, that it would have been surprising if psychology had, to any great extent, escaped it. It is grieving that it has barely thought to question it. We can observe this segmentation of man, both in terms of the way we analyse our personal experience and the ways in which our literature records it. As children we grow rapidly to accept the idea that we are, each of us, two persons—a thinker and a feeler. We learn to speak of our ideas as something distinct from our emotions, we learn to speak of the two as often contrasted and competing.

So deep and continuous has this distinction been in our language, folk lore and philosophy, that literary comments on it achieve the status of platitudinous but inescapable truths. Such "truths" often espouse the rival and crusading causes of thought and feeling. Thus the thought/feeling distinction can be seen, in Kelly's terminology, as a superordinate bipolar construct.

Treasured comments can be found which praise thinking and condemn feeling.

> "All violent feelings . . . produce in us a falseness in all our impressions of external things, which I would generally characterise as the 'pathetic fallacy'." (Ruskin)

Then again there is the kind of pronouncement which, while favouring reason, seems sadly convinced that passion will conquer.

> "The ruling passion, be it what it will,
> The ruling passion conquers reason still" (Pope)

Then there are those comments which are contemptous of the thought aspect of the dichotomy.

> "And thus the native hue of resolution is sicklied o'er with the pale cast of thought" (Shakespeare)

or Keats' cry

> "Oh for a life of sensation rather than of thoughts".

The contrapuntal relationship between the concepts of feeling and thought are further explored in those treasured platitudes which counterpose the two —Pascal's

> "The heart has its reasons, which reason knows nothing of"

or Walpole:

> "This world is a comedy to those that think, a tragedy to those that feel".

Our language is replete with expressions of the dichotomy: rationality *versus* emotion, reason *versus* passion, feeling *versus* thinking, the brain *versus* the heart, cognition *versus* affection, faith *versus* argument, mind *versus* flesh. Whole subcultures, periods and groups have swung the pendulum to those credos which worship the rational man, he who fights the forces of blind instinct, prejudice, chaotic emotion and bestial passion. Conversely the pendulum has swung oft-times the other way, to the Romantic and the apotheosis of emotion as the authentic, sincere and soaring expression of the true nature of man, as opposed to the mercenary practices of intellectualising cleverness and bloodless logic.

The Uses of the Dichotomy

If the construct thinking *versus* feeling has enjoyed such a long history and played so major a role in our ways of delineating ourselves and others, then clearly it must serve many purposes and serve them well. It must reflect and express aspects of experience which we need to express and reflect. Even a cursory consideration brings to light some of the purposes the distinction may serve.

If I am willing to negotiate my position and entertain yours, then I may say "I think that . . ." and proceed to a verbal accounting. If I am unwilling to negotiate my position and do not wish you to challenge it, then I may say "I feel that . . .".

If I want to notify myself or you of some, *as yet*, publically unsupported suspicion I may say that "I know that the evidence is in favour of the view (thought) that . . . nevertheless I have this feeling that . . .".

If I want to picture and represent to myself or to you *conflicts* I am experiencing I can say that "while I *know* (think) that I ought to do that, I *feel* that I ought to do this . . .".

If I want to make some kind of sense out of, excuse, respond to the inexplicable behaviour of myself, my neighbour, lover, friend, I can believe that the puzzling actions are not for this or that *reason* but are caused by mood, passion, overwhelming fear, rage, desire.

Historically, each pole of the thinking *versus* feeling dichotomy has had its implications extensively developed so that we have arrived at an elaborate language of feeling and an elaborate language of thinking. We can make a myriad subtle distinction which stem from and support the basic idea of two occasionally interacting *personae* within each of us—a thinking man and a feeling man. Thus reason takes unto itself the subsets of memory, logic, the accurate and systematic observation of events: we can assess our thoughts as fulfilling the principles of the syllogism, the rules of linguistic definition, the fair weighing of evidence. Feeling has become feelings and we can work with the distinctions between sadness, resentment, exhilaration, tension, grief, triumph, anxiety—the pallet of passion enables us to portray the world and ourselves in many hues. Thus these developed subsystems concerning thinking and feeling are guides to action and movement so that we can *assemble arguments to influence* or *arouse feelings to attack*. Logicians can teach us strategies of thought while encounter group leaders broaden our resources for feeling. Lawyers can weigh evidence and poets evoke emotions, (though be it noted that successful lawyers often plead poetically while great poets emote cogently).

The Limitations of the Construct

Kelly repeatedly made the point that a bipolar construct both liberates and restricts: it brings events within our grasp in one set of terms, while blinding us to other aspects of the same configuration of events.

Thus, in both the informal psychology of our culture and in formal academic psychology, the distinction between thought and feeling has often proved disintegrative and hindering. It is significant that it is precisely in those areas in which the distinction makes least sense that psychologists have spoken to least purpose. *Invention*, *humour*, *art*, *religion*, *meaning*, *infancy*, *love*: all seem areas of particular mystery to psychologists and it may be that they puzzle us because it makes no sense to see them as clearly "cognitive" or clearly "affective". Nor does calling to the rescue that holy ghost of the psychological trinity "conation" seem to help matters.

The separation of feeling from thought seems to have driven psychologists into a barren physiologising in a vain attempt to give substance to the dichotomy. Thus there is a long tradition in psychology which seeks to deal with "emotion" by translating it into a physiological language and redefining it so that it even has a geography (*vide* the "pleasure" centre) or a transporting fluid (*vide* endocrine secretion). Psychologists never seem to have broken entirely free from the kind of concretism, of which even a man as sensitive as William James was guilty, in arguments such as the following (1884).

> "And yet it is even now certain that of two things concerning the emotions, one must be true. Either separate special centres, affected to them alone, are their brain-seat, or else they correspond to processes occurring in the motor and sensory senses, already assigned, or in others like them but not yet mapped out."

Equally in their ponderings on the issue of "thought" psychologists have been driven to that ultimate in hardening of the categories, the notion of the "engram"—the notion that a thought is somehow more real if you think of it as permanently altered state of living tissue.

The central limit set to our understanding by our adherence to the bipolarity of thought and feeling has been that it has prevented us from adequately elaborating the notion of a *person*. Psychologists came close to beginning their study of a person with the concept of "personality" but again they failed because this was turned into a segment, a chapter heading, a branch of psychology. Either personality is psychology or it is not worth the study. Thus a person is not emotions or thoughts, not cognition or drive. To speak of a person is to invent a concept which points to the integrity and uniqueness of your experience and my experience and to the wholeness of your experience of me and my experience of you. The distinctions that such a construct encourages us to make are those of time, the past and the future person and the continuity between them; the distinction between person and object; the distinction between the ways in which the person is free and the ways in which he is determined. None of these distinctions gain and all are obscured by the traditional dichotomy of thought and feeling. Consider the distinction between free with respect to and determined with respect to. The very notion of feeling has developed in such a way as to entail the idea of determinism. Thus we are *overwhelmed* by rage, *seized* by anger, *moved* by joy, *sunk* in grief. It is interesting to note here our failure to develop the idea of feeling in its active sense as meaning exploration, grasp, understanding, as in feeling the surface of a material, feeling our way towards (Mair 1972). Equally, psychologists have followed that intellectual tradition which makes rational thought almost something which is a determining external reality. We must *follow* logic, we are not credited with *inventing* logic. Had we pondered the person rather than the two homunculi of thought and feeling we would have seen man as an active agent rather than the passive object of the environment or his own uncontrollable innards.

Whatever Happened to Emotions?

It is a platitude in personal construct theory, but a very powerful platitude, that to elaborate one's own understanding of oneself and the world it is necessary not only to develop new constructs but to escape from some old constructs. Yet to abandon a construct is to abandon a part of oneself and this is a task not lightly undertaken. However, George Kelly addressed himself formally to just such a task. He lists at one point (Kelly 1955, p. 9) the constructs that do not appear in personal construct theory although they are hallowed by respect and use in traditional psychology.

> "For example, the term learning so honourably embedded in most psychological tests, scarcely appears at all. That is wholly intentional; for we are throwing it overboard altogether. There is no ego, no emotion, no motivation, no reinforcement, no drive, no unconscious, no need".

Clearly, in constructing his theory, George Kelly had a right *not to use* the constructs of other theories, just as each of us have an inalienable right not to use the constructs of another person. However, when this is done, both professions and persons tend to accuse the doer of failing to deal with *the facts*. If I do not deign to categorise people in terms of their "intelligence" then others may respond that I am ignoring the *fact* of intelligence (and thereby being *stupid*). In psychology this strategy most frequently takes the form of transmuting a concept into a "variable" and then arguing that it is something which *must* be taken into account.

There can be no onus on any theory to duplicate the constructs of another. To do so would have the effect of making alternative theories simply co-equivalent sets of different jargons. If my public theory or my private construct system, lacks certain constructions, then you may legitimately ask me with what constructions I intend to deal with the kinds of problems which you handle by using the constructions which I have abandoned. But if you do so it must be a serious inquiry designed to find out how I am handling aspects of my world, what meaning I am giving to them, what usefulness I find in the constructs I am using. It must not be simply an attempt to prove that there are culpable gaps in my system because it does not exactly duplicate yours. To make such an enquiry seriously is no easy undertaking for we all tend to long for familiarity, even in what is new. This is presumably the reason why some vegetarians strangely refer to some forms of vegetable as "nut meat rissole".

Kelly left the great traditional dichotomy of emotion *versus* thought out of construct theory and proposed an alternative way of dealing with the kinds of *issue* which are classically dealt with by the emotion *versus* feeling dichotomy.

Inevitably it was assumed that he had somehow retained the dichotomy but failed to elaborate one end of it, one pole of the construct. In this case the accusation was generally that he had failed to deal with "emotion". Thus, when the two volumes presenting the theory first appeared in 1955, Bruner (1956) reviewed them favourably. But inevitably he saw the theory as a "cognitive" theory—i.e., one which does not deal with "emotion". He commented on what he saw as its limitations as a "mentalistic" theory which failed to deal with issues of emotion. Carl Rogers (1956) went even farther by not only pointing to the same "deficiency" but waxing much more angry and much more concerned about what he saw as a failure to deal with the passions of mankind.

For two decades construct theory has been expounded, discussed and used. One might imagine that by now psychologists would have stopped construing construct theory pre-emptively as "nothing but a cognitive theory". One might hope that they would recognise the novel and adventurous attempt to elaborate a theory of man which did not dichotomise him into a reasoning man and a feeling man. One might hope, that even if they felt that this integrative venture had failed, they would recognise the *deliberate* nature of the venture and understand that it was not simply that Kelly had failed to consider "emotion". But, for the most part, psychologists are not, in philosophical terms, Kellian constructive alternativists—they are naive realists and emotion is apparently a *real thing*, not a construct about the nature of man. Two decades after the presentation of the theory we have exactly equivalent condemnations offered to those propounded by Bruner and Rogers.

Mackay (1975) opens his critical appraisal of personal construct theory as follows.

> "PCT has been widely criticised on the grounds that it is too mentalistic. The ideal rational man, as depicted by Kelly and Bannister, seems more like a counter-programmed robot than a human being who is capable of intense emotional experience."

Peck and Whitlow (1975) comment similarly.

> "Kelly's approach to emotion is deliberately psychological but in order to achieve this position he is forced to ignore a wealth of knowledge from the field of physiology; furthermore some of the definitions seem to fly in the face of common sense. Bannister and Mair (1968, p. 33) state that 'Within this scheme "emotions" lose much of their mystery'; it can be argued that they also lose most of their meaning."

Virtually every textbook over the past two decades that has dealt with personal construct theory has unquestioningly classified it as a "cognitive" theory. Kelly was, in many ways, a man of real patience but even he chafed at the persistent attempt to allot him to constructs whose range of convenience did not span his work.

He used to plan/fantasise a new book which he might write to re-present construct theory. Essentially the content and force of the theory and the nature of its argument would remain the same but it would be stylistically re-presented as "personal construct theory—a theory of the human passions". His dream was to complete the volume and let the people who saw construct theory as a cognitive theory wrestle with the new presentation. Had he completed such a work it seems likely that he would have been open to academic attack for failing entirely to understand the rational aspects of man, the nature of thinking and the degree to which behaviour is a function of cognitive processes.

In summary then, psychologists have failed to take seriously Kelly's attempt to dispense with the thinking–feeling dichotomy. He stated it as explicitly as may be. Thus (Kelly 1969):

> "The reader may have noted that in talking about experience I have been careful not to use either of the terms, "emotional" or "affective". I have been equally careful not to invoke the notion of "cognition". The classic distinction which separates these two constructs has, in the manner of most classic distinctions that once were useful, become a barrier to sensitive pscyhological inquiry. When one so divides the experience of man, it becomes difficult to make the most of the holistic aspirations that may infuse the science of psychology with new life, and may replace the classicism now implicit even in the most "behaviouristic" research".

Alternative Construing

Kelly attempted to deal with the kind of issue normally handled under the rubric of "emotion" by offering constructs which relate to *transition*. The underlying argument is that while a person's interpretation of himself and his world is probably constantly changing, to some degree, there are times when his experience of varying validational fortunes make change or resistance to change a matter of major concern. At such times we try to nail down our psychological furniture to avoid change or we try to lunge forward in answer to some challenge or revelation by forcefully elaborating our experience. Or we may be tumbled into chaos because of over rapid change or move into areas where we cannot fully make sense of our situation and its implications and our system must either change or the experience will become increasingly meaningless. It is at such times that our conventional language most often makes reference to feeling.

Kelly strove to maintain construct theory as an integrated overview of the nature of the person; to deal with all aspects of our experience within the same broad terms. Whether he succeeded or failed the theory is thus essentially grandiose in that it attempts to deal with all aspects of human experience. Thereby it is in contrast with most psychological theories which are essentially theories *of* something. Conventionally, even broadly structured theoretical frameworks such as learning theory, specifically acknowledge that there are areas of human experience and behaviour which are outside their range of convenience—learning theory is a theory *of* learning. Other theories are much more explicit and limited—being theories *of* memory or theories *of* perception or theories *of* sensation and so forth. Most relevant to this argument is that they may be theories *of* cognition or theories *of* emotion.

Constructs Relating to Transition

In naming his constructs relating to transition, Kelly adopted a curious strategy. He chose terms such as guilt, hostility, aggression, anxiety, and so forth, all of which have a traditional lay and formal psychological meanings and then re-defined them in construct theory terms. In each case the new definition is cousin to the traditional definition but the differences are such as to cause some confusion on first inspection. Kelly was never explicit as to why he adopted this strategy, rather than create entirely new names for these constructs. However, one suspects that he did little without malice aforethought and one possibility is that he was trying to draw attention to the *difference* between his proferred definition and the standard one, by using the same term. The suspicion is strengthened when we examine the nature of the difference. In every case it seems that what Kelly is pointing to is the meaning of the situation *for the person* to whom the adjective is applied, as contrasted with the meaning of the situation for those of us who are confronted *by the person* to whom we apply the adjective. Thus standard ways of using terms such as *hostile* are such that the emnity, attack and hatred of the person is seen as directed towards us, almost as if they were traits of the person, almost as if they were unreasoning hatreds. True, we may inquire for what reason a particular person is hostile, but the term hostile itself does not carry with it any kind of causal explanation. In construct theory *hostility* is defined as "the continued effort to extort validational evidence in favour of a type of social prediction which has already been recognised as a failure". Essentially Kelly is pointing, in this definition, to the situation as it exists *for the person* who is being hostile. For such a person some part of his theoretical structure for making sense out of the world is threatened, some central belief is wavering and, because he cannot face imminent chaos, he attempts to bully the evidence in such a way that it will "substantiate" the threatened theory. Similarly, the traditional definitions of *aggression* give it a meaning very much like the meaning we attach to the term hostility, whereas Kelly defines it as the polar opposite of hostility thereby seeking to draw our attention to the nature of aggression from the aggressors point of view as distinct from its discomfort for those of us who confront it. Thus, *aggression*, as defined in construct theory terms is "the active elaboration of one's perceptual field". Aggression is the hallmark of a person who is being adventurous and experimental, who is beginning to make more and more sense out of a wider and wider range of experience and who is leaping into further experience to capitalise on the sense he is making. Truly it can be very unpleasant for us to face aggression of this sort because we may not always wish to be part of the other person's experiment, to be the means whereby he enlarges his understanding.

Kelly's definitions try to make us recognise that we can only understand the persons from within, in terms of the "why" from their point of view. This places construct theory in sharp contrast to trait psychology which sees the person as *caused* from within or stimulus response psychology which sees him as *caused* from without.

Kelly (1955, p. 122) makes this point in the following words.

> "If we are to have a psychology of man's experiences, we must anchor our basic concepts in that personal experience, not in the experiences he causes others to have or which he appears to seek to cause others to have. Thus if we wish to use a concept of hostility at all, we have to ask, what is the experiential nature of hostility from the standpoint of the person who has it. Only by answering this question in some sensible way will we arrive at a concept which makes pure psychological sense, rather than sociological or moral sense, merely."

A few of Kelly's constructs relating to transition are briefly examined in order to give some impression of the way in which the theory handles the issue of "feeling".

Anxiety

Kelly defines anxiety as "the awareness that the events with which one is confronted lie mostly outside the range of convenience of one's construct system". Thus, anxiety is not seen as a kind of psychological ginger pop fizzing around in the system or physiologised into a chemical process or left vague as referring to an uncertain general state of the person. It is given a specific meaning in construct theory terms—it directs our attention to the range of convenience of a person's construct system in relation to the situation which he confronts. Anxiety is our awareness that something has gone bump in the night. The "bump" is within the range of convenience of our construct system in that we can identify it as a "bump" but the implications of the bump lie mostly outside the range of convenience of our construct system. What do things that go bump in the night do next? What can be done about them? A common objection to this definition of anxiety arises from the fact the people often claim to be very familiar with precisely those things which make them anxious. Thus students honestly claim to be familiar with examinations yet feel extremely anxious about them. But here we have to look at the exact meaning of the phrase "lie *mostly* outside the range of convenience of one's construct system". Certainly, as far as examinations are concerned, aspects of them are well within the range of convenience of the student's construct system. He is familiar with the whole business of answering two from section A and not more than one from section B, he is at home with problems of time allotment between questions, he is familiar with all those standard demands to, "compare and contrast", "discuss", "write brief notes on". He may be a positive authority on strategies for revising, guessing likely questions, marking systems and so forth. Yet it is likely that there will be a whole series of questions relating to an examination, the answers to which lie in very misty areas. What will I think of myself if I fail this examination? What will other people think of me if I fail this examination? What will the long term effect be if I fail this examination? It may be these, and a whole range of related questions which run beyond the range of convenience of the construction system of a particular person facing a particular examination.

Not only is the definition cogent but since it is part of a systematic theory, it relates in turn to yet further constructs within the theory. Thus, if we consider ways in which we handle our anxieties, we can observe at least two kinds of strategy which are defineable within the construct theory. We may handle our anxieties by becoming aggressive—that is we actively explore the area which is confronting us to the point at which we can bring it within the range of convenience of our construct system. In Kelly's terms this would involve *dilation*—this occurs when a person broadens his perceptual field and seeks to reorganise it on a more comprehensive level. In contrast we can withdraw from the area altogether. This involves *constriction*—a narrowing of the perceptual field.

Hostility

Kelly's definition of hostility, that it is "the continued effort to extort validational evidence in favour of a type of social prediction which has already been recognised as a failure" can be exemplified by referring back to Kelly's root metaphor, his invitation to consider the proposition that "all men are scientists". We can recognise the plight of the scientist who has made a considerable personal and professional investment in his theory, but who is faced by mounting piles of contradictory evidence. He may well recognise the failure of his *predictions* in an immediate sense i.e., that what he has predicted in a particular experiment has not happened. What he may be unable to recognise and accept is the overall implication that a series of such mis-predictions negates his total theory. His investment may be too great, his lack of an *alternative* theory too oppressive and he may proceed to cook the books, torment his experimental subjects and bully his co-scientists in a desperate attempt to maintain a dying theory. All of us have experienced this situation personally. Clearly hostility is not simply "a bad thing". Sometimes we cannot afford readily to abandon a belief. If the belief is central to our way of viewing ourselves and others and if we have no alternative interpretation available to us, then it may be better to maintain that belief, for a while, by extorting validational evidence, rather than abandon the belief and plunge into chaos. At crucial times the alternative to hostility may be psychosis.

This kind of definition has practical implications for the ways in which we can make change possible. It suggests that we most facilitate change not by assaulting each other's central beliefs but by helping each other to construct alternatives, beginning with areas of peripheral contradiction. Thus, we may gradually replace a central belief without the need for hostility.

We can recognise hostility readily when we witness someone destroying their relationship with someone else in order to "prove" that they are independent. We see it in the teacher who has growing doubts of his own cleverness and therefore begins to bully his students into stupidity so that his superiority as a teacher is demonstrated. It can manifest itself with most brutal clarity in the behaviour of the person who has to physically beat his psychological opponent to his knees in order to prove that he is "best".

The whole conception of the nature of change and resistance to change implied in the idea of hostility recalls the traditional philosopher's model which compares the problem of life to the problem of rebuilding a ship while at sea. If we have to rebuild our ship while sailing it we obviously do not begin by stripping out the keel. We use the strategy of removing one plank at a time and rapidly replacing it so that, given good fortune, we may eventually sail in an entirely new ship. This kind of conception is particularly important in areas of deliberately undertaken change, such as psychotherapy, education, religious and political conversion. We must remember that those whom we seek to change—and it may be ourselves that we seek to change—must *maintain* their lives while change continues.

Aggression

As has already been noted, Kelly defines aggression as "the active of elaboration of one's perceptual field". Thus, aggression is perhaps the centrally triumphant experience for a person. The aggressive poet is one who sees ways of transmuting more of his experience into verse, the aggressive peasant is the one who is grasping ways of making his fields grow more. Aggression is our willingness to risk in order to find out, our passion for truth given embodiment in action. Aggression is the flourishing love affair, the child learning to walk, the conjurer with a new trick.

Again, each construct within the theory links into the total structure. Linked to the idea of aggression is Kelly's notion of "commitment". Morris (1975) makes the point that commitment in construct theory terms, is not a static posture, a clinging to the security of a set position. It means the converse—commitment is to the *frontiers* of one's construct system, a willingness to risk elaboration into what is—at the moment of risk—the unknown.

Kelly discusses at length the nature of the strategies whereby we give force to our aggression and particularly the idea of tightening and loosening. When a construct is used tightly, it leads to unvarying predictions, its relationship to other constructs is fixed: when a construct is used loosely it leads to varying predictions while retaining its identity, its relationships to other constructs is tentative. Loosening is that phase in our inventive cycle when we step back to gain a wider perspective, when we take liberties with the logic of our construct system in such a way that we can examine new possibilities. Loosening is whimsy, humour, creativity, dreaming, a bold extension of argument. Yet to *remain* loose is to deny oneself the opportunity of testing reality, of embodying one's dreams in informative and informed action. Loosening must run into tightening, into operational definition, into concrete forms. When we tighten we give our ideas a form definite enough to yield up the yeas and nays of actual events so that armed with new evidence we can begin again to loosen and re-examine the meaning of what we have concretely found. In relation to the traditional thinking—feeling dichotomy, we can raise here two questions. If you consider your own experience of moving from tight to loose and back again to tight, do you regard this experience as best designated by the notion of "feeling" or best designated by the notion of "thinking" or is it not adequately designated by the construct at all? Equally, it is significant that if we look at the nature of areas such as "problem solving" in Kellian terms, then we are immediately enmeshed in precisely those constructs related to transition, such as tightening and loosening, which Kelly offered as his way into "emotion". We are not safely in the area of "cognitive" psychology as presented in standard textbooks.

Guilt

Guilt is a significant concept, both in theoretical psychology and in social argument, because, at best, it fits awkwardly into the boxes of "feeling" and "thinking". Thus, we often speak of guilt as a tremendously intense and disturbing feeling. At the same time we "*find* ourselves guilty", we argue for or against our guilt, our guilt is presented as a cognitive conclusion.

Kelly defines guilt as "the awareness of dislodgment of self from one's core role structure". Core constructs are those which govern a person's maintenance processes, they are those constructs in terms of which identity is established, the self is pictured and understood. Your core role structure is what you understand yourself to be. For Kelly, self is an element which must be construed as must any other element. Equally, therefore, its unfolding must be anticipated like any other element. You may find yourself doing things which you would not have expected to find yourself doing, had you been the sort of person you thought you were. Indeed you are fortunate if this is not part of your experience. If you find yourself, in terms of those constructs/themes around which your behaviour is centred, behaving as "not yourself", then you will experience guilt. The guilt is experienced not because one has defied and upset social taboos but because you have misread yourself.

Again, this is a construct about transition. Unbeknown to yourself you have changed and guilt comes when you experience your own behaviour as reflecting the change and thereby contradicting that "self" which is now part of history and no longer valid as a contemporary guide.

Words and Constructs

The foregoing sketch was designed merely to indicate the directions which Kelly took in proposing an alternative to the superordinate construct of "thinking–feeling".

The question remains: why has construct theory been so persistently seen as "cognitive". The answer may lie partly in an unrecognised tenet of the theory and partly in the way in which the theory has been received.

A central contention of construct theory is that constructs are not verbal labels. A construct is the actual discrimination a person makes within the elements of his experience. For a given person a particular discrimination may or may not have a verbal label, it may be partly or obscurely labelled with only one pole indicated or it may have been a discrimination which was evolved in infancy before verbal labels became part of operating strategies. Perhaps, because in *discussing* our own and other peoples constructions we have, by the nature of our act, to label them, we too often forget this definitive assertion within construct theory. Thus constructs are most often regarded as verbal labels and thereby denied "emotional" meaning. For it has been a prime characteristic of the way in which the concept of emotion has developed that it should denote those aspects of our experience which are well nigh impossible to verbalise. *Ergo*, by a kind of chop logic, construct theory has been seen as "not dealing with" emotions because it is seen as dealing with words.

Construct theory has been received rather than used. It has been given a modest place in textbooks but it has had only a limited experimental and applied career and this largely in the form of expansions in the use of repertory grid technique. It may be that our failure to argue about *experience* using construct theory has impoverished the theory (for theories live and grow by use) in what it has to say about those aspects about experience conventionally dealt with under the rubric of "emotion". If this is true, then it is only when we seriously undertake explorations of our own and other peoples experience and behaviour in terms of constructs like guilt, aggression, anxiety, hostility, that we will begin to understand their meaning and their content. Till then, construct theory will appear impoverished by contrast with the richness of lay language as a way of talking about "emotional" aspects of experience or the evolved usefulness of, say, psychoanalytic language as a way of delineating interpersonal drama.

The Notion of Personal Test

The crucial and continuing test of any psychological theory is that it should challenge and illuminate the life of the individual from the individual's viewpoint. Traditionally, psychologists have used the construct *objective–subjective* to deny the validity of personal evaluation of psychological theories. It is not only admissable but most appropriate that psychological theories should be examined in the light of personal experience. The reflexivity argument—the notion that psychological theories should account, among other things, for their own construction—has two sides to it. The *nature* of an argument should not, of itself, deny its truth. An argument should be valid for the person by whom it is proposed.

The first demand is rarely made by audiences of and for psychologists. Granted if a speaker were to say "I have proved beyond any possibility of doubt, by carefully controlled experiment, that under no circumstances can a human being utter a sentence of more than four words" we might sense some intrinsic invalidity. Yet we listen solemnly and frequently to psychologists who give us *reasons* for believing that man's behaviour is entirely a matter of *causes* and we rarely protest.

The corollary of the contention that the statement must fit the speaker, is the speaker's demand that the statement must fit him. Personal validity is a necessary basis for consensual validity. Otherwise the speaker is lying.

In terms of personal test, I experienced the need for Kelly's integration/abandonment of the feeling/thinking dichotomy long before he presented personal construct theory or I read of it.

As an adolescent I accepted the distinction and duly thought of myself as a container for two homuncoli—a feeling man and a thinking man. Yet even while I accepted the distinction as reflecting inescapable reality, I found that it served me poorly. The legend seemed to have it that two *personae* were at war within me. If it were not for the harsh discipline of my intellect then, so it seemed, I could have enjoyed a much more intense freedom for and through my feelings. If it were not for the distorting and prejudicial effect of my emotions, then my thoughts would have been so much clearer, more finely formed and truthful. Given this kind of bipolarity I was to choose and re-choose between the demands of reason and the dictates of passion and whichever choice I made seemed somehow to diminish me. I was to be a more miserable lover because I was a better logician, a more muddled philosopher because I was a more sensitive man. In one area after another I was to be intellectual Roundhead or libidinous Cavalier. The choice seemed inescapable.

My then solution for this dilemma was either to alternate or to seek some middle position which made me a compromised representative of both. This seemed less damaging than to take up everlasting residence at one pole or other of the dichotomy but at best I felt it/thought it to be a mean and confusing compromise.

Kelly, by offering notions such as tightening and loosening and above all by proposing the notion of constructs in transition solved for me an ancient problem. I could, in the vision of personal change and resistance to change, account for the intensity of my experience while accepting that the "me" that changed or resisted change was a whole person and not a pair of warring dwarfs. And the "me" that commented on myself I could see as reflexive and superordinate but still entailed in all levels of me.

I no longer see myself as the victim of my "emotion". My "emotions" may torment me but I accept them as an integral part of me, as entailed in all that I have "thought". I now accept that in my "emotions" lie the beginnings, endings and forms of my "thoughts" and it is to what I have "thought" that I "feelingly" respond.

Nor can others so easily use the schism to confuse and condemn me. If my "anger" is rejected because I have no good "reasons" or my "argument" is dismissed because I lack "feeling" then I accept that others experience me as segmented. I do not have to experience myself thus.

Conclusion

The idea of thought and feeling as the two great modes of human experience relates to and bedevils a number of superordinate debates.

A popular vision of Art *versus* Science is arid because it aligns them along this dimension, thus denying the enormous complexity of structure which underlies poetry or music or painting and the intensely personal commitment which informs scientific endeavour. Equally, such a contrasting leads us away from an exploration of the nature of *invention* which is at the heart of both. The blinkering effect of such a contrasting of Art and Science manifests itself through our cultural inheritance and produces a myopia in our educational system so that "artists" are kept ignorant of the creative possibilities of hypothesis while "scientists" are encouraged to see themselves as routine clerks to nature.

The male *versus* female roles which are the root *personae* of society, pivot most often on some version of the belief that man is "by nature" rational and woman emotional. There is no way of calculating how many lives have been constrained, if not crippled, by the attempt to live to such specifications but we can observe the liberating effect of a refusal to bow to the doctrine of the insensitive man and the fearful woman. In work, in relationships, in the very legal arrangements we live by, the thought–feeling dichotomy has been pedestal to man–woman, beginning with the tearful but charming little girl and the tough, capable little boy and elaborating into the adult who cannot find ways of sexually differentiating himself/herself that are not bounded by the poles of this construct.

Even the time line along which we live has been dominated by the exclusivity of thought and feeling so that we seem condemned to move from the enthusiasm and passion of *youth* to the wisdom and calmness of *age*. The range of our choice of style, cause, engagement is arbitrarily limited by what we are socially taught is appropriate to young and old respectively and what we are taught derives much of its content from what are seen as the irreconcilable claims of passion and reason.

A psychological theory cannot be simply a specificaiion of what humanity is. Because it is self-reflecting it must be a tool people use in going about the business of being persons. A psychological theory cannot be a simple representation of a state of affairs—it must be a challenge, a liberating vision, a way of reaching out. If it is not these things then it will be a justification for a personal and social *status quo*, a form of retreat, a prison.

Most psychological theories have not sought to challenge the picture of people as segmented into thought and feeling. Indeed they have not even seen it as *a picture*, they have taken it as "real" and worked within the boundaries thereby set. Kelly was truly adventurous in abandoning the construct and offering alternative ways of interpreting experience. The alternative he offered, the construct of "change", is open to criticism, it is an invitation which we are free to refuse. The least sensible or gracious response to his invitation is not to see that it was being made and to categorise Kelly as a man who did not understand "emotion" and who thereby constructed a merely "cognitive" theory.

References

Bannister, D. and Mair, J. M. M. (1968) "The Evaluation of Personal Constructs", Academic Press, London and New York.

Bruner, J. S. (1956). A Cognitive Theory of Personality, *Contemporary Psychology*, **1**, 12, 355–358.

James, William (1884). Essay "What is an Emotion?"

Kelly, G. A. (1955). "The Psychology of Personal Constructs", Vols. I and II, Norton, New York.

Kelly, G. A. (1969). *Clinical Psychology and Personality: the selected papers of George Kelly*. Ed. B. A. Maher. Wiley, New York.

Mackay, D. (1975). "Clinical Psychology: Theory and Therapy", Methuen, London.

Mair, J. M. M. (1972). *Personal communication.*

Morris, J. Brenda (1975). *Personal communication.*

Peck, D. and Whitlow, D, D. (1975). "Approaches to Personality Theory" Methuen, London.

Rogers, C. R. (1956). Intellectualized Psychotherapy, *Contemporary Psychology*, **1,** 335–358.

The Self and the Stereotype

Fay Fransella

The Background Data

Ten years ago I was intrigued by a clinical finding which arose during an attempt to demonstrate the use of a rank order repertory grid given to the same man on a number of occasions (Fransella and Adams, 1966). While serving a prison sentence for nine acts of arson, this man became depressed and was admitted to a psychiatric hospital for treatment.

In four of the grids, given over a period of four months, there was a consistent finding that would have made one suspect the validity of the grid measures had it not been so persistent. The correlations between the constructs *like me in character* and *the sort of person who is likely to commit arson* were statistically significant but *negatively* so. On the first occasion, rho was −0·59, on the second −0·75 and on the last −0·90. One thing was clear. This man's stay in hospital had resulted in at least one change. When he left he was much more certain than when he came in that he was *not* the sort of person who was likely to commit arson.

I then came across Bannister's report of a woman who had a six-year history of agoraphobia (1965). On the grid, the construct *people who can go anywhere with confidence* was orthogonal to all other constructs, including the self. This woman saw no relation at all between herself and her symptom.

Next there was the case of the stutterers (Fransella, 1968). How could it be that eighteen people with varying degrees of a most public and distressing complaint could see themselves as significantly unlike the group to which they obviously belonged? For here the arsonist finding was replicated. The self and symptom differed significantly and *negatively*, this time on a semantic differential. It is one thing to "pretend" to oneself that he is not an arsonist, because no one can tell whether he is or is not simply by looking at him. But it is quite another thing to hide being a stutterer. But that was not all. Stutterers seemed to share the image of other stutterers with those who did not stutter—and the image was "bad".

However, the same stutterers also completed a rank order grid. This showed that the correlation between the constructs *like me* and *like stutterers* was not significant for the group as a whole. The variation was very considerable. Some were like the arsonist in seeing themselves as definitely *not* like stutterers in general, some like Bannister's phobic woman in seeing no relation at all between herself and other women with her symptom and finally a small group who definitely *did* see themselves as stutterers. Whether or not the stutterer saw himself as one may well be an indication of the degree of difficulty he is likely to experience when he tries to give it up. But that is something for the future. Also unexplained is why the semantic differential failed to show up these individual differences.

In 1968 there also appeared Liam Hudson's book "Frames of Mind". Schoolboys and undergraduates in the arts and science streams completed a semantic differential in which they rated the typical science and arts graduate in relation to thirty characteristics. Stereotypes emerged. But Hudson was led to comment that

> Hidden beneath the relatively smooth surface of the last two chapters, there still lurks an awkward and potentially dangerous finding. Though boys show few inhibitions in reproducing the stereotyped view of their own academic group, they seem unwilling—as Professor Watson—to take the logical step, and apply these stereotypes directly to themselves. (p. 71)

As a final example, there came the finding that alcoholics construe other alcoholics as *weak*, *sexually frustrated*, *lonely*, and *unhappy* but not significantly like themselves as a group (Hoy, 1973). As with the eighteen stutterers on the grid, there was no consistent negative or positive relationship between the self and the construct *alcoholics* for these fourteen men.

There are, of course, many other examples in the literature of problems pertaining to the self. I am simply trying to trace my personal path through the findings during the last ten years that have aroused my curiosity about the relation between the self and the "symptom". This chapter is an attempt to bring together these apparent self and "symptom" or stereotype dichotomies and personal construct theory.

Personal Construct Theory and Stereotypes

Some people seem able to divorce their behaviour from their view of themselves. They light illicit fires, fear open spaces, stutter or drink alcohol to excess, but do not necessarily see themselves as arsonists, agoraphobics, stutterers or alcoholics. The former are simply descriptions of pieces of behaviour—deviant behaviour—whereas the latter are descriptions of people. I commented earlier that the arsonist might be able to "pretend" to himself that he was not one because the desire to commit arson is not obvious from looking at him, but the stutterer could not hide the fact. But, of course, that is not so. The stutterer is only obviously one when he speaks, just as the arsonist is only obviously one when he sets fire to some building. At the time I had fallen for the confusion that the verb is the same as the noun.

As many people have pointed out (e.g. Becker, 1963) deviant behaviour is not construed solely as a quality of a person's acts. In the first place it is construed as deviant by "the public" and also by so-called "experts". But a piece of deviant behaviour becomes more than that—it is construed as the most important, and sometimes the only important, aspect of that individual. Lighting illicit fires makes the person an arsonist; being afraid of going out makes her an agoraphobic; stuttering makes him a stutterer and drinking alcohol to excess makes him an alcoholic.

But nothing is essentially deviant in its own right. It is only deviant if society chooses to construe it as such. No one looks twice at two children, two girls, or a man and a woman walking hand in hand. But many people's attention is alerted when two men indulge in the same behaviour. Similar behaviour will be responded to differently depending, in this instance, on whether it is performed by a man or a woman.

COMMONALITY

This is what the commonality corollary is all about. *To the extent that one person employs a construction of experience which is similar to that employed by another, his psychological processes are similar to those of the other person.*

Within a given culture there will be certain similarities between individuals in the construction of certain events. It is when we take a particular sub-system of constructions for granted and use them in a constellatory or pre-emptive way that we have the stereotype in operation. If society construes a certain group of people as *stutterers*, then the people making up that society will also see them as being *weak*, *unintelligent* and *lacking in ambition* or whatever the stereotype may be for that particular group. If the person is construed as a *stutterer* then he may be seen by many as *nothing but* a stutterer.

Since the commonality corollary stems from the fundamental postulate which states that *a person's processes are psychologically channelised by the ways in which he anticipates events*, and since constructs are the bases of our anticipations about events in our life, then our cultural commonality constructs lead us to have certain common expectations about the people to whom these constructs apply. Arsonists are all basically the same, so are stutterers, so are alcoholics. We characterise certain behaviours deviant in the first place, then make this an identifying characteristic of that person, call him by that name and construe him along a set of specific constructs used in a constellatory or pre-emptive fashion.

A fine example of commonality construing of deviance is to be found in psychiatry. Where would psychiatry be without its classificatory system? A person is pushed into a category which results in her ceasing to be an individual. She is now characterised by her behaviour. This makes life a lot easier for all concerned. There need not be expectations about, say, 100 different depressed people, but rather about two different classes of people, those with endogenous and those with reactive depression. Commonality of construing exists between all psychiatrists so that they can communicate to each other about these stereotypical behaviours. Without such construing the discipline could not exist in its present form.

INDIVIDUALITY COROLLARY

Not all stutterers, alcoholics and so forth see themselves in relation to their particular stereotyped group in the same way. Nor will they be seen by all members of a society in the same stereotyped way. *Persons differ from each other in their construction of events* states the corollary. People can construe others in common terms because each sees the other as an external figure characterised by his deviant or expected behaviour. But they may differ among themselves in how they view their own group since "each experiences a different person as the central figure (namely, himself)" (Kelly, 1955, p. 55). Some will see a relationship between himself and the stereotype while others will not. The individual knows his own experiences. He knows his intentions and feelings. He knows what his behavioural experiments are all in aid of. For understanding to take place between two people we need sociality rather than commonality of construing.

SOCIALITY COROLLARY

Commonality is only a basis for duplicating each other's psychological processes.

> . . . commonality can exist between two people who are in contact with each other without either of them being able to understand the other well enough to engage in a social process with him. The commonality may exist without those perceptions of each other which enable the people to understand each other or to subsume each other's mental processes. . . . For people to be able to understand each other it takes more than a similarity or commonality of thinking. (Kelly, 1955, p. 99.)

For the proper understanding of another person, one needs sociality. *To the extent that one person construes the construction processes of another, he may play a role in a social process involving the other person.* As we attempt to stand in the shoes of another and look at the world through her eyes so we cease to construe her in terms of socially agreed constellatory or pre-emptive constructs. When we construe in relation to the sociality corollary, our constellatory or pre-emptive construing in relation to the commonality corollary ceases to occur.

The All-pervasive Nature of Commonally Applied Constellatory Construct Sub-systems

The all-pervasive nature and the practical consequences of socially-agreed constellatory construing can be seen in the currently popular work on stereotyped sex attitudes.

Rosenkrantz and his co-workers (1968) developed the sex-role questionnaire consisting of 122 bi-polar items that one hundred people had said they thought differentiated the behaviour and attitudes of men and women. In the original study one hundred and fifty-four college students filled in the questionnaire according to the instructions "Imagine that you are going to meet a person for the first time and the only thing you know about him in advance is that the person is an adult male". They then completed the same questionnaire for the "average female" and again as it described themselves. Commonality was seen to be clearly operating by a correlation of 0·96 for male and female response as to what characterises the average man and 0·95 for what typifies the woman. Similar results have since been found after testing around one thousand people ranging across class, age and educational level.

On 41 of the items, the students gave significantly different descriptions of the typical man as opposed to the typical woman. But here comes the sting in the tail. These 41 stereotypic items were also assessed as to their social desirability. Groups of subjects were simply asked to indicate which pole of each item they considered to be the more socially desirable for the population at large (sex unspecified). Twenty-nine poles of the 41 items chosen as being socially desirable described typical masculine behaviour. Only 12 of the socially desirable poles described typical women's behaviour. When these items were factor analysed, they yielded two orthogonal factors, representing male-valued and female-valued items.

Broverman *et al.* (1972) are of the opinion that the male-valued items reflect a "competence" cluster. This includes such attributes as being independent, objective, active, competitive, logical, skilled in business, worldly, adventurous, able to make decisions easily, self confident and ambitious. If we truly take the constructs to be bi-polar, then women are characterised by a relative absence of these characteristics—they are dependent, subjective, passive, non-competitive, illogical and unambitious. The socially desirable characteristics which women are seen to have are gentleness, sensitivity to the feelings of others, tact, religious feelings, neatness, quietness, interest in art and literature and being able to express tender feelings. Men stereotypically have an absence of these characteristics relative to women.

In a further analysis, college men were asked to indicate the poles of the 41 stereotyped items that were most desirable for men and which most desirable for women. Twenty-eight of the 29 male-valued stereotypic items were selected as being desirable for men but only 7 of the 12 female-valued items were considered more desirable for women than for men. These men thus reserved for themselves virtually all the masculine characteristics considered desirable for people in general but they also thought 40% of the female-valued items were also equally applicable for men. Women thus were left with 7 characteristics that were desirable for adults in general leaving 34 for the men.

There is some evidence that these stereotypic items were incorporated into the students' self concepts. When the questionnaire was filled out as it applied to the self, the men and women differed significantly ($p < \cdot 001$) in their use of the items. However, the men differed significantly from the stereotyped "man" and the "women" differed from the stereotyped woman. Men were less male than men in general and women less female then women in general.

In fact, more recent research (Spence *et al.*, 1975) has found that the self and its stereotype are independent on this questionnaire. They suggest that one of the problems is that the two poles are not opposites—the opposite of *masculine* is not *feminine*. (Constantinople in 1973 also came to this conclusion after a survey of the literature.) They go on to argue that the concept of androgyny is a useful one for describing an individual who possesses a high degree of both masculine and feminine characteristics. It indeed seems that such individuals may be the psychologically well-adjusted ones in our society.

But the fact remains that a commonality of construing results in differential expectations of behaviour depending upon whether the individual is a man or a woman. An example of the practical effect of such construing can be seen in the study of male and female clinicians (Broverman *et al.*, 1970). These authors argue that the literature indicates that behaviours regarded as socially desirable are also positively related to ratings of normality/abnormality, health/sickness, and adjustment. If this is the case, and if it is taken together with the finding that men's stereotypic behaviours are rated as more socially desirable than women's, then clinicians may rate women's behaviours differently from men's. What may be considered pathological in one sex may not be seen as such in the other.

To test this, Broverman and her colleagues had seventy-nine medical and non-medical clinicians fill in the sex-stereotype questionnaire so as to describe a mature, healthy, socially competent adult man, or adult woman, or adult person with sex unspecified.

The clinicians were more likely to consider the socially desirable masculine characteristics as healthy for men and not for women. But only 7 out of the 12 socially desirable feminine characteristics were more often seen as healthy for women than for men.

Broverman points out that this is not as innocuous as it seems when one looks at the specific items. Healthy women differed from healthy men in being more submissive, less aggressive, less competitive, more excitable in minor crises, having feelings more easily hurt, being more emotional, more conceited about their appearance and less objective. This powerful negative assessment of women by clinicians "seems a most unusual way of describing any mature, healthy individual". In view of the additional finding that the ideal health concepts differ significantly from those describing the healthy woman, Broverman states that

> ". . . a double standard of health exists for men and women, that is, the general standard of health is actually applied only to men, while healthy women are perceived as significantly less healthy by adult standards". (p. 5)

Both men and women clinicians shared this view. They did also in Fabrikant's study (1974) in which male therapists rated 70% of "female" concepts as negative and 71% of "male" concepts as positive; the figures for female clinicians were 68% and 67% respectively. It thus seems that the sex stereotype affects the expectations of clinicians sufficiently to account, in part at least, for the finding that many more women are diagnosed as psychiatrically disordered. Not really surprising since they are apparently often seen as half way to being neurotic already!

Ringing the women's liberty bell is not appropriate here so I will refrain from a discussion on why these stereotypes are maintained. Instead, I return to stutterers to demonstrate how two changed, conceptually as well as behaviourally, from being stutterers to being fluent speakers and how they dealt with the stereotype.

The Self and Stereotype and Change

As part of a research project on the application of construct theory to the treatment of stuttering (Fransella, 1972), twenty stutterers were given a series of implication grids. The prime purpose of these was to provide measures of the degree to which the construing of fluency was elaborated as the stutterer gave up stuttering. It is suggested that no one gives up some essential feature of himself until he has some idea of the difference this will make to life.

It was found that the meaningfulness of fluency did indeed increase as the stutterer became more fluent. But I want to sketch here the changes in some of the constructs as they relate to the self and the stereotype.

The aim of treatment was to help the stutterer construe what difference it would make to life to be a fluent speaker. The procedure focused on fluency; on those situations in which the stutterer had been fluent (or construed himself as fluent). There was explicit or implicit emphasis on prediction. If he had predicted fluency, what were the bases for his prediction? If he had been invalidated, why was this so? What had gone wrong? What constructs had been up for testing at the time? In this method of controlled elaboration, it was hypothesised that the stutterer would in time have sufficiently elaborated his construing of himself as someone who could be fluent, that he would increasingly be willing to experiment with fluency. The following is a description of some grid results of two men who became virtually fluent.

These were the only two men to be tested on as many as five successive occasions over a period of eighteen months to two years. The possibly indigestible mass of data to follow comes from ten grids per man. On each test occasion there was one grid completed to do with "me as others see me when I am stuttering" and another "me as others see me when I am fluent". Constructs were elicited each time from triads of two photographs of men plus the self as either stutterer or fluent speaker.

Each pole of the construct was presented separately. For example, the person would be presented with *like me in character* and told that all he knew about the person was that he was like him in character. He was then asked to state which, from all the other attributes that had been elicited from him, he would expect to find in a person like him. A tick was placed in the appropriate grid cell for every construct pole he indicated he would expect to find in someone *like him in character* (or *like a stutterer*).

Stutterer number one

The first man was 21 years old and a computer operator. He started stuttering at about the age of five. During the time covered by his 82 attendances for treatment, he married and was promoted in his job three times. Between the first and last treatment occasion his speech improved greatly; his disfluencies decreasing by over 90%. At first testing he had 31·8 disfluencies per 100 words.

The meaning for the *self–not self* and *stutterer-normal speaker* constructs in terms of other constructs elicited can be seen in Fig. 1. The first set of constructs are the relationships on the "me as a fluent speaker" (non-stutterer) grid and the second set for "me as a stutterer" (stutterer) grid.

At the start of treatment this young man saw himself as a stutterer and as a quiet person who was concerned about other people. The only notable feature being the total absence of any significant meaning of being a normal speaker.

In Fig. 2, showing the construct relationships at the second testing, the self and stutterer construct poles are only linked on the "stutterer" grid. But as the stutterer pole has become more clearly defined, so it has become negatively viewed. There is a need to be respected but an awareness that stutterers are not respected. Both the self and stutterers are incomplete people. In the "non-stutterer" grid there are some interesting changes. While the construing of the negative implications of being a stutterer are elaborated on the "stutterer" grid, the self on the "non-stutterer" grid now has a lively and enquiring mind. Also, the not-self pole implies that the self does not mistrust

Non-stutterer Grid 1

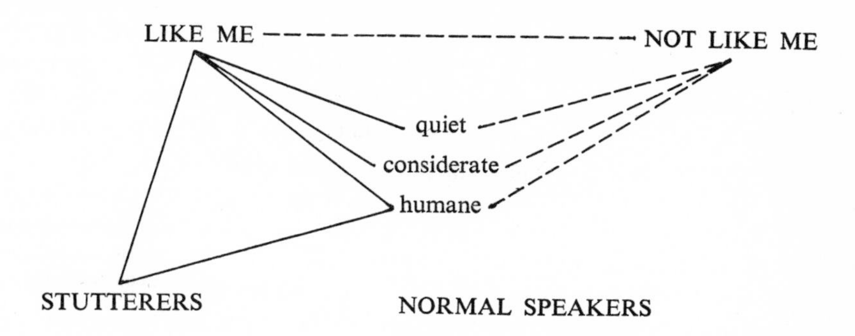

talkative
likely to have their feelings hurt
do not make their presence felt

Stutterer Grid 1

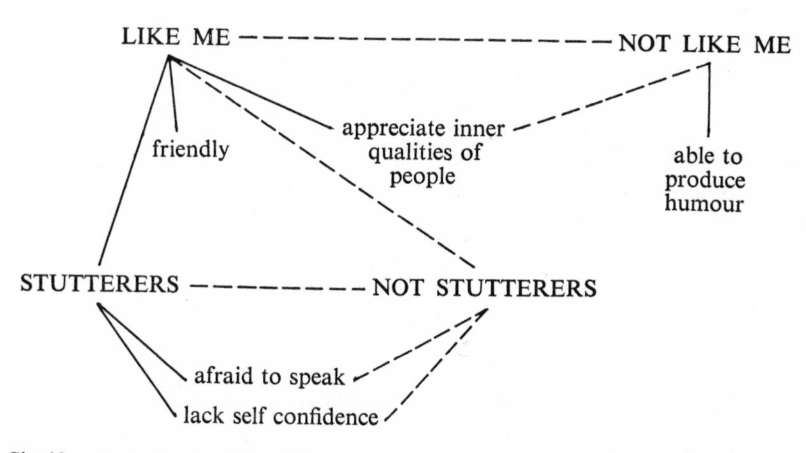

Fig. 1. Significant construct relationships with *like me* and *stutterer* constructs in the first "non-stutterer" and "stutterer" implications grids completed by stutterer no. 1. Positive relationships represented by a solid line and negative relationships by dashes.

Non-stutterer Grid 2

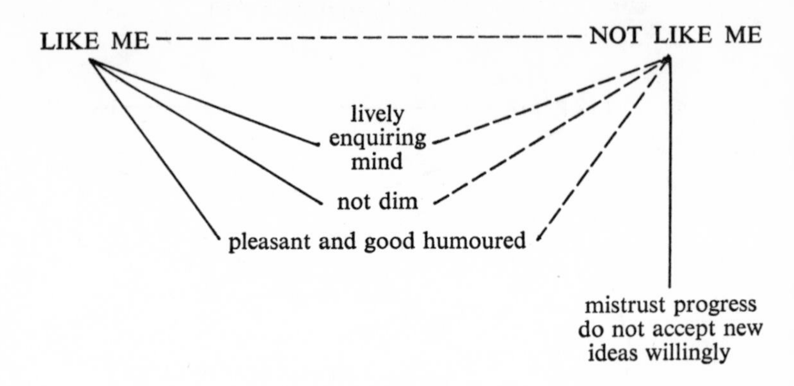

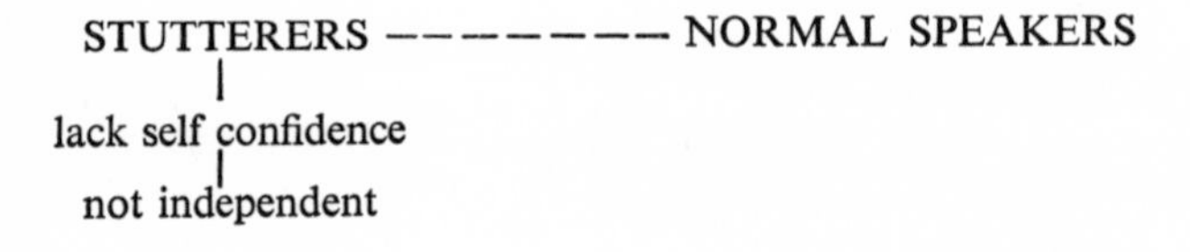

Stutterer Grid 2

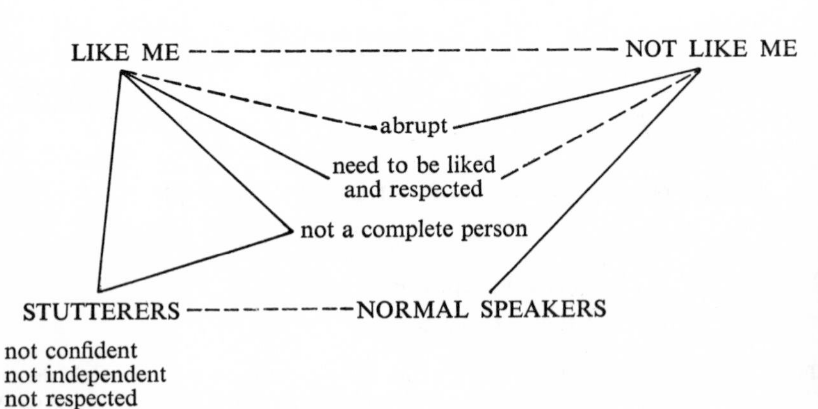

Fig. 2. Significant construct relationships with *like me* and *stutterer* constructs in the second pair of "non-stutterer" and "stutterer" implications grids completed by stutterer no. 1. Positive relationships represented by a solid line and negative relationships by dashes.

Non-stutterer Grid 3

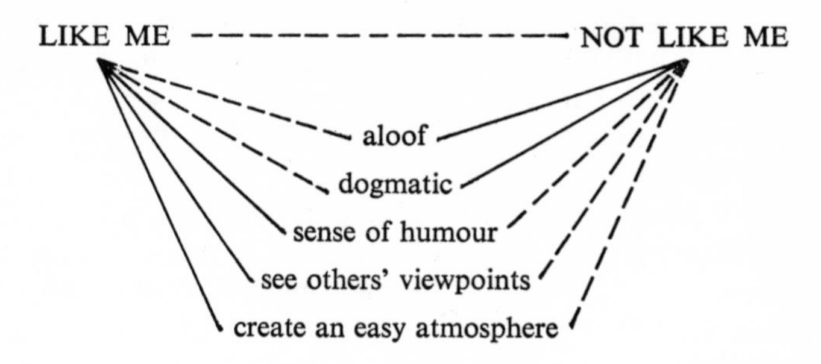

STUTTERERS ———————— NORMAL SPEAKERS

not self confident
not at ease in company
not independent
trampled on and used by others
not aloof
not a complete person
do not command respect

Stutterer Grid 3

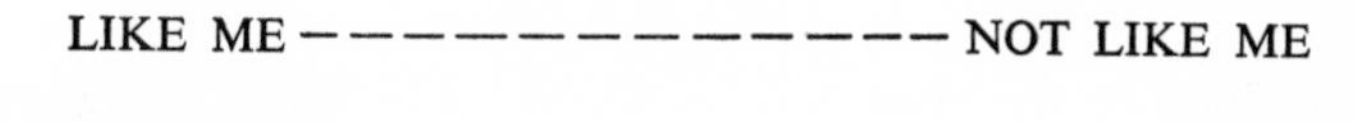

give value for money
not dishonest
do not let others down
efficient

STUTTERERS ——————————— NORMAL SPEAKERS

not confident
incapable of implementing things
not talkative
not capable of imposing own wishes
do not command respect

Fig. 3. Significant construct relationships with *like me* and *stutterer* constructs in the third pair of "non-stutterer" and "stutterer" implications grids completed by stutterer no. 1. Positive relationships represented by a solid line and negative relationships by dashes.

progress. Both these constructs can be seen as very useful superordinate ways of reducing threat and anxiety while, at the same time, acknowledging that change is probably going to occur. By this time his disfluencies had reduced by nearly two-thirds.

The definition and elaboration of being a stutterer continued on the third test occasion (Fig. 3), this time on the "non-stutterer" grid. But on neither grid is the self now seen as a stutterer. The self on both grids has not been elaborated further and the lively mind has now become an ability to see others' points of view. But the most important self change, as was found out later, was the appearance of the construct *efficient*. This proved to be, or became, the most superordinate construct for looking at his experiments of fluency over the following few weeks. His disfluencies were now down to 7·7 per 100 words.

On the fourth test occasion (see Fig. 4), there does not seem to have been a great deal of change, but disfluencies continued to be reduced until they were now only occurring on 2 words per 100. The message from the "non-stutterer" grid is that he is thinking and is not willing to accept things without doing so. Also, not only are stutterers not efficient, they do not even appear to be so.

By the last test occasion (see Fig. 5), his stuttering was down to an average of 2 disfluencies every 100 words. He was in fact fluent, since most fluent speakers have more disfluencies in their speech than that. As you can see, the self is at last well elaborated and stutterers as a stereotype have almost ceased to exist—they are *negative* people.

Figure 6 shows the changes in number of implications of the *self* and *stutterer* constructs on the "stutterer" and "non-stutterer" grids over the five occasions. On both grids, the construct *stutterers* became more clearly defined on the third and fourth occasions before being greatly reduced in meaning on the last occasion. It was only on the "non-stutterer" grid that the self became well defined at the end. A year and a half after discharge, he reported that his speech had continued to improve and that he now rarely thought of stuttering.

Stutterer number two

This man started treatment at the age of twenty-six. He was married and, during the three years in which he attended for treatment 108 times, he had a son and was promoted at work three times.

At the first session for eliciting constructs, he said it was quite impossible for him to imagine himself as a fluent speaker and therefore to say what other people would think of him. His stutter was a moderately severe one as he stuttered on 15·3% of all words, sometimes being held up more than once on

Non-stutterer Grid 4

LIKE ME ------------------→ NOT LIKE ME

thoughtful

no sense of humour

do not accept things without thinking

STUTTERERS

NORMAL SPEAKERS

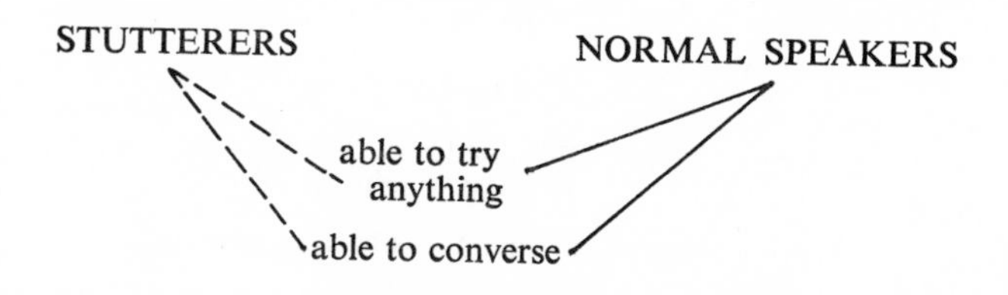

weak, indecisive
stay-at-home
no one cares about them
not respected

intelligent
sophisticated

Stutterer Grid 4

LIKE ME

NOT LIKE ME

easy-going
sense of humour
care about opinions of others

STUTTERERS

NORMAL SPEAKERS

sensitive
disturbed by small upsets
weak
not confident
minds filled with small worries
not able to live life as they wish
do not appear efficient
are not efficient

Fig. 4. Significant construct relationships with *like me* and *stutterer* constructs in the fourth pair of "non-stutterer" and "stutterer" implications grids completed by stutterer no. 1. Positive relationships represented by a solid line and negative relationships by dashes.

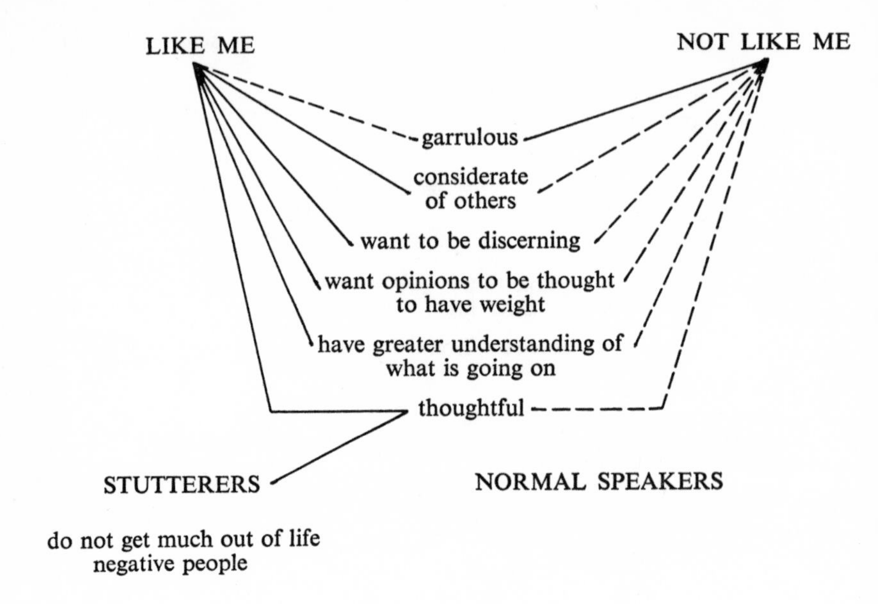

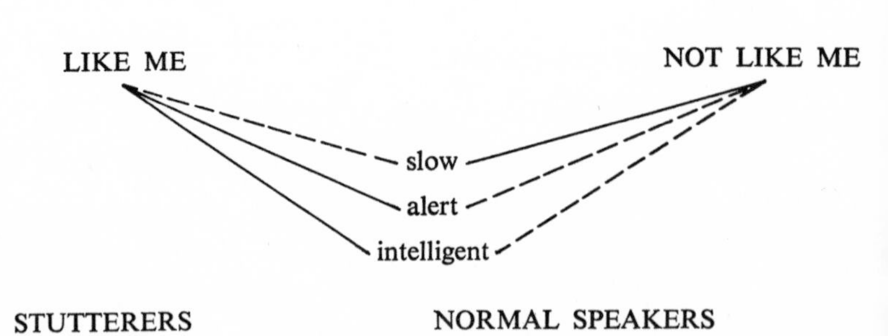

Fig 5. Significant construct relationships with *like me* and *stutterer* constructs in the fifth pair of "non-stutterer" and "stutterer" implications grids completed by stutterer no. 1. Positive relationships represented by a solid line and negative relationships by dashes.

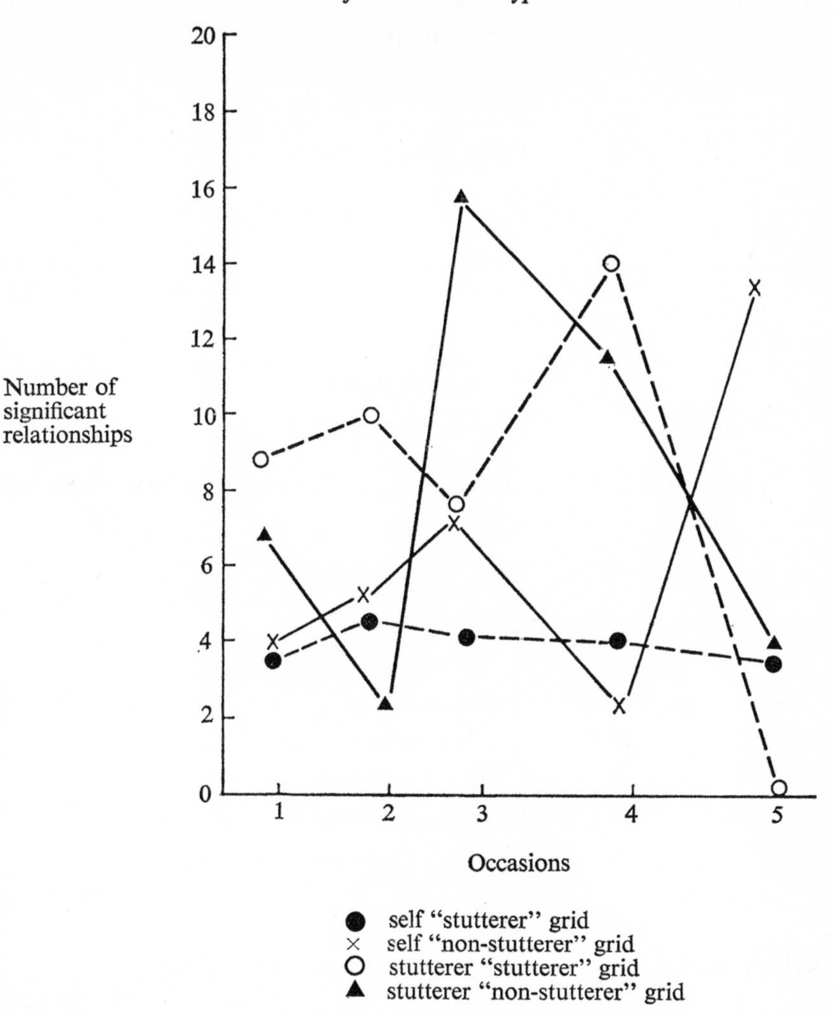

Fig. 6. Significant construct relationships on two implications grids on five test occasions for stutterer no. 1.

a word and often for some considerable time. He had 39·4 disfluencies per 100 words.

As judged by the constructs elicited from how others see him as a stutterer, he seems a nice enough fellow, easy-going and tolerant (Fig. 7). But, unlike the first stutterer example, he definitely does not see stutterers as like himself. Most important here as an indicator of trouble in the future was the construct *antagonistic* which stutterers are and he is not.

Non-stutterer Grid 1

No Constructs Elicited

Stutterer Grid 1

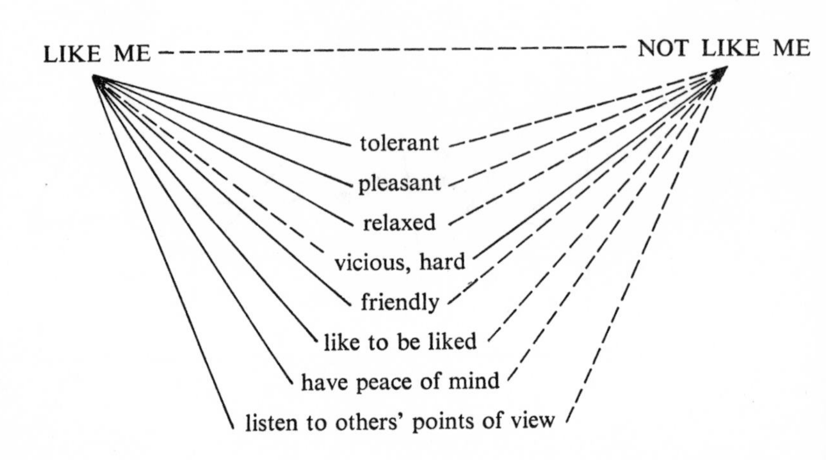

Fig. 7. Significant construct relationships with *like me* and *stutterer* constructs in the first "non-stutterer" and "stutterer" implications grids completed by stutterer no. 2. Positive relationships represented by a solid line and negative relationships by dashes.

On the "non-stutterer" grid, at the second occasion (Fig. 8), the self is well-defined but with an increasing concern about respect and status compared with the first "stutterer" grid. The antagonistic construct has changed into *gentle* as opposed to *aggressive*.

But something strange seems to have happened in the second "stutterer" grid. The self is only defined by what he is *not*. The construct *stutterers* on the

Non-stutterer Grid 2

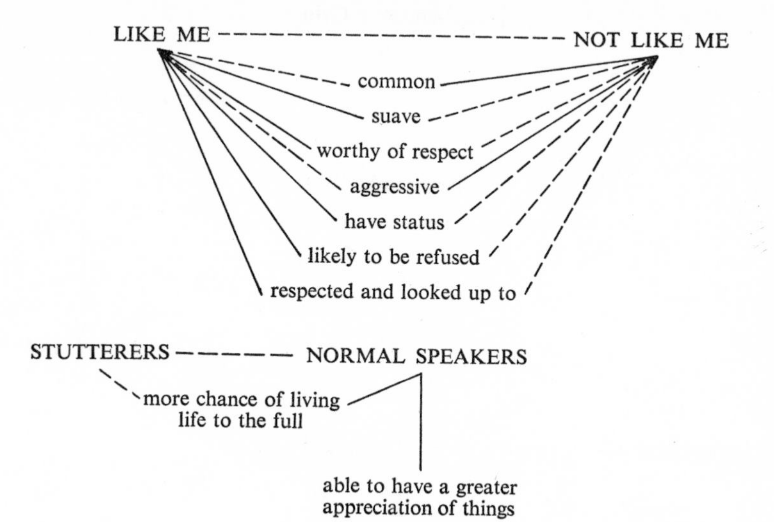

Stutterer Grid 2

LIKE ME ——————————————— NOT LIKE ME

not relaxed
not distinguished
unintelligent
no depth of character
envious
not always respected
concerned with impressing people
no depth of enjoyment of life
no opportunity to live life to the full

STUTTERERS —————— FLUENT SPEAKERS

feel inferior
not confident
sober
not distinguished
do not relax and laugh much
not happy with status in life
do not live life to the full
no depth of enjoyment in life
not always respected

Fig. 8. Significant construct relationships with *like me* and *stutterer* constructs in the second pair of "non-stutterer" and "stutterer" implications grids completed by stutterer no. 2. Positive relationships represented by a solid line and negative relationships by dashes.

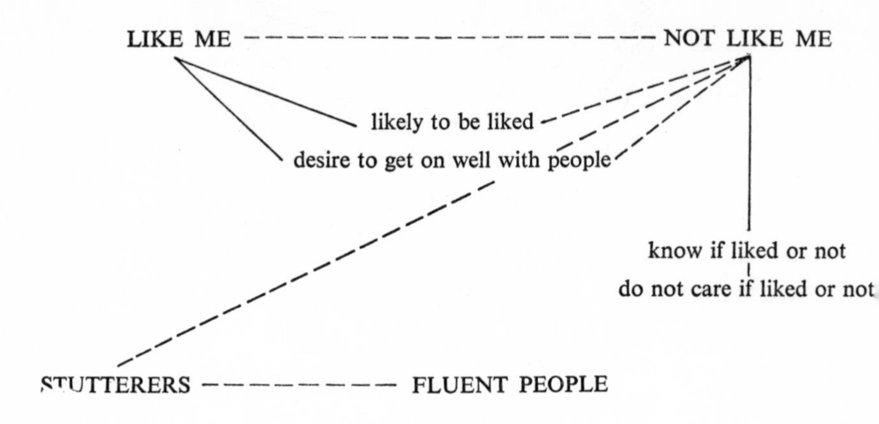

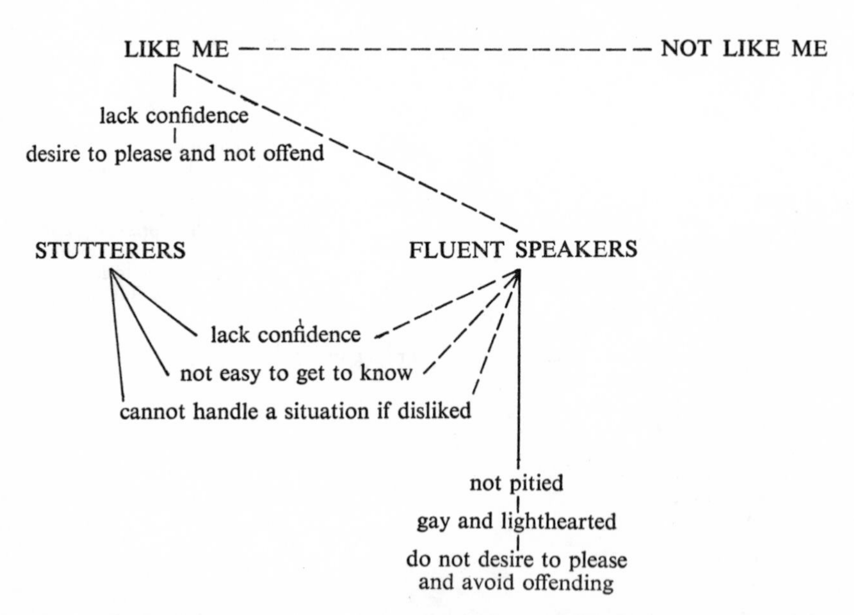

Fig. 9. Significant construct relationships with *like me* and *stutterer* constructs in the third pair of "non-stutterer" and "stutterer" implications grids completed by stutterer no. 2. Positive relationships represented by a solid line and negative relationships by dashes.

Non-stutterer Grid 4

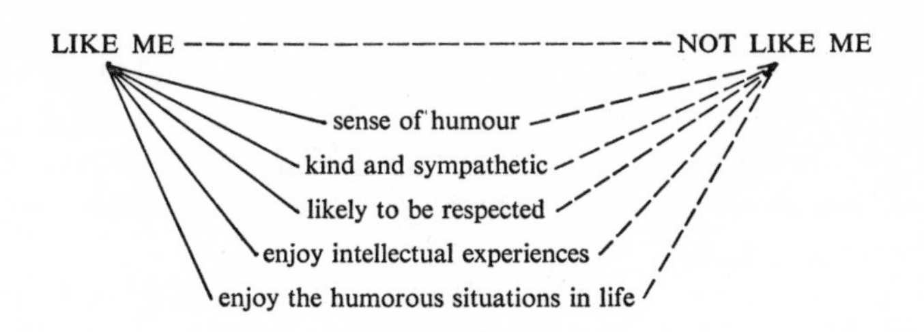

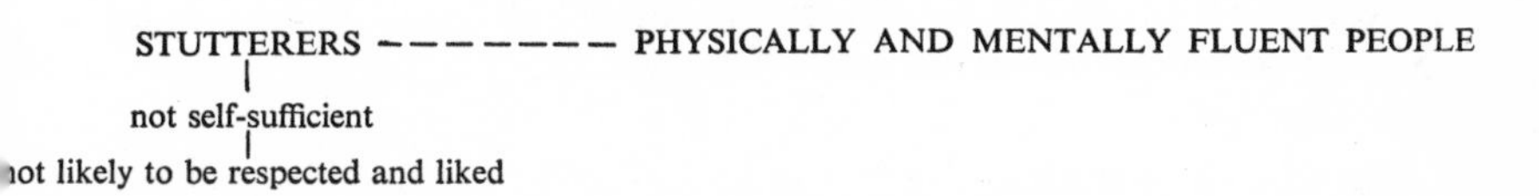

Stutterer Grid 4

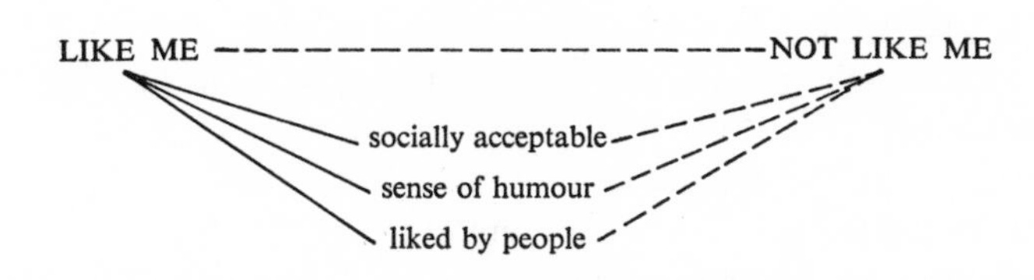

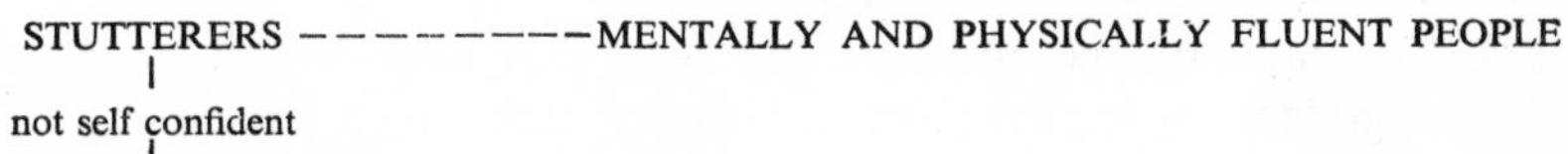

Fig. 10. Significant construct relationships with *like me* and *stutterer* constructs in the fourth pair of "non-stutterer" and "stutterer" implications grids completed by stutterer no. 2. Positive relationships represented by a solid line and negative relationships by dashes.

other hand has become much more clearly defined than it was on the first grid. His disfluencies were now down to 28·2 per 100 words.

On the third occasion his speech was fluctuating wildly and at the time of testing his disfluencies had more than doubled to 71·3 per 100 words. The self has been reduced in definition and, for the first time, there is a significant relationship with the stereotype—people not like him are not stutterers (Fig. 9). This is reflected on the "stutterer" grid—people like him are not like fluent people. What he seems to have done here for the first time is to elaborate his views of what fluent people are, and he has seen for himself that he is not like them. He does not, however, construe himself as a stutterer. The actual difficulties he was experiencing at the time centred on how to deal with being disliked. He had come to realise that this was always a possibility if you were fluent. And it was particularly threatening as he very much wanted always to be liked and respected. An important implication of being disliked is that people may not do what you ask them to and this was a situation with which he felt he could not deal.

But by the next test occasion he had dealt with the problem and his speech had improved to 18·4 disfluencies per 100 words. The self on the "non-stutterer" grid (see Fig. 10) had increased in definition and there was a reduction in the number of constructs relating to the self and stereotype on the "stutterer" grid.

The fifth "non-stutterer" grid speaks for itself (see Fig. 11). Although on measurement he had 11·0 disfluencies per 100 words, most of his speaking time was fluent except for some very brief hesitations. Along with the massive definition of the self on the "non-stutterer" grid, the self and the stereotype in the "stutterer" grid are insignificant in comparison, especially the stereotype.

No more grids were done with this man, but a two-year follow-up indicated that he had continued to improve and sees himself now as having no speech problem.

Figure 12 shows the fluctuations occurring over time in the number of constructs related to the self and the stereotype on both types of grid. In the case of each stutterer, there was an increase in the definition of the stereotype on the "stutterer" grid followed by its virtual demise. On the "non-stutterer" grid there was fluctuation of the self but a final considerable increase in its meaningfulness by the time fluency was well within their grasp.

By focusing on the construing of fluency using the method of controlled elaboration, these two stutterers were able to reduce their disfluencies to a minimum. They were being asked to look at themselves behaving in a way that was not usual for them. And they were being asked to construe what implications this new way of behaving would have for them. In order to get a clearer view of what it really meant for them, they had to have a clear view of what it meant to behave as a stutterer. As the advantages of behaving as a fluent person became increasingly clear, so they were forced to construe their life-long way of behaving as "bad". Having clearly defined what is undesirable, it must then become easier to experiment with the desirable.

The process of change involves invalidation and in the case of the life-long stutterer, invalidation of a most fundamental sort. But if stereotypes are in any way implicated in our core construing, change will not be easy.

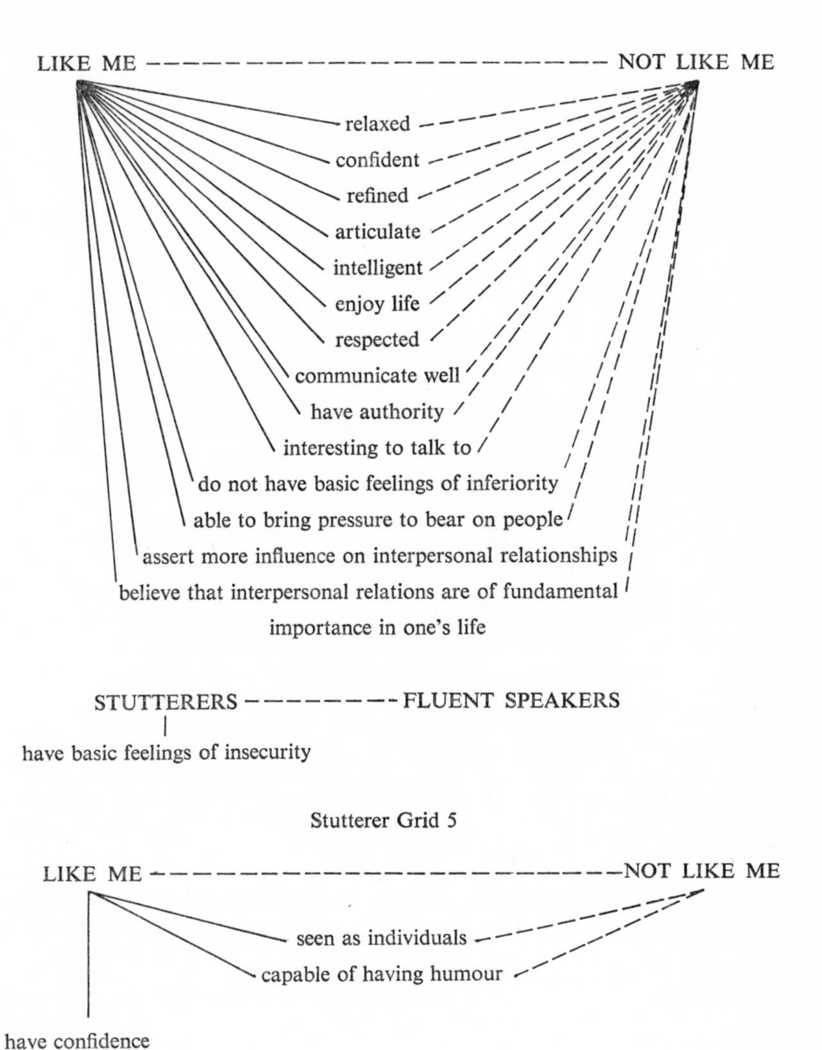

Fig. 11. Significant construct relationships with *like me* and *stutterer* constructs in the fifth pair of "non-stutterer" and "stutterer" implications grids completed by stutterer no. 2 Positive relationships represented by a solid line and negative relationships by dashes.

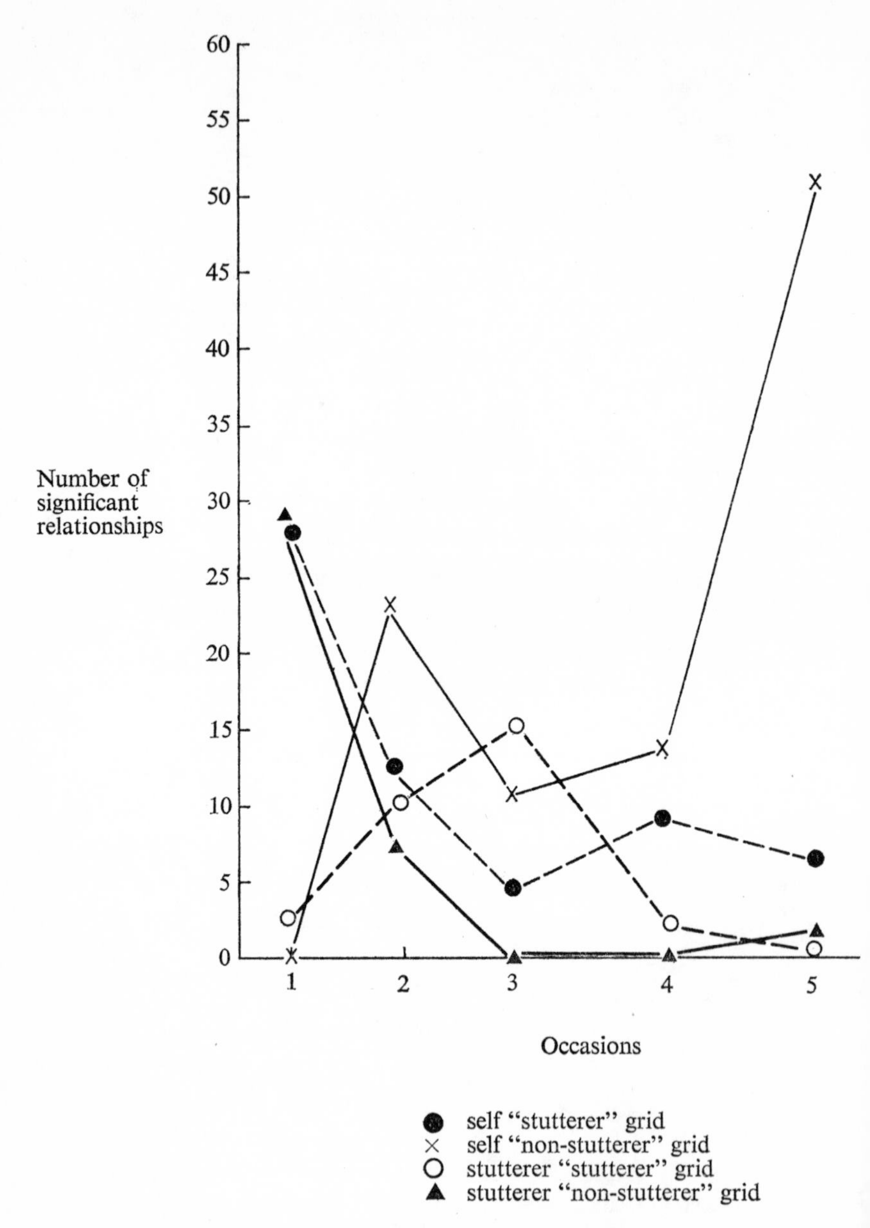

Fig. 12. Significant construct relationships on two implications grids on five test occasions for stutterer no. 2

STEREOTYPED CONSTRUING AND INVALIDATION

"Invalidation represents incompatibility (subjectively construed) between one's prediction and the outcome he observes". (Kelly 1955, p. 158)

If we see stereotyped construing as a set of constructs used in a constellatory or pre-emptive way and about which there is agreement between members of a given culture, then one would expect a certain amount of invalidation to occur in the normal course of events. Individuals will not always conform to our expectations. When this happens there are several courses of action open to one, in construct theory terms.

We can just not "see" examples of invalidation. Ninety-nine women can drive in an exemplary manner and go unnoticed, but the one who fails to immediately make way for the would-be passing male is at once construed by him as providing evidence to validate his socially approved stereotype of "women are rotten drivers".

Or, we can construe contrary events and agree that something is amiss. Our constellatory commonality network does not function as we like our construct system to function. We see the need for change. But how an individual will deal with this situation will depend on the implications such change will have for him. He may be able to loosen this constellatory network of constructs, rearrange them or find out how to use them in a more propositional way ("any roundish mass can be considered, among other things, as a ball").

However, the contemplation of change can give rise to anxiety or threat. Anxiety is the *awareness that the events with which a man is confronted lie mostly outside the range of convenience of his construct system* or, in Hinkle's terms *the awareness of the relative absence of implications with respect to the constructs with which one is confronted.* It is no good the stutterer elaborating his constellatory network of constructs to do with being a stutterer and realising they are "bad" if he has no alternative ways of behaving and experimenting with his world. Anxiety appears when such a construct network has to be abandoned because it is no longer useful and there is nothing to take its place.

Another result of the awareness of the need for change is threat. *The awareness of an imminent comprehensive change in one's core structures.* Or, in Hinkle's terms *the awareness of an imminent comprehensive reduction in the total number of predictive implications of the personal construct system.* Threat occurs when the construct(s) involved in the proposed change are superordinate or else closely linked with superordinate constructs. To be aware of the need to change such constructs means an awareness of having to undergo very profound personal change. For a man to contemplate the need to construe women as individuals and not as belonging to the stereotype group "women", might mean he would have to give up his own "male" stereotype and prove to himself, personally, that he is virile, competent, assertive and so forth. He might have to re-think his whole view of himself if he were to accept that each person has to be assessed as a person rather than as belonging to a specific group. For the use of stereotypes does most of the construing for one.

One reaction to the presence of threat is to show hostility: *the continued effort to extort validational evidence in favour of a type of social prediction which has already been recognised as a failure.* The male driver could be seen as showing hostility by only paying attention to evidence that supports his expectations. Certain clinical constructs can be construed as hostile—a woman who is ambitious or competes with men is doing so because of *penis envy*. The same woman could be construed in a hostile way by the man using another derogatory female term—*neurotic*. The basic sub-system of constructs can remain intact so the individual cannot escape from the category no matter how hard she/he tries.

The Self and its Relation to the Stereotype

It is interesting to speculate how far the differences between commonality and sociality construing can be applied to the individual. One starting point could be Mair's notion of the community of selves. There are some selves we strive to know and to understand. There are others we view as obviously belonging to the community but set apart, not "us". There are enough links between the self and the stutterer stereotype for the stutterer to agree it is part of his community, but not enough for him to acknowledge he is like "them". These sub-systems of constructs defining the "outsider" are always used in the same way. They are never elaborated. The constellatory network remains the same because the questions asked never vary. Whereas *I* am in a constant state of potential movement, *stutterers* are not. Perhaps before the stutterer can be persuaded to give up his stuttering he has to make this network more permeable and settle himself well and truly within its midst so as to see just what it is he is supposed to be like. He can then make a considered judgment about whether he wishes to have this self as part of the inner cabinet or not.

CONSTRAINTS

Does it really matter if a society first agrees on how it wants to categorise people and then construes them in a constellatory or pre-emptive fashion? I would say "yes" it does. It imposes limits on that group's behaviour and on the alternatives open to its members. It becomes a self-fulfilling prophecy.

If we accept the view that people, on the whole, do not wish to be construed as deviant, then they will be inclined to avoid behaviour that they think will lead others to construe them in a negative way.

Cultural construing in terms of commonalities leads to similar experiences. This then has an influence on the types of behavioural experiment we will be prepared to conduct. In most cases we will avoid behaving in such a way that will lead others to perceive us as deviant. For this usually means they will construe us in a way we would not like. So we modify our behaviour accordingly, thus making it a self-fulfilling prophecy.

All the great painters and composers are men. So the expectation of the woman is that she is unlikely to be very successful as a painter or a composer. She therefore limits her activities to succeeding as a pottery painter or piano player. All the great painters and composers are men.

Let us not forget that women as well as men "buy" the stereotype about members of their sex. Women were asked to judge the quality of scholarly essays (Goldberg, 1968), or the quality of contemporary paintings (Pheterson *et al.*, 1971) reputedly written or painted by *either* men *or* women. They judged the work of men to be significantly better than the identical work by women. The same results were found in a replicated study using male college students but the men were *less* strongly prejudiced in their views (Dorros and Follett, 1968).

At least two factors must be involved in determining whether we see ourselves as others see us. It seems reasonable to suppose that we are unlikely to embrace the stereotype to "us" if it is evaluatively "bad". Since most people in a given society relate deviance with undesirability, then we construe our own deviant behaviour or "bad" habits as not part of "us". But there is the example of Hudson's boys having the stereotype of a typical arts or science student but not agreeing that this was like "them". It seems that such self/stereotype divorce can occur when no negative evaluation is involved. Hudson's boys may well not have made up their minds exactly what sort of person they really are and what they are precisely going to be. This means that the stereotypes develop before we decide which pole we are going to call our own. It is of interest to note that Hudson's boys correctly identified themselves as belonging to the arts or science group in terms of *how others saw them*.

Acceptance of the stereotype may occur when we see our own group as "good" and the opposite group as having undesirable implications. How this may operate was shown in a study in which people were asked to construe Conservative, Liberal and Labour voters and to relate these stereotypes to their view of themselves and to their ideal self (Fransella and Bannister, 1967). There was overall common agreement as to what constituted the typical Conservative and Labour person. There was also a close relationship between the two "selves", party preference and actual voting. This was made possible by each person seeing her own political party as "good" and the other as "bad". That is, if a person saw herself as being *like Conservative people*, then

Labour people were *prejudiced*; if *like Labour people*, then Conservative people were *prejudiced*.

Thus at least one way in which people can differ in the application of constellatory construct sub-systems to themselves is in their personal evaluation of them. No doubt evaluation is related to the superordinacy of the constructs involved and to how these relate to core role constructs.

So when society imposes its constellatory construct sub-system on a group of its members, it is not only defining the poles of those constructs on which the members must sit, defining appropriate behaviours, but also influencing core role constructs. The personally restricting nature of these construct sub-systems can be seen in how they deny the individual alternative ways of behaving. An example of this was reported the other day by a woman who had taken part in a "consciousness-raising group" for women. These arise from the women's movement and are designed to help women look at their own attitudes and behaviours, to see how many of these are determined by the negative social stereotypes and how much women themselves are responsible for the continuance of the stereotypes by conforming to them. This woman commented that it can turn out to be a somewhat painful process. First there comes the realisation that one's behaviour *is* submissive, dependent, and generally conformist. Like the two stutterers, there is an elaboration of the stereotype to the point at which its negative elements are clearly defined. But for the stutterer, the path is clear. He has been helped to elaborate an alternative way of dealing with life—as a fluent speaker instead of a stutterer. But what are the alternative ways of behaving for women? To be submissive and dependent is "bad". The only alternative this woman saw was to behave in terms of the masculine stereotype. She can polarise. But what exactly are the implications of being assertive and independent? And indeed, does one necessarily want to be assertive in order not to be submissive? She was distressed to find that there seemed to be no other alternatives available to her. Hence the occurrence of extreme anxiety in some members of the group. A feeling of great confusion. In psychotherapy this sort of situation would be seen as likely to produce relapse. Anything is better than nothingness. Or is it?

PERSONAL DEFINITION

But these socially agreed constellatory sub-systems also give us a defined frame within which to move. They express the limits for our behavioural experiments. We know that we can afford to deviate to a certain extent and then no more. The stereotype therefore needs to be constellatory or pre-emptive and very impermeable otherwise it cannot be of real use. Sitting on those clearly defined poles of constructs we know what we are supposed to be and do within that context. But perhaps more important than that, we constantly remind ourselves about what we are *not*. We know we are not mad by having a very clear idea about what madness is. We resist any attempt to redefine the concept for this will reduce our ability to define ourselves. Knowing I am not a behaviourist helps me understand myself as a personal construct theorist.

Maybe the self is in large part the sum of our construed similarities and differences to our personal stereotypes. The stereotype of a Catholic priest enables me to reflect on the similarities and differences to *priestliness* in myself. "Me" is the extent to which I am like and unlike a host of commonally held and idiosyncratic constellatory or pre-emptive sets of constructs. My stereotypes help to define me.

References

Bannister, D. (1965). The rationale and clinical relevance of repertory grid technique. *Brit. J. Psychiat.* **111,** 977–982.

Becker, H. (1963). "Outsiders: Studies in the Sociology of Deviance", New York Free Press.

Broverman, Inge K., Broverman, D. M., Clarkson, F. E., Rosenkrantz, P. S. and Vogel, Susan R. (1970). Sex-role stereotypes and clinical judgments of mental health. *J. Consult. Clin. Psychol.* **34,** 1–7.

Broverman, Inge K., Vogel, Susan R., Broverman, D. M., Clarkson, F. E. and Rosenkrantz, P. S. (1972). Sex-role stereotypes: a current appraisal. *J. Soc. Issues.* **28,** 58–78.

Constantinople, Anne (1973). Masculinity–femininity: an exception to a famous dictum? *Psychol. Bull.* **80,** 389–407.

Dorros, K. and Follet, J. (1969). Prejudice towards women as revealed by male college students. Unpublished manuscript. Cited in F. Denmark (ed) 1974 "Who discriminates against women?" Sage Publications, London.

Fabrikant, B. (1974). The psychotherapist and the female patient: perceptions, misperceptions and change. *In* "Women in Therapy" (Violet Franks and Vasanti Burtle, eds.) Brunner/Mazel, New York.

Fransella, Fay. (1968). Self concepts and the stutterer. *Brit. J. Psychiat.* **114,** 1531–1535.

Fransella, Fay (1972). "Personal Change and Reconstruction", Academic Press, London and New York.

Fransella, Fay and Adams, B. (1966). An illustration of the use of repertory grid technique in a clinical setting. *Brit. J. Soc. Clin. Psychol.* **5,** 51–62.

Fransella, Fay and Bannister, D. (1967). A validation of repertory grid technique as a measure of political construing. *Acta Psychologica.* **26,** 97–106.

Goldberg, P. A. (1968). Are women prejudiced against women? *Trans-action.* **5,** 28–30.

Hoy, R. M. (1973). The meaning of alcoholism for alcoholics: a repertory grid study. *Brit. J. Soc. Clin. Psychol.* **12,** 98–99.

Hudson, L. (1968). "Frames of Mind", Penguin Books, England.

Kelly, G. A. (1955). "The Psychology of Personal Constructs", Vols. I and II. Norton, New York.

Pheterson, G. I., Kiesler, S. B. and Goldberg, P. A. (1971). Evaluation of the performance of women as a function of their sex, achievement, and personal history. *J. Person. Soc. Psychol.* **19,** 114–118.

Rosenkrantz, P., Vogel, Susan, Bee, Helen, Broverman, Inge and Broverman, D. M. (1968). Sex-role stereotypes and self-concepts in college students. *J. Consult. Clin. Psychol.* **32,** 287–295.

Spence, Janet T., Helmreich, R. and Stapp, Joy (1975). Ratings of self and peers on sex-role attributes and their relation to self-esteem and conceptions of masculinity and femininity. *J. Person. Soc. Psychol.* **32,** 29–39.

Inside Mystical Heads: Shared and Personal Constructs in a Commune with some Implications for a Personal Construct Theory Social Psychology*

Thomas O. Karst and John W. Groutt

Like psychoanalysis, Kelly's Personal Construct Theory (PCT) was developed by a practicing clinician with the expectancy that others would find it useful in describing, ordering, and explaining clinically significant processes. The expectation was that it might be useful in helping troubled people who come to a psychologist for consultation: "For Kelly the focus of convenience of personal construct theory, however wide its range of convenience might be, was psychotherapy." (Bannister and Fransella, 1971, p. 125). PCT, if nothing else, is essentially a *psychological* theory, described by Rychlak (1973, p. viii) as "a marvelous application of . . . a Kantian theoretical model", accepting the primacy of interaction with the world and the resulting experience over a totally independent, factual "reality". All of this is to say that, at least at first blush, one would not necessarily see PCT as the most relevant tool to put in your back-pack as you leave for a social-psychological, quasi-anthropological, participant-observer-type study of the new and "foreign" cultures represented by the contemporary communes of the late 1960's.

But as psychoanalysis was taken to other cultures by psychologically oriented anthropologists (e.g., Sapir, 1938), this chapter describes an effort by the authors to use PCT in the field (see also, Groutt, 1973; Groutt and Karst, 1974). It is addressed to the question of the range of convenience of PCT (see Bannister, 1970; this volume; Bannister and Fransella, 1972). Could it be a useful tool in understanding the people who were living in a relatively large, mystically-oriented, contemporary, rural New England community called the Brotherhood of the Spirit?

* This chapter is an extended version of a paper presented by the authors at the 1972 meeting of the Society for the Scientific Study of Religion, Boston, Massachusetts, entitled "Inside mystical heads: The personal construct theory of George A. Kelly as applied to a religious–mystical commune."

PCT never purported to be a systematic social psychology, although Kelly laid down a base for a social psychology (1955) and has presented a social psychological analysis of Cold-war Europe (Kelly, 1962). Others have explored its applicability to a social setting, but the general field of social psychology, with the exception of the person-perception literature (e.g., Rosenberg, 1975), has largely ignored it.*

The phenomenological analysis of society and social experience is a method of interest to sociologists, and Holland (1970) discusses the relationship between Kelly and Alfred Schutz (1953). Berger and Luckmann (1967) draw heavily on Schutz in their presentation of a sociology of knowledge, *The Social Construction of Reality*. And in anthropology constructual aspects are found as illustrated by these lines by Goldschmidt taken from a forward to a book by Castaneda: "The central importance of entering into worlds other than your own—and hence of anthropology itself—lies in the fact that the experience leads us to understand that our world is also a cultural construct" (Castaneda, 1969, p. viii). PCT, in any case, does encourage adventure and the attempt to look at any phenomena from a different point of view is certainly in the spirit of Kelly's thought.

The Brotherhood of the Spirit

The mid 1960's were a time of intense social and personal upheaval for American society. Two phenomena most responsible for the social iconoclasm of the era were the escalating Vietnam War and the fast-growing popularity of mind-altering drugs among youth. Old structures, traditional values, and former ways of establishing social relationships were rejected by many and a search for alternatives was begun. Communal living was one class of institutions that evolved, and the Brotherhood Community was one of the specific groups.

In 1968 in Massachusetts, a 17-year old youth experienced a vision in his tree house that began a series of events resulting in the growth of a relatively large community (see Groutt and Karst, 1974). In 1972, when our study took place, the group was four years old, had some 240 members including 120 men, 95 women, 25 infants and children, and included four black members. The median age of the adult members was 22, and excluding one man well past retirement age, the range of ages was from 17 to 42. The group owned or was purchasing four sizeable parcels of land with substantial buildings, an elaborate sound system worth thousands of dollars for a rock band, and a tractor-trailer to transport the band equipment. The group surrounded a 21-year old rock-music "superstar" and "guide", and their

* See also J. Frank's (1973) social–psychological and comparative analysis of psychotherapy which uses the notion of "assumptive world" and acknowledges Kelly's PCT.

main objective was to attain "spirituality". One of the important sources of spiritual knowledge for the group was the trance-lectures of a 60-year old neighbour who was a psychic but not a member. The rock band was a central facet of this community, treated somewhat like a sacrament, which they believed served to channel spiritual energy into the world. It was also the community's main contact with the outside, non-believing world and the *Spirit in Flesh* recorded and marketed at least two LP records during the early 1970's.

The three mile drive off the main road to get to the commune's central location takes the traveller into a world of distinctive greetings, clothes, hair styles, architecture, music, dancing, stories, and tales of reincarnation. A sign greets visitors with three rules: "No drugs, no alcohol, no promiscuity". However, more than distance must be covered and more than three rules are needed to enter into the world that is this community.

The Brotherhood is a group of people who feel they were together in other lifetimes and are now joined together in working out the eternal laws of karma. Individual past lifetimes were spent in Atlantis, the Judean desert, Pharaonic Egypt, medieval monasteries, colonial America, and elsewhere. Now, in the wooded countryside of New England, live together reincarnations of John the Baptist, Peter the Disciple, ancient Pharaohs, English kings, Vikings, Robert E. Lee, Thomas Paine, and dozens of common folk from all past eras of man's history. Discussions about reincarnation occupy a great deal of time. The members of this group use their understanding of personal reincarnations to define and explain their current behaviour to themselves and others. Trance-lectures, auras, visions, staring in silence, dreams, flashes of insight, listening to the band, and close contact with feelings are the common means of coming to knowledge of former lifetimes. One does not use intellectualising and logical arguments which are functions of the "brain". Those only serve to hide "truth" which is attained by "letting go" of old ways and becoming attuned to "feelings" and the world of "spirit".

The importance of communicating with the spirit world is central. These vibrant and friendly young people believe that the world has entered a New Renaissance and will soon be subject to catastrophic "earth changes". The majority of mankind, who live in negativity, will be destroyed. Those who, like the Brotherhood, live in positive energy will form the new beginning and balance will once again be re-established. The "most pure channel of creative energy in the Universe" is the rock band, *Spirit in Flesh*, and the world exists today only because of the high positive spiritual energy generated by the community and its band. The band is the heart of the group and one can grow in spiritual awareness by listening to its rock music. Even getting ready for a concert brings about "high" feelings.

A monistic unity provided by the spirit mythology (all persons are spirit, differences are illusion, etc.) is central to the belief system. However, like many gnostics before them, the concept of the spirit in the Brotherhood is a monistic model that operates simultaneously with a strong dualism (Jonas, 1963): all men are brothers, but there are the sheep and the goats. Positivity is drained by negativity, community is set against the world, reality is hidden by illusion. The coexistent monistic and dualistic models are also evident within the polity of the community. Insistently one is told there is no leader nor is anyone better than or above anyone else. A sign directed to community "kid watchers" (babysitters) admonishes that even the children are "old spirits and have a lot to teach us if we listen". On the other hand, there is a clear identification of levels of influence and power: the Band, the Twenty, the Forty, Members, Prospective Members, Visitors, and Outsiders. Clearly there are in-groups and out-groups in the fashion of high dualism.

Method

This report focuses on one of four contemporary communes that were ultimately studied in a similar fashion. One of the authors (JG) visited frequently, established relationships with many of its members, and lived with the group for a continuous period of two months as a participant-observer during the summer of 1972. Week-end contacts with the Brotherhood continued for a period of time. One occasion, when both authors spent a few days with the Brotherhood is illustrative of the quality of the interaction with the community. One of the purposes of this visit was to discuss a paper that was a general, ethnological description of the community (Groutt and Karst, 1974). A draft had been submitted to the group for their reaction and this led to many hours of discussion, some negotiation, and a generally better paper than otherwise might have been written. The relationship, then, was fairly open, cooperative, two-way, and allowed for mutual respect (Mair, 1970).

During the initial period the participant-observer gradually experienced the group members, its organisation, rituals, and processes. As he became known and accepted, a series of semi-structured interviews were conducted with a number of community members. Included as interviewees were the apparent leaders of the group, new and older members, and marginal or "drop out" members. Interviews were tape recorded and usually conducted in two sessions lasting up to an hour and a half each.

The interview's structure was inspired by PCT and to some extent the REP Grid (Kelly, 1955), but no rigorous attempt was made to standardise the experience. It was explained that the questions were an attempt to understand the communal experience through the eyes of the people living it, and the first question usually dealt with communes; e.g., How did you find and come to this group? Have you lived in any others? Do you know any other communes fairly well?

The member was then asked to select two communes which he knew well in addition to his own. The instructions were explained: "I would like you to make some comparisons, but in a special way. We will do the same thing with people later, but right now I would like you to think of the three communes—A, B, and C. How are two of them like each other and how do they differ from the third? You can list as many ways as you can think of". Sometimes these groupings were listed and charted on paper to aid the person in making comparisons. Next he or she was asked to rearrange the three, so that if A and B were initially grouped, the person was asked to group A and C as different from B; then B and C as different from A.

The interviewee was then asked to select any three persons (other commune members, family, friends, etc.) whom he knew well. These selections were grouped as before, comparing two in contrast with the third, then regrouping the three. This was done with at least four different sets of people. If members of the person's family had not been selected, it was suggested that a further set be done which would include them.

Our procedure departs from Kelly's in several ways. First, allowing free choice of elements presents the possibility of missing some important roles. Second, the interview is generally less formal, systematic, and complete than would be the traditional REP test. Third, each set was compared and contrasted in as many different ways as the person could think of, whereas the typical procedure asks the responder to compare and contrast only on one bi-polar dimension. The advantage of these departures was that the interview was not construed as a psychological test which most members of this subculture despise. While the presented task was seen as difficult and "artificial" by some respondents, none rejected the interview outright. Most found the procedure an interesting challenge (one member construed it as giving "readings" on others) and some commented on new insights they had attained through the process. The free selection of elements itself gives some evidence and feeling for the respondent's present existential world. This can be observed, for example, by comparing the person who freely selects only persons from his own commune with the person who selects only people from his earlier experience; or by noting how some construe three persons in a set in many ways while others have a more limited repertoire.

Since the task required rather intense concentration and was somewhat structured, other questions were often inserted in order to keep the interaction lively. These questions were suggested by Kelly (1955) to uncover other significant constructs or test how far reaching those already uncovered were. Here were included questions such as: What were you like when you were a child? Where do you see yourself in five years? At age 65? Describe yourself now, but in the third person ("he" or "she") rather than the first person ("I"). What do you like most about communal living? Least? If there was time left, questions were asked about significant books, music, movies and comparisons and contrasts were made.

Tapes of these interviews were analysed by the authors and lists of constructs were set out. Such lists were collated, grouped, and hierarchically arranged in a clinical manner depending primarily on procedural evidence (Rychlack, 1968). This analysis should yield the sets of reference axes through which the person channels his own behaviour and understands the behaviour of those around him. This is Kelly's notion of "psychological space" (1955, p. 279).

A basic expectation of the interviews was that they would reveal some of the channels through which new experiences of the person, as well as old, may run. This should lead to a better understanding of the alternatives available to the person which led him to select a communal life-style and allow the prediction of which direction he might move if his current setting should prove unacceptable to him. It also provides a method of entering the perceptual world of another and seeing that world through the personal filters or constructs that person employs in organising his rational world. Kelly points out, "the subject's measure of understanding of other people may actually be inadequate or preposterous; but, if it is the basis of a real social interaction with them, it is indeed related to his role construct system" (1955, p. 230).

This method results in a large data pool consisting of field notes, interview notes, printed materials, taped interviews, and, of course, the memories of the participant-observer experience which can be analysed and reported in a number of ways, none of which tend to be statistical. Certain aspects of the group or its ideology can be brought into focus (e.g., Groutt, 1973; 1974; Groutt and Karst, 1972), ethnological descriptions can be provided (Groutt and Karst, 1974), and individual personal construct systems (case studies) of group members can be outlined. In what follows, we have chosen to try to demonstrate some of this variety. First, a construct analysis will be presented of one of the Brotherhood members—the central figure—to illustrate an individual analysis. The next section illustrates what might be done with eight individual protocols when an attempt is made to construe what these people might have in common. Finally we will conclude with some observations about PCT as a social psychology that are based in what we have learned in this experience. Kelly's Fundamental Postulate is, "A person's processes are psychologically channelised by the ways in which he anticipates events". Given that the community we are dealing with sees itself as being designed for the *coming* renaissance society, this method seems well fitted to uncover some shared and personal anticipations of its members.

Mark

The charismatic leader of the Brotherhood community was Mark*, 21 years old at the time of our study. He relates stories of childhood mystical experiences which pointed to his future role as a spiritual guide. At 16 he hitch-hiked to California to join the Hell's Angels motorcycle gang because a magazine story about their "brotherhood" deeply impressed him. "They would die for each other . . . I said I'm going to do anything for that". He was quickly disillusioned by their violence and returned home in less than a year. Meanwhile he had tried drugs and when he returned home he became deeply involved with the drug culture. A vision and 35-foot fall from a tree house he had built and was living in brought him the realisation that he did not need drugs to attain spiritual consciousness and he determined to discontinue using them. Meanwhile he had met an older man, considered by many to be a mystic, and attended his trance-lectures. Elwood (Hapgood, 1973) introduced his young pupil to a seven-levelled world of spiritual consciousness and to his own spiritual writings. These greatly influenced the community's mythology.

Shortly after our extended visit to the commune, Mark and the group produced a profession of their beliefs in a local newspaper. Mark wrote:

> "I, Mark . . ., truly believe that I stand as the foundation of the new Church of Christ for the New Age . . . I have the natural ability to perceive Life-Force in its full manifestations of color, vibration, and light. There are many who have witnessed this upon occasion—I experience it continuously. This evolvement has been gifted from the life when I stood as well as I could with Jesus the Christ as his beloved disciple Peter [in another incarnation], and my life is dedicated in all its motivation to restoring the world to the brotherhood Jesus lived . . ."

Mark's picture and proclamation were followed by a full page containing photos and confessions of faith in him and the Brotherhood community by seventeen other members. The act of calling for a public commitment to Mark as a person is often done in the history of the Brotherhood. It has served to separate the truly faithful from those who waver and is a result of one of the ways in which Mark channels his own experience as well as those around him.

In response to one of our questions, Mark believes that by the time he is 65, "People will have come into a total acceptance of this spirituality as the way. We'll be into a spiritual revival or Renaissance where everything we don't know now will be part of it. It'll be heaven on earth." He himself will have brought in this power and will "probably be able to dematerialise and materialise so I can get to different places in this world immediately and be teaching [in a] no-limelight way," (i.e., he won't have to be in the limelight in order to teach).

* All names of persons are fictitious.

Given his choice, Mark would now prefer others to see and hear him in his role as lead singer with the band: singing, gyrating, screaming, throwing his body with pelvic thrusts in all directions, waving his arms, twisting and contorting his facial muscles, throwing his shoulder length blond hair into the air with twists of his head. Then one experiences the physical Mark. However, what is seen and heard on the physical level is the world of illusion: what one experiences at a spiritual level is the real. That is what the Brotherhood believes it is all about.

Personal construct analysis

An analysis of Mark's interviews demonstrates the following to be the personal constructs he uses most frequently to channelise his perceived world:

Emotion (Sexual)	*v.* Sensitive (Sympathetic response to sexual approach)
Active (Initiative)	*v.* Reactive
Strong	*v.* Weak
Male	Female
Hard hat	Gentle
Strong conviction	Influenced
Firm	Maleable
Superior (Powerful)	*v.* Inferior (Subservient)
High	Low status
Rich leader	Poor people
Stand With Me	*v.* Deceit
Total belief	Leave the group
Guilt	*v.* Balance

We shall now examine the ways in which this young man uses these bi-polar categories.

The most consistent construct Mark uses is one that appeared immediately when he began to compare and contrast. In Mark's words this is "emotional" as opposed to "sensitive". When comparing Brotherhood members Charles and Tim as different from Mary he said the two men were

> "emotional in a situation where a woman is involved" whereas Mary is "totally sensitive. She isn't into those worldly things. She would feel the emotion of Charles when he's getting emotional and expressing it, but not because she is really emotional. The same . . . with Tim; if he felt emotion toward her this way, she'll be emotional only because of that too."

Obviously the word "emotional" in this context expresses erotic feelings or sexual attraction and "sensitive" means a sympathetic response to that initiative. This construct could be more generally labelled "Active *versus* Reactive". In another sort comparing and contrasting two different men and another woman the latter was seen as reacting to the sexual and other material needs of the men (money and power) whereas the men actively initiated the relationship.

Strong *v*. Weak is another construct Mark employs. Subsumed under its poles are descriptions like:

"a hard-hat"	*opposed to*	"gentleness, can adapt"
"don't bend"		"balance of sympathy and firmness"
"a man channel . . . fanatical bravery"		"weak"

He used this dimension variously through several sorts with the result that a strength of will is an alternative to weakness and/or gentleness. Weakness and gentleness are considered equivalent and are positively valued. While comparing and contrasting, Mark frequently used karmic (causal) reasoning to explain these "facts" about persons. In one sort two persons had experienced many past incarnations as women which accounted for their "weakness". The contrasting person "stood only in a man channel" in past lives and had "exhibited fanatical bravery". He was so firm he was considered "out of balance" in this regard.

Building somewhat along these same lines, Mark revealed another channel which we labelled Superior (Powerful) *v*. Inferior (Subservient). Comparing his mother and sister he described how in other lifetimes the two had a master–servant relationship in which the present mother "totally blew up and hacked her [the daughter's, then a servant] head off or something". His father who was supposed to be contrasted against them "must have been something to my mother before [in another lifetime] that he totally respected her, maybe a servant but not that either, endlessly respected her . . ."

Obviously there are many ways to understand behaviour. Kelly's Fundamental Postulate states that we interpret the world based on the explanations and predictive systems (i.e., personal constructs) which we build and then maintain through validation. This is demonstrated by Mark's use of this construct (Superior *v*. Inferior) and how he sees it validated by his interpretation of events. It came into even sharper focus as he continued to compare and contrast his parents and sister. His mother, he said, "has been in a position of secondary power a lot, like rich, but never actually the person who saw somebody else make it and had some of the benefits.

> "She was always trying to make it, but always cut off. This [current] incarnation's been totally successful because she's busted out of that . . . For instance

[in the past] she might want to tell somebody off, confront them, but if their house or car were bigger and newer than her's, she wouldn't do it, just because of that. She is scared of the status."

In this same sort his father was

"so influenced by karma that he's never really lived his life. He was a Mohawk Indian in his last reincarnation. He just totally relates to that peace and tranquility of the Indian. If I could impress you I would impress you with this man living in the middle of the woods, where things were much more peaceful: no roads, no highways, living at ease with nature, the deer, hunting and stalking the deer, talking to the trees. Like you'll read in Hiawatha."

It should be noted that the "superior" pole of the construct carries negative feelings of fear and frustration while the opposite pole ("subservient") brings peace but a truncated fulfilment in life through being murdered, taken advantage of, or "not really living life". Thus Superior *v.* Subservient are two alternatives in a system which Mark uses to organise his world but neither end of the construct appears any too desirable. This may help account for the fact that he refuses to construe his role in the commune along this dimension and chooses instead to see himself as "guide" rather than leader. By side-stepping the problems this dimension would present him, he is free to interpret his role along other lines which allow him to perceive greater self-fulfilment.

By examining these last three constructs (Active *v.* Reactive, Strong *v.* Weak, Superior *v.* Subservient) we note how they often intertwine. His perceptions of persons shifts back and forth from one pole to the other highlighting perceived alternatives: power to subservience, strong conviction to not really living life, failure as opposed to total success. All this is set in other lifetimes so that karma provides causal explanations others may seek in drives, impulses, or needs.

An important element in the life of the Brotherhood is the parallel Mark draws between himself as the "guide" of the commune and Peter the Disciple, whom he was in another lifetime. The construct which subsumes this identity is labelled Stand with me *v.* Deceit. Centuries ago the church resulted from persons who stood with Peter; the community came about today because of people who stood with Mark. "The thing that stands out most in Peter's life is his denial of Christ. But I know I totally denied him . . . so that lives in me too." In his lifetime as Peter he was deceitful for not standing with his leader just as today many are "deceitful" toward him.

In the experience with the community Mark considers an incident at one of the earlier locations of the commune to be one of the most significant in the history of the group. The problem arose as to who could command the loyalty of the group, Mark or the owner of the property and cabin the group was living in at the time. Mark demanded a blind faith in himself and asked those who believed in him to leave the warm house and good food. He could promise them none of this. Most chose to follow him and not the landlady who could provide the physical comforts.

This particular construct of total commitment to Mark as opposed to "deceit" has been adopted by the entire commune as an effective way to build its boundaries (Siegal, 1970; Kanter, 1972; Mead, 1970; Klapp, 1969). Operationally, it translated into a construct we labelled Belief *v.* Leaving. Not infrequently members are asked at general meetings to raise their hands if they have "any doubts about what we're doing here or about Mark". On at least one occasion, which we witnessed, the several people who did have doubts were told to leave *that day*, work out their doubts, do what they felt they had to do, and then come back. Everyone said, as is always said whenever anyone leaves, "they'll all be back". It operated effectively to push doubts out of conscious consideration for most members, to divide Brotherhood members from people in the world, and to divide the "totally committed" core groups (the Band, "the Twenty," "the Forty") from the rest of the members who "are trying . . . growing". Mark described his experience relating to the core groups as opposed to ordinary members as being the difference between "talking to aliens as opposed to . . . these people who stand close to me". He understands that his job as guide is to separate out the people who want to stop growing in spiritual awareness and have them leave the community. After they "balance out" their lives they can return. This may take twenty years or more. Meanwhile "the people in the community must continue to expand with me".

This is undoubtedly one of the constructs Mark used earlier in life when he was attracted to Hell's Angels. We have already seen how he found himself attracted by their ideas of close brotherhood—they stood together. Since he could not *totally* accept their way of life he left.

Guilt and the functions of group constructs

In George Kelly's theory, guilt is the perception that a person is dislodged from core role structures, that is, dislodged from the networks of constructs that serve to predict and control essential interactions with others. Obviously an individual has much at stake in maintaining perception of self in line with core roles, for invalidation or loss of one's perceived social role brings guilt. In this perspective the concept of guilt is broadened from its commonly used meaning dealing with transgression of a moral order to describe the consequences which follow acting out of role. The material we have been considering provides occasion to examine how this can operate. For example, when Mark interprets behaviour through the filter of "power" or "active" he associates it with guilt. Commenting on the fact that his mother was always trying to get ahead as well as be able to confront others in past lifetimes as well as this one, he noted that she felt guilty and that was why she over-compensated when relating to his sister in this life. After describing Charles and Tim as "totally dedicated to what they believe in" and labelling Charles' total dedication "hard hat" he observes that "the only problem is they're guilty about how they express it". Guilt, for Mark, is experienced when he interprets behaviour as firmness or active pursuit of power or a goal. Self assurance too is usually seen as an ego trip where something needs to be "balanced".

There is, however, an apparent contradiction here, for Mark and the band work tirelessly to project the image of a "superstar" and perfect their music. They proudly relate that the Krishna voice has predicted the group will become more famous than the Beatles. The guilt this blatantly worldly activity might engender is filtered through a group construct Real *v.* Illusion, so that what the outsider observes as ego involvement is labelled "illusion" and what Mark and the Brotherhood members see, hear, and experience is "real" and spiritual. Just what the group constructs mean in the communal setting will be discussed shortly. Presently we can begin to understand one way shared constructs function. They offer alternate channels to interpret selected activities. In this case a core role will not be violated and result in guilt. The extent to which these external group constructs can be internalised and used by this individual will undoubtedly aid in the removal of guilt feelings. But it is interesting to note that neither of the constructs used here (Real *v.* Illusion, Spiritual *v.* Material) were elicited in Mark's individual protocol. Du Preez (1972) has noted that constructs of key political leaders in a society may become public constructs. If they are shared, they may define alternatives for the population. Given what we have presented, these alternatives may more readily find acceptance for personal use if they serve special personal needs such as help avoid guilt, i.e., help one perceive congruence with one's core role. It is an interesting area for further investigation.

This has special significance for clinicians and those working in subcultures. Isaacson and Landfield (1965) found that one's personal language held greater meaningfulness for self understanding than other kinds of language. The present material suggests that if the person adheres to an ideological (e.g., political) or mythical (e.g., religious) explanatory system, such a system might offer alternate constructs for a therapist to employ in attempting to help clients experiment with alternatives. The obvious problem is that these alternatives may seem even more "God-given" than the original personal constructs (Berger and Luckman, 1967; Du Preez, 1972) and make extension of a constructural repertoire even more difficult.

Mind altering drugs and the development of personal constructs

An hypothesis to consider concerns Mark's drug experiences in California. When he was tripping there he found that he was becoming paranoid, so much so that after entering a room "the whole place was down on me". We can see one end of the construct (deceit, against me) becoming dominant. This provided a powerful negative experience for this teenage youth who was away from home and disillusioned with his dream. Consider for a moment the poles of this construct: Stand with me *v.* Deceit. His ideal, the Hell's Angels, had not fulfilled his hopes to stand together in brotherhood and he was forced to the opposite end of his construct where people are really against you. The drug experience magnified this.

Those who use mind altering drugs or are familiar with their effects say that a person should never use them when depressed or "down". The drugs tend to emphasise the mood, good or bad, at the time they are taken. In terms of PCT we can hypothesise that the drug experience might be seen as highlighting submerged poles of the personal constructs which the person has chosen to bury. If the experience is powerful, as Mark's was, it could serve to strongly validate a particular construct as a channel for perception and behaviour and make it more difficult for the person to experiment with other constructs and/or develop new ones. Thus the drug experience, like other traumatic experiences, could serve to stifle the development of a large repertory of personal constructs helpful in facing the varieties of life.

On the other hand the drug experience can be seen from another perspective. When one begins to plot shifts in the construction processes a dimension labelled *looseness–tightness* is relevant. When explaining how it feels to think loosely, Kelly explains that one thinks loosely in dreams. The loosening releases "self-evident" facts from their rigid conceptual moorings. This opens up the possibility of realising experiences in new patterns and obtaining wholly new insights.

One of the important sequences of constructions which persons use in order to meet everyday situations Kelly labels the "Creativity Cycle". When this is employed in the reconstruction of a person's life, the person starts with loosened construction and terminates with tightened and validated construction. He begins by showing a shifting approach to problems, experimenting minimally with transient variations, then seizes upon one of the more likely ones, tightens it up, and subjects it to a major validational test. In fixed-role-therapy, the therapist tries to help the client work through creativity cycles by releasing his imagination and then harnessing it. Each new construct becomes the blueprint for a novel experiment which the creative person tests. Extensive use of psychedelics will produce dream-like experiences which suggest infinite alternate possibilities for restructuring personal worlds much like

Don Juan (Castaneda, 1969). For some this experience also invalidates previously significant constructural elements and validates new ones. For example, the illegality of these drugs alone makes users experience, possibly for the first time, the alternatives of Criminal *v.* Lawabiding, and see these in a very different light.

Once the constructs were loosened for many youth of the 60's there was no fixed-role therapist to help create experiments with alternate constructions. However when these "new criminals" compared experiences, it was religion, expecially Eastern religions which seemed to provide the most appropriate frameworks to construe the strange new feelings and experiences (Needleman, 1970; Robbins and Anthony, 1972; Zaretsky and Leone, 1974). Consensual validation of a group helped to structure the new and often exciting experiences. Canopies of meaning (Berger and Luckman, 1967) were provided into which experiences could be fitted. Such was provided by the myth evolved by the Brotherhood. Mark believes his experiences and spiritual beliefs to be true for three reasons: (a) external consensual validation, (b) they proved out and (c) internal personal experience. Mark spoke of his mystical experiences as "real" because they influenced people who were considered by everyone to be sane (Groutt, 1973, p. 281). These persons were both outsiders and members of the community. Secondly, events foretold in his visions were validated in how things had worked out for him and the group. The visions and myth explained why and how things were happening. Thirdly, his personal spiritual experiences were profound.

William James (1902) noted that "mystical states . . . are absolutely authoritative over the individuals to whom they come". They do not contain sufficient authority to force others to accept their revelations uncritically. They do, James writes, open up the possibility of other orders of truth. Certainly for Mark and the Brotherhood members these experiences provide the most important pathway to truth. They have their own criteria for validation and offer an alternative approach to living and understanding in the world.

Thus we have entered into the construed inner world of a young man who claims the spiritual allegiance of many persons. His view of the world affects their lives as well as his own (and his neighbours!). When we began to analyse the function of the imagery we find a person validating and testing his hypotheses just as other folk do. The social results are not greatly unlike what happens when one calls himself "President", "King" or "Pope". They too receive consensual validation for the role, and a national, world, or religious society is organised which uses these constructs in an operational way. Each of these roles evidence an alternative way to construct "reality" and each in turn *becomes* the reality by means of personal and social constructions.

Common Understandings?

The Brotherhood's world view, translated into construct form, yields the following list of shared contrast terms:

Spiritual	*v.* Material
Brotherhood	*v.* The world
Real and natural	*v.* Illusion
Positivity	*v.* Negativity
Feeling high	*v.* Feeling low
Mind	*v.* Brain/Intellectual
Letting go	*v.* Holding on

These terms and contrasts are culled from the language, conversation, writings, stories, etc. that were observed and experienced by the authors. They are shared in the sense that the whole community used the terms constantly, but no precise frequency counts were made. They are called shared contrast terms because technically they are not elicited, personal or group constructs. What this listing does that a simple Brotherhood dictionary would not do is to display the terms with their contrasts and suggests not only what the group is choosing but also what it is rejecting when making those choices. A brief example of this is the general subject of drugs. The Brotherhood tends to construe drugs as part of the "illusory, material world". It is the logic of the group to "Let go" and "Feel high" without drugs. Recall the signs posted at the entrance mentioned in the general description of the Brotherhood. Using drugs, including alcohol and tobacco are grounds for expulsion.

These terms have some particular functions within the group. Hierarchically, Spiritual *v.* Material is a most important contrast and the idea that in the spirit one finds the essential, real, unified experience is reflected in the group's names for the community, the band, and their newspaper: *Brotherhood of the Spirit*, *Spirit in Flesh*, and *Free Spirit*, respectively. Real *v.* Illusion is a common analytic tool. Most of the outsider's common sense reality is seen by the Brotherhood as illusory, but it is through such "illusions" as the rock-band that the Brotherhood can relate to (and possibly convert) outsiders who will eventually see that the rock-band is not a rock-band at all but the "most pure channel of creative energy" i.e., almost completely spiritual, natural, and real. In conversations among group members, feeling high or low is frequently mentioned to describe the person's general state. The exhortation to "let go!" (or the accusation of "holding on") is most often heard during interactions between established members and initiates, perspective members, or sympathetic visitors who are all considered potential converts.

Positivity *v.* Negativity is a kind of "weather" term used to describe the general atmosphere of the day or setting (the Band is seen as exuding positivity), but the level of positivity is much more important to the community than the barometric pressure, precipitation, or cloudiness, This particular contrast is somewhat elusive. The words are heard often in daily conversation, but in fact not once did the terms positivity or negativity appear as eight members made their individual comparisons and contrasts. A possible explanation is that the terms are so abstract and general that the contrast is difficult to use in describing people. Thus it is a type of meta-construct at a second or third level of abstraction appropriate to use when talking about the world in general or "vibes".

When the listing is taken as a whole, and taking into account the strong positive and negative values the community associates with these contrasts, it can be construed as representing the major elements of a single evaluative dimension (Osgood, Suci and Tannenbaum, 1957). Rosabeth Kanter (1972) notes that successful nineteenth-century communities tended to emphasise a clearly negative view of the outside while at the same time conceptualising the community in positive images. One is reminded of the strong group identifications made during early adolescence in the process of identity formation (Erikson, 1968). Self definition is usually accompanied by defining what one is (like the in-group), and what one is not (not like the out-group; Newman and Newman, 1975). Siegal (1970) sees this type of phenomenon as characteristic of any group that is proposing deep, cultural transformations as opposed to superficial corrections or changes. In some ways, then, this listing can be considered a psychic boundary language which tends to reify limits and provide members with a distinctive new identity. In any case, much activity within the Brotherhood community is channelised by an evaluative construct.

Studying this matrix of the group world (Kelly, 1962) leads to the following understandings: the materiality of the world is to be avoided in preference to spirituality which is found in the community. In the Brotherhood one lets go of old ways, especially an analytic, intellectual approach to life; one follows feelings. The true reality will be gained in this way and the material world will be seen for what it really is—illusion. Members grow in positivity as they increase their spiritual awareness and they know this is happening when they feel high.

It is important to note again that the above listing was not inductively constructed from the personal constructs of individual members. In what follows an attempt is made to take three of the dimensions from this table and evaluate the degree of sharing evident in eight members' individual constructions. These groupings are, in a sense, our constructions of their constructs. Bannister and Fransella (1971, pp. 155–56) have noted that ". . . in practice it would be extremely difficult to make comparisons across group members without standardising the verbal labels". Yet this is what we have tried to do.

There are some difficulties in detecting shared meanings with this method; the personal aspects continue to stand out. Hans Toch (1965, p. 161) has noted in regard to social movements that "... *each person joins a somewhat different movement*. Every person's perception of the appeals of his movement is partly a reflection of his private concerns and interests."

Brotherhood community v. The world. As a first example, Table I is a complete list of the personal constructs of eight individuals which we have subsumed under the shared contrast of Brotherhood *v.* The world.

TABLE I. *Individual Constructions of a Shared Contrast.*

Member	Brotherhood Community	*v.*	The World
A	Those who stand with me		Those who deceive
B	Insider		Outsider
C	Close friendship		Merely together
D	Evidence of total belief		Holding back
E	Companionship		Solitude
F[a]	—		—
G	Communicate with easily		Communication only with difficulty
H	Companionship		Sexual relationship

[a] No personal construct was elicited from member F which could be coordinated with this grouping.

The general interpretation of the Table would follow the form: Member E sees the Brotherhood Community as a place where companionship was found, whereas on the outside he perceived loneliness.

As a group these individual constructions, elicited by comparing and contrasting many very different persons and living situations, denote a strong desire for *gemeinschaft* rather than *gesellschaft* (community as opposed to society; Schmalenbach, 1961). While the objection might be made that our population biased the "results" in this direction (i.e., all had already chosen to live in a group seeking to establish primary relationships), it was never suggested to the member that this was an expected category. The fact that seven out of eight members used some form of this construct to compare and contrast friends from their past life, family, and commune members allows us to infer that for them the greater society was failing to provide for this need. Phillip Slater notes that the fear of radical movements in the larger society arises from the attraction they offer in three need-areas so often frustrated by the larger society: *community*, *engagement*, and *dependence*. While he sees these as "secondary needs" his thesis is that when society fails to meet them they can and will become primary, and social reorganisation (e.g., the communal movement) will be necessary to satisfy them. Thus, channels which are personally important to the individual for organising his life are built up into shared channels for organising lives in a community.

A further observation concerning these constructs used to distinguish the group member from the world is the fact that while it does indicate a sense of alienation it does not carry overtones of rebellion. This stands in contrast to some of the other groups studied where rebellion and revolution are central aspects and also in contrast to the view of many outsiders who see all communes harbouring a lot of rebellious kids. This group, especially, impressed the observers as a case of old wine in new wineskins. That is, the personal constructs of the members were those which parents and other "normal" people could understand and relate to in spite of the new form of religion so strange and upsetting (especially to the parents) that they embraced.

In the above listing the first three members (A, B, and C) are more tenured in the community and are in fact part of the leadership. Taken as a subgroup it is of interest to note that while two of these people did have a leadership-type construct available (see Groutt, 1973, p. 403–5) the fellowship or Brotherhood construct seemed to carry more weight. The founder-leader, it will be recalled, sees himself as a reincarnation of Peter the Disciple. The way in which he interprets his spiritual connection with Peter is not so much in the leadership role, although that is one aspect for him. More importantly, he sees people standing with him or treating him with deceit. And put into operation this construct culls the wheat from the chaff.

Note that while on the side of the Brotherhood Community there is some similarity among the poles of the individual constructs there is considerable variety at the other end. What each of the people was moving away from in joining the community is quite different. A is avoiding those who deceive; C those superficial groupings that occur simply by virtue of being in the same place at the same time. E rejects being alone, G people he has difficulty communicating with, and H relationships based only on carnal sex.

Spiritual v. Material. A second group dimension, together with the individual constructions related to it, is outlined in Table II.

The member who did not express any construct built directly around "spiritual" subsumed it within his construct of Insider *v.* Outsider (see Table I.) For him a sense of spiritual ("something more") was felt when people were close together in the Brotherhood community and this could not be felt when one was outside or perceived oneself as outside of the group.

Member H felt that the Brotherhood's notions about the Spiritual gave people a false sense of security. In other words she saw this as a group defensive manoeevre, a way to run away from reality. It may not be surprising that she finally withdrew from the community with a certain contempt. For her it was not a place that she could find "real stability".

One can see the distinct ways that each individual experiences what he or she calls "spiritual". There is some agreement that it is important to get beyond the phenomenal and surface elements of life into the true reality of the unseen. Beyond that there is little agreement. For D spirituality is to "feel as though you were with people" and happiness. To A it means that he can view his participation as a band-leader and rock-singer as a spiritual enterprise in

TABLE II. *Individual Constructions of a Second Shared Contrast.*

Member	Spiritual	*v.*	Material
A	Spiritual		Illusion
B[a]	—		—
C	Depth		Superficial/Surface
D	To be with people/Happy		Shattered/Unhappy
E	Change/Learning		Stability/Naivete
F	Knowing/Certainty		Intellectual/Doubt
G	Spiritual		Need for comfort
H	False security		Real stability

[a] No personal construct was elicited from member B which could be coordinated with this grouping.

spite of the fact that all other bands and rock-singers are considered to be wasting time with illusions. (A prospective member "caught" listening to a Joni Mitchell record was told to turn it off because it "wasn't relevant to what we're doing here"). To C spirituality means getting to know himself and the world better (Depth), whereas to E it carries with it the idea of change, learning, and growing beyond the naivete of formerly held beliefs. For F it means certainty, and for G it connotes that one is beyond the need for material comfort; that he did not attain that level of spirituality might be one of the reasons that he too left the group a short time after the participant-observer phase was completed.

The fact that the word "spiritual" is believed to mean the same thing to all members hides the greatly differing personal meanings which this central element of the Brotherhood world-view connotes to individual members. Almost invariably, this paradox—that a supposedly shared idea had a different meaning for each individual—was not recognised by the individuals. As such this central construct takes on the quality of a myth, as described by Rollo May (1960, p. 24) ". . . which lends form and unity to the culture" and is a means of transcending the immediate situation. Following the thought of Berger and Luckman (1967, p. 98) we can say that it is in this myth that "the symbolic universe provides the highest level of integration for the discrepant meanings actualised *within* every day life in society. . . ." At the level of myth "all reality appears as made of one cloth" (Berger and Luckman, 1967, p. 114). The PCT methodology provides a way of determining how this dialectic between the specific and general operates (see Groutt, 1973).

Letting go v. Holding on. Six of the eight members have a construct dealing with the Brotherhood's idea of change (Howard and Kelly, 1954). We have subsumed these under the shared contrast terms, Let go *v.* Holding on in Table III.

TABLE III. *Individual Constructions of a Third Shared Contrast.*

Member	Let Go	*v.* Hold On
A	Reactive/Passive	Initiate/Active
B	Let go and form bonds	Hold on to past and remain outside group
C	By natural means	By drugs
D	Free/Unconventional	Inhibited/Traditional
E	Learning	Naivete
F	New consciousness	Intellectual
G[a]	—	—
H[a]	—	—

[a] No personal constructs were elicited from members G or H which could be coordinated with this grouping.

We have already indicated that two of these members left the group and that these members were G and H. Neither had an elicited personal construct that was compatible to the group's idea of letting go. Most of G's constructs (Groutt, 1973, pp. 368–81) are built around the ease or difficulty in relating and he did not seem to conceptualise himself as able to change and undergo the resocialisation that the Brotherhood requires.

H's perspective was somewhat different (Groutt, 1973, pp. 312–96). She alligned herself with an active, problem-solving end of a personal construct and rejected the other end which includes, for her, ideas like "tranquility, peace of mind, and boring". In the Brotherhood's ethos, (and in Mark's personal construct system), "Letting go" has a very submissive aspect to it—you let go in order that you can experience (have done to you) what is out there (and in there) trying to happen. You submit to the spirit world and take the attitude of "do with me what you will". The Brotherhood views problems as illusions—when one has attained perfect spirituality there are no problems—and H sees such a state of affairs as dull and boring. (In fact, however, there are indications of slot rattling along this continuum in H's protocol.) Also in regard to the idea of change is the fact that H exhibited one of the more extensive repertoires of personal constructs. Within a PCT framework, therefore, she had more alternatives open to her; but not in the Brotherhood's way. She viewed the Brotherhood as limiting freedom, "Being swallowed up". The Brotherhood saw as illusion that which she considered creative and exciting, a set of problems to solve. Thus we are able to see, in terms of personal constructs, why two persons (G and H) were unable or unwilling to be resocialised (changed) and adapt to the Brotherhood Community. One could not see himself changing, the other's ideas of change were not compatible with the community's notions.

Conclusion

What does Kelly's system add to the study of a community like the one we've considered here? First of all the typical kind of ethnological description is not lost (Groutt, 1973; Groutt and Karst, 1974) and, at the same time, given a facet it might otherwise not have. And secondly, the communal movement seems a most appropriate social phenomenon for PCT analyses. Joining such a group is a rather dramatic behavioural statement and a prediction that a significant change in life style will somehow serve to elaborate one's world-view. The notions of "alternatives, search, and discovery" are an integral part of the general movement and of PCT. Paul Tillich (1967, p. xxi), commenting on the functions of Utopias, points out that the fruitfulness of utopian ideas is their ". . . ability to open up possibilities". And, indeed, the communal movement might be seen as a type of fixed-role-therapy for society—not too threatening since society as a whole does not have to change all at once. Within the movement, alternative ways to view and organise social relationships are made available. Communes do in fact operate as fixed-role-therapy for certain individuals and one might say that this is one of their latent functions (Merton, 1957). For some participants, the new, emerging constructs are found to be elaborative and functional while for others, like the two members who left the Brotherhood, the new roles are found unhelpful, at least in their entirety, and are discarded.

Thirdly, the study of individual community members provides insight into the psychological processes that lead to such actions as starting or joining a new social group while "dropping out" of the dominant society. What is illustrated through the PCT analysis is how active this process is, that it is not simply a "dropping out" but a movement toward experiences predicted to be more meaningful. We have also seen how one individual's constructions (Mark's) strongly influence the group's public ideology, while a good deal of variability is still present at the level of personal constructs.

The latter point is of some significance. We set out to make some comparisons between individual group members' personal construct systems. The few studies of groups that have made a similar attempt have always provided standardised verbal labels for such comparisons thus assuring a great deal of surface "commonality" (see Bannister and Fransella, 1971, pp. 99–124; 155–58). While it is, at least partially, a function of our method, we were still impressed by the uniqueness of the personal constructs elicited from a group of people living together in a highly ideological community ostensibly striving for a great deal of conformity in personal beliefs.

A common finding was an overlap on one pole of a construct but not the other. How much does person A have in common with B if person A's construing of people is on a dimension of Warm and loving *v.* Cold and indifferent, while person B uses the dimension Warm and loving *v.* Out to get me? In this situation communication and even understanding is possible when A and B are talking about people they love and feel close to. But when they are talking together about people they don't like they are really talking about much different experiences. When the eight Brotherhood members talk together about the comradeship they feel, they are probably communicating meaningfully; but when they talk among themselves about the world outside, the depth of understanding and communication is in doubt. Groutt's (1973) analysis of this phenomenon within the Brotherhood and another communal group leads to the conclusion that this type of "one-sided commonality" is the stuff that makes up a group's ideology or the group myth. But this is an interesting general issue pointing to degrees of commonality and will need to be faced both at a theoretical and methodological level as a PCT social psychology is developed.

Lastly, and in our case after the fact, the study of a community from this point of view brings to awareness the complex issues surrounding Kelly's views of the relationship of the person to the group. His most detailed analysis of the group process was in connection with therapy groups (1955, pp. 176–79; 428–51; 1158–78), and an individualistic theme, as opposed to a social perspective, is apparent. As Bannister and Fransella (1971) have noted, for Kelly the fullest development of a therapy group was for an individual group member "to stand on his own feet without group support (p. 155)". This individualism, possibly influenced by Kelly's own rural, midwestern, conservative-religious upbringing (Rychlak, 1973, pp. 471–73), forms a background of his critique of communal groups such as college fraternities, monasteries, and collectives. While such groups offer secure and highly specific role relationships they do not point to nor tend to allow this last step—independence from the group (Kelly, 1955, pp. 1174–75).

In a more technical sense, the study puts into focus the relationship between the commonality and sociality corollaries.* People can be construed as

* See Salmon (1970, pp. 205–07) for a discussion of commonality and sociality as these relate to issues in developmental psychology.

psychologically similar to the extent that they use the same constructions of experience (Commonality). A person can play a role in a social process (relate to another person) to the extent that he is able to construe the construction process of the person he is relating to (Sociality). Kelly (1955, p. 95) brought these two aspects together in a short discussion of social psychology in which he stated "... social psychology must be a psychology of interpersonal understandings, not merely a psychology of common understandings".

A PCT analysis of social relationships, groups, or communities, then, is a function of both commonality and sociality. One's first impulse is to categorise groups (from the point of view of an outside observer) as to "how much" commonality and sociality are observed in the groups' operation. Such a description would be a related but a more systematic alternative to such traditional sociological categories as *gemeinschaft*, *gesellschaft*, and communion (Schmalenbach, 1961). A community, such as the one considered here, might be considered a group showing a high degree of both commonality and sociality whereas other groups might show high degrees of commonality but low degrees of sociality, etc.

Yet such a categorisation assumes that commonality and sociality are to be construed as independent. Indeed, commonality can occur completely independent from sociality. Person A can be construed as similar to person B to the extent that A and B use the same constructs, even though A lives in England, B the United States, and they've never met. To what extent can sociality occur independent from commonality? In the special instance of a therapist we have a person, supposedly skilled in role relationships, able to empathise with (construe the constructions of) a wide variety of people including people much different than himself; and this would be a case of sociality somewhat independent of commonality. To carry this point further, a person can relate to (play a role relative to) his dog; e.g., the owner construes the dog as hungry and feeds it: the dog then eats, indicating that its construction of the world, at that point in time, had been correctly interpreted. The dog, in turn, relates to its owner, for example when it construes a small part of its owner's world and responds to a command to "sit" or "heel". Over a long period of time a dog and its owner might develop a great deal of sociality in their relationship. But can the owner develop a communal relationship with his dog? We think not—partially because of the very little commonality, in the PCT sense, that is present in this relationship. In a PCT social psychology, however, the nature of the *interaction* of these factors will need to be elaborated. One would explore how commonality potentiates sociality and vice versa.

We are left, then, with some uncertainty as to the relationship between commonality and sociality and a construction of social psychological phenomenon that is, perhaps, a bit too psychological or individualistic (i.e., the "role" from the point of view of one individual at a time). In retrospect, our work with communes puts these difficulties into sharp focus, and future work with groups will be aided by a more systematic theoretical analysis than we undertook. What seems to be needed is a system to construe not just a single role relationship but a mutual role playing process—person A's role taking relative to person B and person B's role taking relative to person A—*at the same time*. Such reciprocal role relationships (cross-constructions) would be an interpersonal relationship in PCT terms, would be a function of both commonality and sociality, and would be focused on the processes *between* people (see Leman, 1970, cited in Bannister and Fransella, 1971, p. 123).

Two very recent reports that also deal with this problem are known to the authors. Bonarius (1975) has worked out an intricate and successful methodology for dealing with diadic interpersonal relationships within a PCT context, and it seems possible, in theory, to expand this to larger groups. The resulting complexity, however, is a bit frightening. Sarbin's (1975) recent critique of PCT implies an alternate and more radical approach: he suggests that the construction of man as "Actor" is superordinate to the construction of man as "Scientist". Such a reconstruction would seem to have a certain utility in the analysis of social psychological phenomenon. Kelly (1955, pp. 588–89) notes that a therapist needs to be able to accept a client's constructs, but ". . . the psychotherapist should be prepared to fixate those constructs within a more comprehensive frame, a frame which he [the therapist] constructs within a more comprehensive frame, a frame which he himself ought to be able to provide. . . ." The personal construct theorist interested in social psychological phenomena will, like the therapist, need to find a comprehensive frame to construe such phenomena.

References

Bannister, D. (ed.), (1970). "Perspectives in Personal Construct Theory", Academic Press, New York and London.

Bannister, D. and Fransella, F. (1971). "Inquiring Man", Penguin Books Ltd., Middlesex, England.

Berger, P. L. and Luckmann, T. (1967). "The Social Construction of Reality", Anchor Books, Garden City, New York.

Bonarius, H. (1975). "The Interaction Model of Communication: from Experimental Research to Existential Relevance", Paper presented at the 1975–76 Nebraska Symposium on Motivation, Lincoln, Nebraska.

Castaneda, C. (1969). "The Teachings of Don Juan: A Yaqui Way of Knowledge", Ballentine Books, New York.

Du Preez, P. (1972). The construction of alternatives in parliamentary debate: psychological theory and political analysis. *S. Afri. J. Psychol.*, 23–40.

Erickson, E. H. (1968). "Identity, Youth and Crisis", Norton, New York.
Frank, J. (1973). "Persuasion and Healing", Johns Hopkins University Press, Baltimore and London.
Groutt, J. W. (1973). "Communal Ideology and Myth: Interpreted Within the Framework of Personal Construct Theory", (Doctoral dissertation, Temple University), University Microfilms, Ann Arbor, Michigan, #73–30, 155.
Groutt, J. W. (1974). "Reincarnation Beliefs Considered as Pathways of Personal Growth", Paper presented at the 1974 meeting of the Society for the Scientific Study of Religion, Washington, D.C.
Groutt, J. W. and Karst, T. O. (1972). "Inside Mystical Heads: the personal construct theory of George A. Kelly as applied to a religious–mystical commune", Paper presented at the 1973 meeting of the Society for the Scientific Study of Religion, Boston, Massachusetts.
Groutt, J. W. and Karst, T. O. (1974). "Brotherhood of the Spirit", unpublished MS.
Hapgood, C. (1973). "Voices of Spirit", Putnam, New York.
Holland, R. (1970). "George Kelly: Constructive Innocent and Reluctant Existentialist", *In* "Perspectives in Personal Construct Theory" (D. Bannister, ed.), pp. 111–132. Academic Press, New York and London.
Howard, A. R. and Kelly, G. A. (1954). A theoretical approach to psychological movement. *J. abnorm. soc. Psychol.* **49,** 399–404.
Isaacson, G. I. and Landfield, A. W. (1965). Meaningfulness of personal and common constructs. *J. Indiv. Psychol.* **21,** 160–166.
James, W. (1902). "The Varieties of Religious Experience", The Modern Library, New York.
Jonas, H. (1963). "The Gnostic Religion", 2nd ed. Beacon Press, Boston, Massachusetts.
Kanter, R. M. (1972). "Commitment and Community: Communes and Utopias in Sociological Perspective", Harvard University Press, Cambridge, Massachusetts.
Kelly, G. A. (1955). "The Psychology of Personal Constructs", 2 Vols., Norton, New York.
Kelly, G. A. (1962). "Europe's Matrix of Decision", *In* "Nebraska Symposium on Motivation" (M. R. Jones, ed.), Vol. X, pp. 83–125. University of Nebraska Press, Lincoln, Nebraska.
Klapp, O. E. (1969). "Collective Search for Identity", Holt, Rinehart, and Winston, New York and London.
Leman, G. (1970). "Psychology as science fiction", unpublished MS.
Mair, J. M. (1970). "Psychologists are Human Too", *In* "Perspectives in Personal Construct Theory' (D. Bannister, ed.), pp. 157–184. Academic Press, New York and London.
May, R. (1960). "Symbolism in Religion and Literature", George Braziller, New York.
Mead, M. (1970). "Culture and Commitment: Study of the Generation Gap", Doubleday, New York.
Merton, R. K. (1957). "Social Theory and Social Structure", The Free Press of Glencoe, Glencoe, Illinois.
Needleman, J. (1970). "The New Religions", Doubleday and Co. Inc., Garden City, New York.
Newman, B. M. and Newman, P. R. (1975). "Development Through Life: A Psychosocial Approach", The Dorsey Press, Homewood, Illinois.

Osgood, C. E., Suci, G. J., and Tannenbaum, P. M. (1957). "The Measurement of Meaning", University of Illinois Press, Urbanna, Illinois.

Robbins, T. and Anthony, D. (1972). Getting straight with Meher Baba. *J. Sci. Study Rel.* **11,** 122–140.

Rychlak, J. F. (1968). "A Philosophy of Science for Personality Theory", Houghton Mifflin, Boston.

Rychlak, J. F. (1973). "Introduction to Personality and Psychotherapy", Houghton Mifflin, Boston.

Rosenberg, S. (1975). "New Approaches to the Analysis of Personal Constructs in Person Perception", Paper presented at the 1975–1976 Nebraska Symposium on Motivation, Lincoln, Nebraska.

Salmon, P. (1970). "A Psychology of Personal Growth", *In* "Perspectives in Personal Construct Theory" (D. Bannister, ed.), pp. 197–221. Academic Press, New York and London.

Sapir, E. (1938). Why cultural anthropology needs the psychiatrist. *Psychiat.* **1,** 7–13.

Sarbin, T. (1975). "Contextualism: The Root Metaphor for Modern Psychology", Paper presented at the 1975–76 Nebraska Symposium on Motivation, Lincoln, Nebraska.

Schmalenbach, H. (1961). "The Sociological Categories of Communion", *In* "Theories of Society, I" (T. Parson *et al.*, eds), Free Press, New York.

Schutz, A. (1953). Common sense and scientific interpretation of human action. *Phil. Phenomenol. Res.* **XIV,** 1.

Schutz, A. (1967). "The Phenomenology of the Social World". (Translated by Walsh, G. & Lehnert, F.), Northwestern University Press, Evanston, Illinois.

Siegal, B. (1970). Defensive structuring and environmental stress. *Amer. J. Soc.*, **76,** 11–32.

Tillich, P. (1967). "Paradise and Utopia: Mythical Geography and Escatology", *In* "Utopias and Utopian Thought" (F. E. Manuel, ed.), pp. 296–309. "The Daedalus Library"; Beacon Press, Boston, 1967.

Toch, H. (1965). "The Social Psychology of Social Movements", Bobbs-Merrill, Indianapolis, Indiana.

Zaretsky, I., and Leone, M. P. (eds), (1974). "Religious Movements in Contemporary America", Princeton University Press, Princeton, New Jersey.

A Reconstruction of Emotion

Mildred M. McCoy*

Man the Scientist—his behaviour an experiment, a prediction about reality, his processes psychologically channelised by the ways he anticipates events; or *Homo Patiens*—passionate, suffering, caring, feeling Man. Must we choose one or the other of these constructions? Do these models lurk in incompatible sub-systems of our construction of the human condition? I propose that they are compatible and complementary; that the view of the scientist as objective, cold and unfeeling is our own construction. It imposes a boundary which we should transcend.

Kelly has reminded us that

> "a good psychological theory should be expressed in terms of abstractions which are of a sufficiently high order to be traced through nearly all of the phenomena with which psychology must deal. . . . If the abstractions are well taken they will possess a generality which will make them useful in dealing with a great variety of practical problems." (1955, p. 27)

In developing his own theory, Kelly saw the task as involving three objectives. He wished to enunciate a principle so basic that it would apply to each and all of the psychological processes, yet, it must have practical significance. At the same time, he wished to "alternatively construe" (i.e. observe recurring regularities from a fresh point of view) human behaviour. He found previous efforts, although initially having given good service, were no longer sufficiently useful. They were too static and unfruitful. Kelly's emphasis on a sufficiently high level of abstraction so that his theory could embrace the totality of the human experience of each human individual was one way to circumvent the dead ends into which Psychology's past efforts at dealing with practical problems had led. Let us briefly first construe his theory along the "sufficiently high order of abstraction" dimension he himself enunciates as being a way to discriminate Psychological theories.

* The author gratefully acknowledges the guidance throughout the development of this paper of Erik Kvan, Senior Lecturer and Head of the Department of Psychology, University of Hong Kong.

A High Level of Abstraction

The statement Kelly makes about human behaviour, i.e., that man is best understood in terms of an abstraction called *Man-the-scientist* whose ultimate aim is to predict and control his universe, and that he does this through the noting of replications in his experiences called "construing", offers us a wide open vista. There are certain landmarks and natural demarcations called the fundamental postulate and the eleven corollaries which amplify it, but the vista itself is an holistic approach to the understanding of man based on only one process, *construing*. Construing in its minimum form, involves making a distinction between two experiences on the basis of a dimension or aspect that is relevant to both as well as some third thing (event or experience) which can be said to be like one of the original ones. While construing can be tightly defined, it nevertheless represents a rather high-level abstraction of human behaviour. According to Kelly it can be applied to all psychological events and even physiological events, although it does not fit quite so well in the physiological realm as it does in the psychological realm, and particularly the psychotherapeutic realm, which is its focus of convenience.

It was Kelly's desire to devise a principle of sufficient generality that it would be useful in dealing with a wide range of human events and at the same time to back off from previous approaches which were no longer sufficiently productive. These two objectives account for the position he took with regard to previous efforts at psychological theorising. He did not find it necessary to repudiate other theories in order to begin from a fresh viewpoint. Without stating much more than his dissatisfactions, he chose *not* to use the traditional conceptualisation of experience which involves terms such as motivation, emotion and cognition. He had found these terms so loaded with a kind of subject-predicate reification that they got in his way. For too long, he claimed, they led to labelling, and labelling led to rigid thinking about them. He suggested that we need only look at the long lists of traits or needs or emotions which have been compiled in order to realise how little has been gained in terms of predictive efficiency from this approach.

Apart from finding the traditional terminology a stumbling block in the path of progress, Kelly also found the trio, motivation, cognition and emotion, did not measure up to the task they were given. He says, "Not only did it seem that the words man uses give and hold the structure of his thought, but, more particularly, the names by which he calls himself give and hold the structure of his personality". (1963, p. 56) Kelly believes that a human being is *more* than an assemblage of motivations, emotions and cognitions. His approach is rightly described as holistic. He said,

> ". . . in talking about experience I have been careful not to use either of the terms, 'emotional' or 'affective'. I have been equally careful not to invoke the

notion of 'cognition'. The classic distinction that separates these two constructs has, in the manner of most classic distinctions that once were useful, become a barrier to sensitive psychological inquiry. When one so divides the experience of man it becomes difficult to make the most of the holistic aspirations that may infuse the science of psychology with new life and may replace the classicism now implicit even in the most 'behaviouristic' research."* (1966, p. 140)

In essence, Kelly consciously abandoned the classical point of view, i.e., the trichotomy, called variously cognition, conation and affection, or intellect, will and emotion, or the more modern terms, thought, action and feeling. His reasons for this were twofold. He desired to take a view of man which would be of a sufficiently high order of abstraction to encompass all of human behaviour, avoiding both the pitfalls of labelling reification and the reductionistic man-is-a-sum-of-his-parts mentality. Secondly, he desired to take a fresh point of view, to alternatively construe the human condition. For example, he wished to deal with both what has been called the rational (cognitions) and the irrational (affection) in the same psychological terms. To do so, he needed to stop talking about each of these as separate entities and discuss what they had in common. The resulting abstraction gives a decidedly intellectual flavour to Personal Construct Psychology. In fact, the cognitive emphasis perceived in Kelly's theory is often singled out for criticism or as a defect which leads to placing too much emphasis on the human intellect. This may seem to be a misunderstanding of Kelly's intentions in view of the passage just quoted, but it is based not only on his statement of principle but on the multitude of illustrations of his therapeutic approach.

The "Too Cognitive" Criticism

Kelly's approach leads to PCP generally being classed as a cognitive theory of personality (Southwell and Merbaum, 1971; Mehrabian, 1968; Patterson, 1973, etc.). Mehrabian (1968) notes, "Kelly (1955) elaborated an almost completely original personality theory within a cognitive perspective". (p. 121). Patterson says, "Kelly's approach is . . . highly rational and intellectual in nature" (p. 323). He seems to criticise Kelly's psychotherapeutic approach for being too much like a scientific experiment, although he notes that in practice it may differ from its description. Rogers (1956) describes a PCP therapist, "He is continually thinking about the client, and about his own procedures, in ways so complex that there seems no room for entering into an emotional relationship with the client" (p. 375). Bruner (1956), despite

* Kelly is using classicism here in contrast to humanism. He says, "Classicism reveres the past; humanism is inspired by it. The classicist engrosses himself in the thought of other times; the humanist uses history as a fulcrum to pry himself loose from the prejudices of his own." (1966, pp. 133–134)

an enthusiastic overall acceptance of Kelly's point of view, expresses a vague dissatisfaction with what seems to him to be an over intellectualised approach. Reviewing *The Psychology of Personal Constructs*, he says, "The book fails signally, I think, in dealing convincingly with the human passions" (p. 371). Kelly says,

> "The psychology of personal constructs is built upon an intellectual model, to be sure, but its application is not intended to be limited to that which is ordinarily called intellectual or cognitive. It is also taken to apply to that which is commonly called emotional or affective and to that which has to do with action or connation. The classical threefold division of psychology into cognition, affection and connation has been completely abandoned in the psychology of personal constructs." (1955, p. 130)

Emotion in Personal Construct Psychology

The terms and the division they concretise are abandoned but the facts that the common term "emotion" refers to have not been banished from Kelly's scheme. It is clear he did not banish *events* which others might call emotional. In one example, he observes that,

> ". . . if we apply the scientist paradigm to man, we someday are going to catch ourselves saying, in the midst of a heated family discussion, that our child's temper tantrum is best understood as a form of scientific inquiry." (1965, p. 293)

Post-PCP children still have temper tantrums even though they are construed in a new perspective. It is important to make this point because events do have a way of being lost when one takes a new viewpoint. Perspective and degree of abstraction tend to cause some highlights to appear and other features to recede, even vanish into an undifferentiated background. This is not the case with temper tantrums and other emotional behaviour in PCP. They are still with us. But it is true they can be seen as having much more in common with other human behaviour than often is the case. Kelly seems to say they simply are not so special as to deserve all that rhetoric and attention from psychologists over the years.

Another argument for not considering PCP as a purely cognitive theory is that within the intellectual model based on the process of construing and the integration of constructs into a system, Kelly has redefined certain experiences which in other theories would be called emotional or antecedents of emotion. In other words, he deals with a set of emotions by name, albeit by giving them very special definitions within his system as well as assigning them a special therapeutic role. I am referring to what Kelly calls "the professional constructs" of threat, guilt, fear, anxiety, aggression and hostility.

Professional constructs are the framework and the dimensions which are appropriate for the psychologist/clinician to use in construing phenomena of the client's situation in which he is interested. They represent one of several ways in which Kelly has demonstrated the practical significance of PCP and its emphasis on construing as the basic psychological process. It is having nd using this set of professional constructs which distinguishes the PCP therapist from others of different persuasions. Kelly elaborated a substantial list of professional constructs under two categories. The *general diagnostic onstructs* describe the nature of constructs and their relations to one another. The category includes terms like "preverbal constructs", "submergence", "suspension", and "level of cognitive awareness", all of which refer to events that are commonly labelled "unconscious". It also includes terms such as "superordinate" and "subordinate constructs" and "loose" and "tight onstruction". The category of *constructs relating to transition* includes the six already mentioned as professional constructs, as well as a few others descriptive of regularly occurring cycles of the construing process.

In defining commonly used terminology in a special way, Kelly noted,

> "The commoner definitions of these terms are not abrogated by their being given limited meanings within our system. It is not our intent to preempt the words, and it is important for the reader not to assume that we have. . . . What we have said is that, within our psychological system—which is only a part of our total personal construction system—these terms are assigned restricted meanings." (1955, p. 489)

Additionally, Fransella (1972) points out that despite Kelly's special definitions it does not mean that what he called, for example, anxiety, was thereby different from what everyone else calls anxiety. Kelly's special definition is intended to enable a therapist to construe another's behaviour in a thera-peutically useful manner. It focuses on the construing process and relates to he reason for the experiences in every day life which one commonly calls anxiety without actually being very concerned about the precision of the decision and whether psychologists have reached a consensus about its essence.

Characteristically for Kelly, his definition is based on the phenomenological experience. For example, the immediate cause of anxiety is not an external stimulus such as the knock on the door by the bill collector or a "Dear John" letter. Rather it is the internal phenomenon, the recognition of the impact of a prediction one makes regarding the self in these circumstances. n actual practice, a PCP therapist observes a set of events in a client's behaviour commonly designated as anxiety. The PCP therapist construes this behaviour as a sign that the client's construction system does not apply to the events at hand. Kelly states, anxiety

> ". . . is, therefore a precondition for making revisions. Now it does not follow that the more anxious one is, the more likely he is to make an effective revision. Sometimes a person is so generally anxious that he spends all his time running around putting out small fires and has no time to design any fireproof structures.

> A therapist usually has to spend considerable time repairing minor anxieties s
> that the client can do something about the chaos represented by the major are
> of anxiety. Sometimes a therapist has to create anxiety in a certain area by bringin
> invalidating evidence to bear upon the defenses used there. Sometimes a therapi
> wastes time trying to bring all the anxieties under control in order to make h
> client comfortable." (1955, p. 498)

This does not sound so very different from many other discussions of anxiet in therapy. Yet, viewing anxiety in terms of an awareness of the inadequac of a person's construction system suggests various ways the therapist ma deal with it. It enables him to design a therapeutic programme around series of validating and invalidating events which will normally lead t construct system revision.

The emotion terms as redefined by Kelly enable the therapist to constru behaviour associated with construct system transition. While admittedl ignoring controversies over distinctions between various emotions whic have occupied the efforts of so many psychologists, Kelly's set of speci definitions not only attests to the relative comprehensiveness of PCP, it als indicates that PCP should not be regarded as exclusively intellectual.

Of the six definitions of emotion-related terms which we have singled o for discussion, four, anxiety, threat, fear and guilt seem to describe "emotion states" or to be the names of "emotions". The other two, aggression an hostility, describe behaviour which results from emotional states.* It shoul be noted that each of the emotion definitions has in common with the othe an aspect of an awareness state or the recognition of some fate of the constru system. There are two features of interest in that last observation which PC construction can subsume conveniently.

In other schemes there might be some difficulty with trying to communicat the notion of a state of awareness which does not necessarily involve co sciousness as that term is used to mean "in the forefront of attention" Consciousness is not an essential feature of construing. Discriminatior outside the range of convenience of the concept of consciousness are als construing if they meet the other criteria. This entire range of behaviours ca be subsumed as gradations along a construct dimension "level of awareness" This construct includes at one extreme such out-of-awareness events as th physiological discriminations involved in digestion. The automatic decisio of an experienced driver on his daily journey over a familiar route to wor represent a slightly higher level of awareness. The conscious choice of the mo informative word in a telegram is a sample of behaviour at a still higher lev of awareness. At the highest level might be the acute and heightened sense o awareness associated with mystical and some psychedelic states.

* These terms are all described in some detail later in this paper. See the glossary (p. 12 for a short definition.

The second feature of interest in the definition which is easily handled ithin PCP is the implication of two distinct stages, and a successive, cause-nd-effect-order occurring as the emotion experience. There is a state of wareness distinguished from the attempt at construing. However, awareness f construing is via yet another construction. It is not a sort of behaviour of class apart from other construing. As soon as there is an awareness-event hich is noted to be like some other and different from a third, it is quite gitimate to regard this also as an act of construing, and until then it is not ifferentiated.

The notion of a *succession* of construing is also natural to the PCP system. elly says, "One's construction system is never completely at rest. Even the hanges which take place in it must themselves be construed. The successive ates of one's construction system need to be treated as elements, and this alls for superordinate construction" (1955, p. 488). Professional constructs re superordinate dimensions which the psychologist can use to construe onstruct systems. We are dealing with a two phase situation. This is an ssential distinction. Being in a state of awareness of some fate of the construct ystem is the essential aspect which distinguishes some behaviour as emotional om other behaviour which is non-emotional. In so far as awareness of the ate of the construct system is construed, we can be said to be construing ersonal emotional events or experiences. So-called emotional behaviour is a ymptom that one construes one's own construct system in the process of eing used.

Kelly did not abandon emotional events; in defining in a special manner ertain events commonly called emotions, he has identified a basis for istinguishing them from other human events. It is an interesting challenge try to expand the PCP range of professional constructs to include additional motional events which we commonly distinguish.

But first, some justification seems in order. Kelly's expressed wish to bandon emotion as a separate category of human behaviour deserves some spect from his loyal beneficiaries. A glance at the plethora of confusions, ifficulties and disagreements abounding in psychological studies of emotion ems to confirm his wisdom. Yet beyond the "tendency for closure", a desire complete a pattern which alone could inspire an effort at expansion of the st of professional constructs subsuming emotions, there is a feeling expressed y the "too cognitive" critics that there is something of value missing in the CP model of human beings. Contrast Kelly's fundamental postulate "A erson's processes are psychologically channelised by the ways in which he nticipates events", with for example, a statement which appeared in a recently xposition of a theory of emotion, "The theme here is homo patiens—assionate, suffering, caring, feeling Man" (Tomkins, 1971). No doubt much f the difficulty lies in a false assumption that construing is primarily a cognitive process, yet the tactic of playing down other schemes of categorising behaviour without more completely reconstruing those events called feelings contributes to that false impression.

There have been developments in the last quarter century of work on emotion which justify a reconsideration of the desirability of reconstruing it within the PCP framework.

Sources of Confusion

There is no doubt that vast conceptual difficulties have been encountered in trying to deal with emotions. Chief among them is the question of which academic discipline has the best approach to their explication. Psychology hasn't been particularly successful. Explanations of this are rampant. Peters (1969) attributes the difficulty to behaviourism. He says,

> "It was the concept of 'palpability' that went with this particular brand of dogmatic methodism, with all its conceptual confusions and antiquated notions of scientific method, that both restricted the questions that psychologists felt they could respectably raise about emotions and that occasioned them to ignore the obvious point that we cannot even identify the emotions we are talking about unless account is taken of how a person is appraising a situation . . . the investigation of emotional phenomena has been hamstrung by a widely influential methodological dogma." (1969, p. 154)

Izard *et al.* (1966) see a source of confusion in the abundance of terms such as affect, emotion, feeling, mood, sentiment etc. which are sometimes used interchangeably and sometimes to identify separate and distinct concepts. Ekman *et al.* (1972) point to the difficulties in deciding which aspects of the experiences commonly called emotion are essential. The different factors which some authors have considered central to their definitions include:

(a) *A special class of stimuli* which usually elicit emotional behaviour, both those to which the response may be innate and those for which it is learned.
(b) *Physiological responses* such as visceral activity, other signs of autonomic activation, and hypothalamic activity have been viewed as relevant to some definitions of emotion.
(c) Some *motor responses* also have been regarded as central to the definition of emotional behaviour, chief among them facial expressions, tics, and restlessness and both voluntary and involuntary actions such as flight and immobility.
(d) Certain kinds of *verbal responses* have been part of some definitions. There include referential verbal behaviour describing internal states and the vocabulary of emotion names.
(e) Finally, it is noted that the *interactive consequence* of certain behaviour has been considered by some to be the necessary criterion for defining

emotional behaviour. This seems to arise in the quest for scientific precision and objectivity, particularly in studies of nonhuman primates where there is a sequence of events between two or more organisms in which a behaviour on the part of one is immediately followed by a particular kind of response by another.

The lack of clarity about what to regard as definitive of emotion probably accounts for some of the resistance to accepting any one definition. There seem to be too many unsettled issues, too much contradiction in the research results, and perhaps, faulty conclusions deduced on the basis of weak experimental design.

In addition to the difficulty psychology has had in deciding what is essential within the concept of emotion, there has been great confusion over how emotional experience should be categorised. Are individual emotions to be distinguished as discrete from each other, perhaps having the property of being additive or should emotions be regarded as comprising a continuum in which one emotion shades into another with the boundary between them remaining somewhat arbitrary?

In the face of such lavish disorder the relative simplicity of Schacter's cognitive-physiological theory (Schacter and Singer, 1962; Schacter, 1964) which boils down to "the interpretation of arousal = emotion" seems to have been extremely attractive and to have been widely accepted. This is the ultimate extension of the dimensional approach (Schlosberg, 1954) in emotion studies and puts one in the position of having to defend the merits of talking about emotion as anything other than learned labels for arousal. We must face the question of whether it is worth defending the idea that there are distinct emotions which are immediately recognisable or measurable responses common to mankind across cultures, races, sexes and other variables which might be pertinent. Can we talk about any one emotion as being innately different from another as a universal human experience? This is the area of fundamental disagreement between those who favour a "learning" theory of emotion and the adherents of a "biological" theory.

Innate Emotion

The question of whether there is a set of emotions, or a set of different expressions of different emotions which is *innate* is crucial to this discussion. Is it reasonable to regard a set of emotions as innate, or at least biologically based rather than completely learned? Is there a set of emotions, each of which has, or is, a specific pattern of response distinguishing it from others in the set? The approach does not preclude combinations of emotions. For example, Izard has identified a set of "fundamental emotions" and uses the term "patterns of emotions" to refer to two or more fundamental emotions in combination and interaction (1972, p. 24). The point is that if we now can confidently identify as distinct from each other a set of some fundamental emotions, and if now we have a clearer picture of the process we mean when we use an emotion-name, it makes good sense for a psychological theory to be able to talk about separate emotions.

Proponents of theories positing a set of discrete emotions, such as Izard (1972), argue that each emotion has separate adaptive functions and each adds a special quality to life experiences. Each emotion has relationships to certain non-emotional processes as well as (a) a specific, innately determined neural substrate; (b) a characteristic neuromuscular-expressive pattern; and (c) a distinct subjective or phenomenological quality; without any one of these three facets alone constituting emotion. A complete emotion process is said to require all three. Thus we have three areas of exploration in which to look for evidence of distinctions between emotions. Izard (1971) and Ekman, Sorenson and Friesen (1969) present evidence that some fundamental emotions have already been identified and defined empirically at the expressive and phenomenological levels, that is levels (b) and (c).

Research into the neuro-physiological aspect, level (a), is less conclusive at this time but some theoretical and empirical evidence supports the assumption that each fundamental emotion has its own characteristic nervous and biochemical substrate. For example, the two functional divisions of the autonomic nervous system, the sympathetic and the parasympathetic systems have anatomical differences which appear to contribute to some of their different functional characteristics. According to Izard,

> ". . . the parasympathetic system is relatively more segmental in character, being under less direct immediate influence of the CNS. This would mean that emotions involving predominantly the sympathetic system (e.g. fear) might be expected to respond more readily to a brain function like cognition (appraisal, memory, reasoning) than would an emotion (e.g. distress) involving predominantly the parasympathetic system. Indeed, our everyday experience tells us that fear usually passes more readily than distress. Distress, as seen in intense sadness or grief, changes relatively slowly. Similarly, clinical evidence indicates that fear-related emotional disturbances (anxieties, phobias) yield much more readily to relationship or predominantly verbal (cognitive) psychotherapy and conditioning–learning techniques than do distress-related disorders (e.g. depressions)." (1972, p. 10)

However, most of the experiments on the theme of neuro-physiological and biochemical differences have not had sufficiently careful controls to assure that the evidence was a function of a single emotion. Thus at present, the sturdiest argument for emotions being both distinct and innate is based on evidence of innate facial expressions of emotion and evidence that such facial expressions are not a consequence of emotion but actually an integral aspect of the emotion.

Izard (1971) provides a comprehensive survey of research on the question of innate facial expression of emotion. Most adherents of this position trace its origins to Darwin's (1872) attempt to provide evidence that facial expressions of the emotions are the result of a genetic element common to all races of man. This is the approach of much current research such as Izard (1971); Ekman, Sorenson and Friesen (1969); Boucher (1969). Darwin's approach seems to be vindicated. There are several emotions, the expressions of which and the interpretation of which are pan-cultural. Evidence seems to support seven such emotions but the addition of others to the set is not precluded.

An extensive body of research by Ekman and associates (1967, 1969) has focused on determining whether there are separate facial expressions associated with specific emotions. The unequivocal conclusion is that, "There is one fundamental aspect of the relationship between facial behaviour and emotion which is universal for man: the association between the movements of specific facial muscles and specific emotions. This has been found true for the facial appearances associated with anger, sadness, happiness, disgust, surprise and fear". (Ekman, Friesen and Ellsworth, 1972). Recently, interest has been added to the list making a total of seven pan-culturally distinguishable emotions.

This conclusion is contrary to the general trend in psychological thinking of the past 50 or so years. Watson was able to speak confidently about a set of innate emotions, but as research on emotions multiplied and confusions increased geometrically, it became tempting to assume that all research on the identification of facial expressions of emotion was a waste of time. The idea that the expression of emotion was learned and therefore culturally determined seemed to account for much of the confusion. Izard (1971) notes that psychology's struggle to become a respectable science has been a factor in the weaknesses of studies of emotion. Subjective experience was banished along with the technique of introspection. He believes that the lack of progress for so long was determined by three things: (a) the tendency of behaviour scientists to view emotion as a global, unitary concept, which makes operational definitions so difficult as to preclude its consideration as a researchable problem; (b) the dominance of logical positivism as a philosophy of science and of s-r drive-reduction principles in psychological theory; (c) the lack of an adequate theory dealing with separate and distinct emotions, each capable of definition as a construct that could be studied by specified and repeatable operations.

> "The net effect of these three influences was that the term emotion, while falling into scientific disrepute, was at the same time being forced into the semantic position of a vague catch-all for anything not covered by currently sanctioned constructs." (Izard, 1971, p. 3)

If any theory of emotion had any respectability at all, it was a learning account of emotional expression, of which there are various versions. Izard shows the serious theoretical and methodological faults of much emotion research in the 1930s through to about 1960, and points out that the conclusions drawn from this tainted data fitted beautifully into current beliefs of mainstream psychology. Small wonder that when Kelly surveyed that scene he wished to take a fresh point of view!

Why Change Tactics?

The emotion scene is changing. Although much of the chaos which Kelly faced remains, hopeful trends have appeared. Research in physiological psychology has contributed to a resurgence of interest in emotion (P. T. Young, 1961; Pribram, 1967). Likewise, cognitive aspects of the emotion experience have been reintroduced in various theoretical approaches (Arnold, 1960; Leeper, 1970). Efforts have been made to preserve the complexity of the human experience rather than deal with it reductionistically (Tomkins, 1962, 1963). Finally, there is the research indicating that with confidence we can accept a set of biologically based, discrete emotions.

Not only has the emotion scene become more amenable to PCP construing but the need for an expanded set of professional constructs has become apparent. Psychotherapists deal with clients' behaviour which is not adequately subsumed with the present set as offered by Kelly. What can we make of a smile or a smirk, of passion, dedication and a sigh of relief? Yes, in the face of our construct system's inadequacy, anxiety is the result. It provides the impetus for an extension of the psychologist's system of superordinate constructs. Given that Kelly has set the stage by especially defining some emotions within the PCP systems, that at least some emotion theory and research is at a more advanced and acceptable stage today than it was during the gestation of PCP, and that greater completeness of the set of constructs is both possible and will fill a need, the effort to expand this aspect of PCP appears worthwhile.

New Professional Constructs for Emotional Events

The four PCP emotion definitions: fear, threat, guilt and anxiety, have certain common aspects. All are defined as experiential phenomena. They are each defined in terminology which conveys the idea that there is a certain awareness of the fate of part or all of the construct system being at stake. In each case it might be said that the experience is towards sensing a certain problem with one's construct system for a prediction made or about to be made.

The four Kellian emotions involve variations of several dimensions including validation *v.* invalidation, comprehensive *v.* incidental core structure change, dislodgement of self from one core role structure, and outside *v.*

within the range-of-convenience of the construct system. It would seem that each dimension should be regarded as a construct. Therefore, it is bipolar. For some unexplained reason Kelly has not given symbol labels to both poles of these constructs. He is aware of this incompleteness (e.g., p. 489, 1955) but probably regarded as either pedantic or too peripheral to his main interest this particular carrying out of his conviction that a construct is always best interpreted as a discriminating dichotomy. It may also be that in focusing on change he failed to explore construct system stability.

The expanded set of emotions based on models offered by Kelly in defining threat, fear, anxiety and guilt stems mainly from variations in the dimension categories inherent in defining those four terms. But, I was also interested to have the list include the fundamental emotions identified in current research as biologically based and universal. Tomkins (1970, 1971), Izard (1972), and Ekman and associates (1969, 1972) present convincing evidence for a basic set of seven such emotion categories. The list sometimes has eight or nine items. The Disgust-Contempt category appears to be separable and Shame is often added to the more generally agreed upon basic seven. The items on this list have appeared consistently in work of many psychologists who were interested in basic emotion categories. The Ekman, Friesen and Ellsworth (1972) analysis of research studies* on judgments of facial expressions of emotion found remarkable unanimity for a set of seven categories which are basically the same seven appearing in the "biologically-based, universal" lists: Happiness (Joy), Surprise (Startle), Fear, Anger (Rage), Disgust–Contempt, Sadness (Distress), and Interest. In a different tradition, Shand (1914) developed a set of primary emotions which also seems to be identical if his Curiosity category equates reasonably with the Interest category.

It is important to bear in mind that these terms, while commonly used to refer to a single emotion, here represent a group of related or somewhat similar experiences. Davitz (1969) compiled a list of over 400 terms likely to be used as labels of emotion states. One hundred and thirty-seven of them were used by more than half of his 40 subjects, so obviously, there is a well developed common vocabulary in English to refer to emotional states. A variety of schemes has been used to reduce the field to a representative selection. The wide agreement on the seven categories is based on including a variety of differently labelled experiences under a common emotion name. For example, "Happiness" includes complacency, enjoyment, glee, love, quiet pleasure, mirth etc. "Surprise" encompasses at least amazement, astonishment, awe, bewilderment and startle as well as surprise itself.

In developing the expanded list of professional constructs to deal with emotional events, I started with the permutations of the constructs Kelly has

* See Ekman *et al.*, 1972, for complete references.

already used and chose names for these based on general use of the terms rather than get embroiled in the technical distinctions and operational definitions of the psychological literature in which little agreement is apparent. As that exercise did not provide definitions for all the basic emotion categories, it was necessary to work within the scheme to describe the remaining states. Sadness and Surprise were quite easily fitted into the construct system transition model. Anger appeared to be related to two Kelly definitions which I had not considered as emotions. Hostility and aggression are PCP terms which describe behaviour resulting from an emotion state and the definition of anger was, accordingly, related to them.

To a certain unavoidable extent all the names reflect a somewhat arbitrary choice and are an attempt to match usually imprecise descriptive language describing various human states with the rather technical language of PCP describing possible variations in the fate of the construct system.

I wish to add the same disclaimer as Kelly about these special definitions. "The commoner definitions of these terms are not abrogated by their being given limited meanings within our system. It is not our intent to preempt the words . . . " (1955, p. 489).

It is not my intention to claim that these definitions are the only way certain emotions may or should be construed. Book length treatises on each emotion may not say all that could be said. The intent is to provide the psychologist with a fresh way of construing certain events, a way that focuses usefully on their function in therapy. This is particularly the case for those psychologists who subscribe to the PCP view that therapy is "the psychological process which changes one's outlook on some aspect of life. It involves construing or more particularly, reconstruing. That which is reconstrued is usually one's own life or the role he envisions for himself through his understanding of other's outlooks" (Kelly, 1955, pp. 186–187).

As is generally true of the PCP approach, the attempt has been to achieve a relatively high level of abstraction. In this case, it is properly subordinate to the basic process, construing, itself. At the same time, practical application is the *raison d'être* of any theorising.

The presentation of these new definitions is in an order based on their derivation from Kelly's original definitions and discusses each briefly. For completeness, I will include in the listing the four emotions and two emotional behaviours already defined by Kelly. A briefer version of the list is presented as the Glossary (p. 121).

CHANGE IN CORE STRUCTURE

(a) Comprehensive

Threat is the awareness of imminent comprehensive change in one's core structures. Kelly notes that for a threat to be significant, the change must be substantial and imminent and involve a multifaceted alternative core structure. There are many examples of threat. A patient hospitalised for an extended period may be eager to be discharged but at the same time threatened by the imminence of having total sole responsibility for a difficult-to-care-for body. In marriage counselling, as a couple progressively reconstrue their relationship it may be threatening for one partner to see that the other needs a different version of the self than formerly . . . that a relationship based on an infantile dependency is no longer satisfying and that now it is recognition of competence and respect which is called for. Threats occur in the shape of human beings, people like a now banished former self (Landfield's exemplification hypothesis) or people who expect an older, and now rejected, type of behaviour (Landfield's expectancy hypothesis, Landfield, 1954). But not all threats stem from other human beings. Some are realisations of many sorts including the realisation of death, the meaning of praise, punishment, responsibility etc.

(b) Incidental

Fear is like threat except that it involves an awareness of an imminent incidental change in one's core structure. A new incidental construct, rather than a comprehensive construct, seems about to take over. One tends to be threatened by that which he construes comprehensively and is only made fearful by that which he construes simply. One may be seen to flirt with what he fears in the attempt to elaborate a part of the core structure which is experienced as inadequate. But what is threatening is too dangerous to the core structure to be toyed with. When core structure is pervasively involved, experimentation is not possible.

The distinction between threat and fear based on degree of pervasiveness is not common in studies of emotion, but it seems particularly useful in therapy. Threats are elements which tend to elicit a construct which is basically incompatible with the system upon which a person has come to rely for his living. In psychotherapy it is the very plausibility of certain elements which makes them threatening. The closer a client comes to having to construe himself in some new and alien (to the present system) manner, the more he is "threatened". If the behaviour or element were totally implausible, it would not be threatening.

In other approaches to psychotherapy, threat is seen as coming from the therapist's interpretations of client material (e.g., Rotter's Social Learning Theory). These interpretations may be ordered on a continuum in terms of the degree to which they might be threatening or provoke defensive behaviour. Assuming that the client is able to construe his situation in the therapist's interpretation, this is completely consistent with the PCP interpretation. The Rogerian interpretation of threat is also somewhat similar to Kelly's. Rogers

(1959) sees threat as the state which exists when an experience is perceived or anticipated (subceived) as incongruent with the structure of the self. Threat eventuates in anxiety. The process of defence prevents this either through distortion of the experience or denial of it to awareness. Threat is a psychotherapist's term whereas it rarely is used by those studying emotion *per se*. In such studies, the term "fear" tends to encompass both the pervasive and more contained experiences. Where distinctions are discussed, it is usually a question of whether fear and anxiety should be separated rather than fear and threat.

CHANGE IN NON-CORE STRUCTURE

(a) Comprehensive

Bewilderment is an awareness of imminent comprehensive change in non-core structure. When all the world seems topsyturvey, when things are "curiouser and curiouser", but self constructs have not been invalidated, bewilderment describes the resulting sense of precariousness.

Bewilderment is the momentary experience of waking after a short nap to find that morning things aren't going on. The sun isn't where it should be; the smells aren't "breakfasty"; and there's too much traffic noise outside. Bewilderment can also be more than momentary as tourists abroad have often experienced and immortalised in the "If this is Tuesday, it must be Belgium" quip. The external world has been so consistently unfamiliar that its only meaning is derived from a name on the schedule. One's construction system devised to handle familiar architecture, landscapes, transportation, cuisine etc. is about to be changed. This will most likely be through increased permeability of present constructs but eventually, perhaps, there will be a development of some new or different networks of superordinate and subordinate constructs. Some are likely to have foreign, untranslatable, names. For those who have committed themselves to live abroad for an extended period, bewilderment may be a pre-cursor to culture shock in that external events become unpredictable using the familiar construct system. However, culture shock seems to involve core-structures, hence threat, as well.

Although bewilderment is a common human experience it is not commonly discussed in approaches to therapy or studies of emotion. Like doubt, I suspect that it is regarded as a cognitive confusion of little significance. In its more momentary versions or as a single experience it has its charm, but I believe that with this new construing of it, bewilderment will be a more widely useful professional construct, especially in this age of *anomie*.

(b) Incidental

Just as fear is a more contained experience than threat, *doubt* is a more constricted experience than bewilderment. It involves an awareness that an incidental change in a non-core structure is about to take over. Doubt can involve the inadequacy of a single construct or a small subsection of a system although there may be a sizable number of elements for which that construct is useful. For example, ethnologists may doubt the usefulness of the concept of race without their entire professional construct system being demolished and I may doubt whether that is a good example without invalidating much of my construing of the experience called doubt. Clients often express doubt when the PCP approach is explained to them. Former ways of construing the therapy situation must be forsaken and new constructions gradually developed. Their expression of doubt is evidence that the process has begun.

VALIDATION OF CORE STRUCTURE

(a) Comprehensive

Love is a state of awareness of the validation of one's core structure. The loved one is everything needed to be one's whole and true self. Love is feeling accepted for the self you know you are. Needless to say, there is more to say about love and I have not defined it in terms of what one gives to the loved one, which obviously is a large part of the common experience. Rather, I have focused on why one loves another. This definition accounts for unrequited love as well as the reciprocated. In short, in love, one sees oneself completed by the loved person and core structure constructs are validated. Incidentally, it is probable that another *person* is loved since core structure evolves out of interpersonal role construing.

This particular definition has been very useful to me in work loosely described as "marriage counselling". Perhaps the title should be modernised to "relationship counselling", but I am referring mainly to situations where there seem to be no extraneous reasons (legal/formal contracts, dependents etc.) for persisting in a relationship which is relatively unsatisfying. Often the most illogical, painful, unproductive behaviour is regarded as unassailable because the client attributes it to love. Love is a sacred cow.

Obviously core structure must be at risk to be defended so vigorously. In this construction, possession of the loved person is in some way essential to core structure. In most instances, the structure will be preverbal and explication of it will present a rather difficult detective job to the client and therapist. Apart from this PCP point of view, when dealing with such clients, one is reminded of the neurotic paradox. Why does the client persist in such unrewarding behaviour? It appears in many of these cases that love is an element construed constellatorily as the ultimate good as well as the name of a pre-emptive construct. The reinterpretation of the phenomena related to love suggested by this new definition is more fertile than previous explications, e.g., Fromm (1956), and provides a framework for understanding both counterfeit and mature love within which certain deductions may be made and future events anticipated.

(b) Partial

Happiness has many names: joy, pleasure, delight, mirth, whatever state leads to the smiling response. In PCP terminology I have defined it as awareness of validation of a portion of one's core structure. Thus it is part of love, but love involves a more massive core structure validation. Happiness comes only with some core involvement, but it need not be the total core structure which is validated. Happiness *can* be a warm puppy.

The relationship between love and happiness which is reflected in these PCP definitions has been substantiated by research as well as common experience. Davitz (1969) found that the definitions of happiness, joy, love and affection contained many of the same descriptive items although each had distinctive items as well. Tomkins' (1962) very interesting discussion of the smiling response as the affect of enjoyment emphasises the relationship between happiness and an individual's social dependencies. Smiling creates dyadic interaction seen first between the mother and child. Later in life, the adult recaptures this type of communion when he smiles at another person and that one smiles at him. At the same time the eyes of each are arrested in a stare at the eyes of the other. Tomkins says that under such conditions one can "fall in love". Happiness begins in love, but one soon learns to broaden the spectrum of objects and activities to a wide variety of social and non-social events which validate at least a portion of one's core structure. Hence, we are happy with the familiar, the comfortable, the reaffirming. New events are interesting, but newness makes us happy only when it signifies something about ourselves. Perhaps the new event allows us to construe ourself as open minded, up-to-date, affluent and successful. These are all old familiar core constructs which are being validated, but they are only a portion of the self-construing system.

VALIDATION OF NON-CORE STRUCTURE

(a) Comprehensive

Satisfaction is an awareness of validation of a non-core structure. It has a more externalised locus than love but it too is a positive feeling, a sense of the general orderliness of events and their predictability as well as one's own ability to predict correctly.

Common use of the term "satisfaction" includes the connotation of completion; a prediction we made has been validated. "Are you satisfied?" one is asked solicitously or sarcastically. And the answer is "Yes", if indeed we received what we expected. We can be satisfied with a commission paid for our services although the amount had not been agreed upon in advance. The compensation was at least what we estimated the task was worth. Likewise, because of the uncertainty always present with a living thing, we can be satisfied when a wine, chosen with care, complements our food. Our prediction that it would has been borne out.

One can experience satisfaction even in the face of undesirable events as evinced by many a groaning or gloating, "I told you so!" Our prediction, though unhappy, was correct.

(b) Partial

Despite the negative connotations of *complacency* in an achievement oriented culture, it is the name I have chosen for a positive affect, an awareness of validation of a small portion of some non-core structure. It is a state of no dissatisfaction, but since only a small and relatively peripheral construct structure was at stake, there is no great sense of satisfaction either. It is a positive affect in the sense that it does not call for change. I see validation as rewarding because it represents at least a partial satisfaction of the basic human motive, the attempt to predict and control the events of our world.

Complacency differs from boredom in that in the latter state no bets have been placed at all; little or no construing has been attempted; few opportunities for either validation or invalidation have arisen. Although Kelly says man is in motion just by virtue of being alive and that we don't require a push or pull to behave, I suspect we experience ourselves as alive only because we construe this motion and our efforts to make sense of our world. (Shades of Descartes' "Cogito, ergo sum".) The more actively we strive to construe our world the less chance there is that we experience boredom.

INVALIDATION OF CORE STRUCTURE IMPLICATIONS

Sadness is an awareness of the invalidation of implications of a portion or all of the core structure. Note that this does not involve the dimensions of the core structure themselves, but rather of the links these dimensions have with other dimensions including both core and peripheral structure.

Sadness is an emotional state which admits of a wide degree of variation in intensity. It has many names including grief, anguish, distress, sorrow, melancholy, pensiveness etc. Variation in intensity is perhaps a reflection of the extensiveness of the implications of the structures which were involved in the invalidation. Sadness does not have the urgency of an anxiety propelled state but is rather a feeling of loss.

Loss, of a loved person or object, is the prototypical trigger of sadness. The loss may be permanent as through death or destruction, or it may be temporary as when lovers must part. Clearly, no construct dimensions must be abandoned because of this, but what has happened is that some elements no longer are within the range of convenience of some constructs. The relationship between constructs comes from their joint applicability to certain events. Through loss, some implications no longer have any basis. Particularly in the case of the loss of a lover, the deprivation is of some of the connotative meaning of self, for there are some core dimensions which particularly arise to construe events in which the lover has amplified the self structure. Connotative meaning can be considered to arise from the implications of a construct dimension and this is what is lost, including some of the meaning of self. Perhaps sadness is the opposite of "basking in reflected glory", reflected glory being implications linking self structures to the construction of the gloried event.

FIT OF SELF AND CORE ROLE STRUCTURE

(a) Dislodgement

Guilt is "the perception of one's apparent dislodgement from his core role structure". Core role has been defined as one's deepest understanding of being maintained as a social being.

> "Within one's core structure there are those frames which enable one to predict and control the essential interactions of himself with other persons and with societal groups of persons. Altogether these constitute his conceptualization of his *core role*." (Kelly, 1955, p. 502)

This basic definition takes into account both the conventional notion of guilt as man's awareness of the evil within him and the psychoanalytic notion of guilt as a reprimand of the superego. It also is consistent with Erikson's (1950) concept of a developmental stage in which guilt emerges in tandem with initiative during a stage of transition from attachment to the parents towards becoming a parent. This parental set "supports and increases self-observation, self-guidance, and self-punishment" (p. 248). It is the emphasis on an emerging sense of self which is pertinent. While core role involves that part of a person's role structure by which he maintains himself as an integral being, basic maintenance is not only a self-centred matter. PCP emphasises the importance of social construction. Role is structured in relationship to the significant people in one's life. Guilt involves an awareness of a violation of basic identifications and Kelly's definition can be usefully extended towards understanding a wide variety of guilt related situations, from the lack of guilt of the psychopathic personality to the pervasive guilt of involutional melancholia.

(b) Good Fit

Self-confidence is based on an awareness of the goodness of fit of the self in one's core role structure. This definition concerns the opposite of guilt. It is important to re-emphasise that core role structure is both a social interaction ame of reference and the basis of identity. In naming this state I chose self-confidence" rather than Erikson's term "initiative" because it seems to ıe to direct attention to the underlying psychological state which enables ıitiative to emerge. This definition also permits an explanation of the avering of self-confidence under varying circumstances. The core role ructure, for example, may be a better fit within the family circle where elf-confidence is high, but not so adequately trimmed to the facts of an mployment environment. Or, perhaps more typically, core role structure ıay be adequate for situations which require only superficial involvement of ıe self but be quite inadequate where commitment and involvement demands re greater. Such structure is probably unrevised from the pre-verbal mother–hild interaction days and the predictions available based on that unelaborated ole may not fit the demands of a mature relationship. The ensuing lack of onfidence leads to a great variety of behaviours commonly described as efensive. They often hobble intimacy. Self-confidence in this frame of eference comes very near to being the same as "ego-strength" as used by)ollard and Miller (1950).

:) Dislodgement of Other's Construing of Self

'hame differs from guilt in that the locus is rather more external than internal. oth however, involve a phenomenological assessment of the self in a role.

> "The unpleasant feelings called shame are elicited by an expectation that *other people* will be disappointed in the fact that a standard has been violated. The unpleasant feelings called guilt are caused by expectations that the *self* will disapprove." (Kagen, 1969, p. 471)

Vhile aiming in our definition for more generality than Kagen offers, the uotation highlights the main departure in the definition of shame in PCP tyle. Shame is an awareness of the dislodgement of the self from another's onstruing of your role. In PCP theory, the subsuming of other people's onstruing efforts is the basis for social interaction. The concept of role is entral to an understanding of the definition. In Kelly's explication of the ociality Corollary he says,

> "In less precise but more familiar language, a role is an ongoing pattern of behaviour that follows from a person's understanding of how the others who are associated with him in his task think." (1955, pp. 97–98)

Shame involves an awareness of another's construing of your own role and hat your behaviour has not been as the other predicted. Shame is the result of artial invalidation of core role structure, i.e., that which makes a relationship neaningful to you, but it does not involve the actual abandonment of the onstructs involved.

In some conceptualisations shyness is closely linked with shame (Tomkins,

1963; Izard, 1971). A PCP description of *shyness* would define it as a prediction that shame will be experienced, i.e., a prediction that there will be a dislodgement in another's construing of the self. Shyness, as such, is not an emotion in the sense of being an awareness state but is actual behaviour or construing. Darwin (1872) discussed shyness and shame together because blushing was common to both.

> "Shyness seems to depend on sensitiveness to the opinion, whether good or bad, of others, more especially with respect to external appearance. . . . Persons who are exceedingly shy are rarely shy in the presence of those with whom they are quite familiar, and of whose good opinion and sympathy they are perfectly assured." (Darwin, 1901, pp. 349–350)

Darwin's description accords quite well with the PCP position relating shyness to shame, with shame having an external locus, rather than the internal one associated with guilt.

FIT BETWEEN OWN CORE STRUCTURE AND OTHER'S

Contempt and *disgust* are somewhat similar emotions which involve an awareness that the core role of another is comprehensively different from one's own. They may also involve a realisation that the other is, as a consequence, suffering a dislodgement of self from his own core structure (as in guilt). Contempt and disgust can be distinguished in that contempt is a term reserved for human social prediction but disgust may involve the role of any person or object. One is contemptuous of people who do not meet the norms of social expectation but disgust is experienced when things are socially unacceptable without holding them responsible for their state. Contempt can be founded on the simple difference between the self's core role structure and that of another, such as a foreigner, or it can arise when we construe social situations so that we expect that another should experience guilt.

Current work by Izard (1971) and unpublished studies by Ekman and Friesen suggest that disgust and contempt may be separable but the evidence is still inconclusive. Should a separation be confirmed this distinction provides a rationale for it.

RECOGNITION OF CONSTRUCT SYSTEM FUNCTIONALITY

(a) Inadequate

"*Anxiety* is the recognition that the events with which one is confronted lie outside the range of convenience of one's construct system" (Kelly, 1955, p. 495). Considering all the possible professional definitions available for anxiety this may seem like a rather unlikely choice for a clinician. However, it follows Kelly's general plan of trying to achieve a sufficiently abstract level of theorising to both encompass most of the previous positions and provide a ıew perspective. For example, he observes, in reference to the term "free-loating anxiety" that it "is the inability to construe certain impending events neaningfully which gives anxiety its characteristically ambiguous quality" 1955, p. 496). It is not merely the invalidation of a construct that produces inxiety. Anxiety results from invalidation only if the construct is abandoned ecause it is no longer relevant and there is nothing to take its place. Anxiety s the result not only of invalidation but also of construing the outcome of a rediction so that there is an awareness of a discrepancy. It represents an wareness that one's construction system does not apply to the events at and. Therefore, it is a pre-condition for making revisions in the construct ystem.

This is akin to Kierkegaard's view that it is necessary for the individual to ay the price of anxiety in order to expand his range of awareness and dequacy (May, 1950). McDougall (1923) has a parallel definition:

> ". . . anxiety is the name by which we denote our state when the means we are taking towards the desired end begin to seem inadequate, when we cast about for possible alternatives and begin to anticipate the pains of failure."

Anxiety plays a part in construct system revision which leads to it often eing mistaken for Kelly's account of motivation. While it is true that Kelly, lespite his protestations, does offer an implicit theory of motivation, this is ıot, as characterised, simply another version of a tension-reduction point of iew (Maddi, 1972, p. 160) or, quite similarly, minimising the disruptive urprises that the world can wreak on us (Foulds, 1973; Brunner, 1956). This nterpretation seems to be the result of excessive focusing on only one aspect of Kelly's account of human activity. A number of authors, (Sechrest, 1963; Fransella, 1972; Maddi, 1972; McCoy, 1975) have noted that Kelly's theory of motivation as expressed in the Choice Corollary is bi-directional, an spect he had not elaborated sufficiently. Choice is made to enhance one's redictions of the nature of reality and this can be through either (1) extending he predictive range of the system or by (2) constricting it in order to more learly define the system. Kelly is ambiguous about how one can predict which of these strategies will be employed although he recognises differing ircumstances likely to lead in one direction or the other. What he does state learly is that anxiety is a precondition for making revisions in the construct ystem (1955, p. 498). Thus anxiety (as defined by Kelly) must assume a ignificant role in any discussion of Kelly's implicit theory of motivation. However, not all construing involves a construct system revision so it cannot be claimed that anxiety is *the* implicit motivational principle in PCP. Rather anxiety is a symptom, a state present when a person recognises being nescapably confronted with events to which one's constructs do not adequate-y apply. At the same time that it is not the primary motivator, which is the

need to make sense of the world, anxiety provides the impetus for construc
system development.

(b) Adequate

Contentment is the result of the opposite outcome of a prediction from tha
which leads to anxiety. It is an awareness that the events with which one is
confronted lie within the range of convenience of the construct system.

A construct system may or may not fit events. Just as both lack of relevance
and invalidation can lead to anxiety, their opposites must be construed to
experience contentment. It requires an awareness that the construction
system is relevant but such an awareness only can follow upon the system
being put into use successfully. Invalidation of a construct does not abort
contentment provided there is an alternative construction available which wil
serve to discriminate events adequately. In order to construe the adequacy of
the range of convenience of one's system, construction must have been
validated. For this reason contentment is likely to be blended with other
states which are the awareness of validation, i.e., love, happiness, satisfaction
and complacency. Common parlance validates this notion of blend as it is
almost impossible to separate contentment from any of those. The perceived
incompatibility of discontent with each of them is the quickest test of that
proposition. Davitz (1969) has not included complacency and satisfaction or
their equivalents in his operational dictionary of emotions but the relationship between contentment and both love and happiness, which he does
include, can be demonstrated by the fact that they all load on four of the same
factors. Contentment differs from the other two in that they load additionally
on others, especially a factor called an "hyperactivation". (Incidentally, love
loads on an "inadequacy" factor as well which would be expected in the PCP
definition.) Clearly the activity which is part of both love and happiness is not
part of contentment. This is what would be predicted from the successive
construing model. It requires placing one's bets to see if you have real, not
counterfeit, money. The only way to win is to bet with coin of the realm
but you may lose if either your money was counterfeit or your prediction was
not borne out. Contentment is awareness that your money was good although
you would not have known it without winning.

Sudden need to construe

Surprise is often included in lists of innate emotions (e.g., Darwin, 1872
Shand, 1914; Tomkins, 1962; Ekman *et al.*, 1972). All recognise that this is
an extremely transitory state and is often compounded with the state which
follows it. Ekman (1973) distinguishes startle from surprise, viewing the
latter as relatively slower and less negatively toned than startle. Also he notes
that surprise can be viewed as a response to a misexpected event rather than

to a sudden, intense, unexpected event as in the case of startle. This distinction appears to be largely one of degree rather than kind. I doubt that it is necessary for the PCP definition which says that startle or surprise is the sudden awareness of a need to construe events. In fact, there may be an implicit time continuum in which startle depicts the most sudden need, surprise a less sudden but still unexpectedly urgent need and interest a need which arises at a normal pace. At any rate, surprise is often quickly followed by fear where the combination is called terror or by pleasure which we call delight, a happy surprise or simply good news. Tomkins (1962) view of surprise as a general interrupter of ongoing activity is consistent with the PCP definition as is the combination with other emotions. Following awareness of a need to construe, the ensuing construction may be satisfactory or unsatisfactory and the transitory state of surprise is superceded by some other awareness of the adequacy or fate of the construction system.

Behaviour associated with emotion

The definitions which have been offered up to this point all involve permutations of the dimensions involved in Kelly's four emotions: fear, threat, guilt, anxiety. However, definitions of a number of other emotional states flow rather naturally from the PCP outlook I have been developing, especially in consideration of other PCP professional constructs which I have not treated as emotions. They are aggression and hostility which Kelly conceptualises in a particular way. I include a brief discussion of the PCP meanings of these two terms as background before a consideration of yet another new definition.

(a) Active elaboration of one's perceptual field

"*Aggressiveness* is the active elaboration of one's perceptual field." Kelly's basic motivational statement involves the tendency of a person to choose that one of a pair of alternatives which promises the greater possibility of extending his predictive system without endangering it. There are some people who characteristically seem to set up choice points in their lives and make their elaborative choices with greater frequency than do others. These people are always putting themselves "on the line" and occasionally, they force others into uncomfortable stances as well, by precipitating situations which call for decision and action. These people are called "aggressive". As for the relationship between aggression and anxiety, Kelly notes,

> "When a person is aggressive, he seeks out bits of confusion. He fusses over them, he tests out constructs which might possibly fit and he rapidly abandons those which appear to be irrelevant. Indeed, one might say that the areas of one's aggression are those in which there are anxieties he can face." (1955, p. 509)

(b) Effort at validation without change

"*Hostility* is the continued effort to extort validational evidence in favour of a type of social prediction which has already proved itself a failure" (Kelly, 1955, p. 510). When constructs are invalidated remedial action can be directed either towards the construct or towards the invalidating evidence. The construct can be abandoned and/or replaced. The evidence can be doubted or the construing person can try to alter the events in an effort to make them conform to his original expectations. This later method, the method of Procrustes who either stretched his guests or cut them down to size to fit his guest bed, is hostility. Kelly discusses hostility in relation to threat and guilt and also elaborates as "loving hostility" that form of control within a relationship which keeps another individual from maturing, i.e., the parent who treats a child as a doll or the husband who treats his wife as an incompetent child.

In PCP, aggression is more nearly synonymous with adventuresomeness, especially adventuresomeness involving high stakes and persistent action along a single chosen line. In contrast, hostility is a last-ditch effort to achieve validation by extortion or by misleading manipulation of the data rather than by revision of the constructs that have proven themselves faulty (Kelly, 1961). Definitions of aggression seem to provoke a fair amount of hostility among their adherents (Montagu, 1968).

A newspaper reported recently that the answer to the question, "What is aggression?" has been sought by the United Nations for over 23 years. A special committee of 35 members has been meeting regularly for six years to discuss this topic. Recently it concluded a six week session at Geneva no nearer to a definition than at its inception.

Anger

While I have not considered hostility as defined in PCP as an emotion because it does not necessarily involve awareness of the fate of the construct system, there is an emotion associated with hostility. This is *anger* which I define as an awareness of the invalidation which leads to hostile behaviour, i.e., an attempt to force events into conformity so that the prediction should not have been a failure, and the construction should not have been invalidated.

There is nothing already worked out in PCP theory which would predict that anger rather than anxiety and/or other states such as fear would follow invalidation. They are often found intermingled (Izard, 1972) in emotion studies. Invalidation should theoretically lead to anxiety, and from there to either greater extension or better definition of the construct system. Kelly (1961) suggests that a person may have explored both types of revision and landed on a difficult choice. The necessary revision may involve excavating and overturning the very foundations of his construct system. It may seem easier to extort confirming evidence from the events. This later choice is called hostility. In this scheme I propose that anger is the awareness of it and the preceding invalidation is the precipitating factor.

Anger can involve core constructs or more peripheral ones but it is more likely that awareness of the invalidation leading to hostile behaviour will involve the self structure. In efforts to understand anger, it is wise to explore this possibility even when the anger appears to have been triggered by events which seem to lack self involvement. Likewise, responses regarded as passive in situations normally eliciting anger may unexpectedly not involve core structure.

One interesting question about anger has been suggested by Boucher's (1974) description of a Malaysian, Ahmad, who is angry when his newly purchased car breaks down. This anger is clearly identifiable by his facial expression and other behaviour. However, when later Ahmad's father berated him for so foolishly spending money on such a useless vehicle, there was no sign of the anger which an American friend who was with him would have expected. Ahmad says in response to the American's query that he was indeed angry with his father but that it would have been improper to have shown it in that situation. Maybe Ahmad was angry at his father and it is only a cultural constraint which keeps the display of it in check as Boucher seems to suggest. Alternatively, it may be that Ahmad is only being polite and validating his friend's construction. For it is also possible that Ahmad construes his father's judgment of him in such a way that it does not invalidate any of Ahmad's own construction, elicit anger and lead to hostility. For example, he may regard being berated as part of a father–son ritual which by now is merely boring. Or it may reflect an affirmation of Ahmad's self-concept which is defined in terms of his father's consistently low opinion of Ahmad's shrewdness in matters mechanical and financial.

In therapy, clients who are not angry when it seems "normal" are always puzzling. I am constantly reminded of Kelly's discussion of insight, which implies that a client's achieving "insight" may be little more than his adopting the therapist's viewpoint and language. It may be that a culture forbids many displays of anger and therefore anger is experienced, but not communicated. Another possibility is that anger may well be construed only with preverbal portions of the construct system and therefore the construction is at a very low level of awareness. But it may also be that to experience anger in "normal" situations requires a certain elaboration of core role which particular individuals (or many individuals in certain cultures) may never have developed. In such a case, it is not the therapist's task to teach the client to construe his anger. Rather, the task should be seen in terms of an elaboration of the client's core structure if that is desirable.

The whole question of whether one can be in an emotion state without seeming to experience it needs much more explication. The PCP definitions have all hinged on awareness of a state of one's construct system. But the basic process, construing, is the medium of awareness and it applies over the widest possible range of levels of consciousness. The recent research on the innate basis of a set of emotions seems to be leading to contemporary approaches to the question of whether there are at least some innate triggers of emotion. Should it be found that there are, at best this will be an avenue to only a small portion of emotion situations. In the meantime, and perhaps beyond, it seems that the PCP approach provides a *modus operandi*. Psychologists can construe what others might call emotion events in a way which focuses on the basic psychological process and its potential for change. At the same time, it is not necessary to assume that certain people or cultures "repress" emotion either generally or under specific circumstances. The events in those cases can be construed by them and by those trying to understand their construing in a wide variety of ways. Why resort to hostility? To force someone to have a "normal" emotional reaction is counter-productive.

Not wishing to be guilty of cultural chauvinism, this approach is particularly useful to me in exploring the implications of psychotherapy in non-Western cultures. For a number of reasons, emotion events have different meanings and implications in different cultures. When therapist and client do not share a common cultural heritage, emotions seem to pose more puzzles than any other kind of behaviour. Until one experiences the cross-cultural perspective it is almost impossible to construe just how culture-bound one is. Such awareness depends upon evolving both an original and a second construction system for delineating reality and then a set of constructs which subsume both.

The pervasiveness of Western philosophy and values throughout psychology, its theories and practice, is most difficult to grasp from within the system. It is rather like trying to grasp the concept of time while we are in the time frame. Trying to conjure up non-time leaves one with a dimly perceived awareness of the inadequacy of our construction system for the task at hand. Trying to construe emotion events in non-Western clients is one area where I became aware of the discrepancy between reality and my system for construing it. Not having any other construction possibility, most Western social scientists had tried to force the non-Western experience, various outlooks on human nature and society, normality and abnormality, success, virtue, morality etc. to fit into their Western construct systems. Rather than acting with such hostility, once the inadequacy of the usual Western construction is dimly perceived, recognising the possibilities for hostility should permit a choice of behaviour. Anxiety need not culminate in hostility. It can impel extension of the system so that construction can be a closer, improved proximation of "raw reality" than at present. The relatively high level of abstraction of PCP provides a way for backing off from the usual perspective. It is hoped that the proposed expanded set of professional constructs will enrich the PCP system for construing human behaviour both within and outside the therapy situation. Our construing man, *Homo Construens*, can now be seen as happy or sad, in love, angry, bewildered, contemptuous or contented. In short, we now have an expanded PCP system for construing the exciting and interesting diversity of experience that is so typically human.

Glossary of PCP-defined Emotions

Threat*	Awareness of imminent comprehensive change in one's core structure
Fear*	Awareness of imminent incidental change in one's core structure
Bewilderment	Awareness of imminent comprehensive change in non-core structure
Doubt	Awareness of imminent incidental change in a non-core structure
Love	Awareness of validation of one's core structure
Happiness	Awareness of validation of a portion of one's core structure
Satisfaction	Awareness of validation of a non-core structure
Complacency	Awareness of validation of a small portion of some non-core structure
Sadness	Awareness of the invalidation of *implications* of a portion or all of the core structure
Guilt*	Awareness of dislodgement of the self from one's core role structure
Self-confidence	Awareness of the goodness of fit of the self in one's core role structure
Shame	Awareness of dislodgement of the self from another's construing of your role
Contempt (or Disgust)	Awareness that the core role of another is comprehensively different from one's own and/or does not meet the norms of social expectation
Anxiety*	Awareness that the events with which one is confronted lie outside the range of convenience of the construct system
Contentment	Awareness that the events with which one is confronted lie within the range of convenience of the construct system
Startle (or Surprise)	Sudden awareness of a need to construe events
Anger	Awareness of invalidation of constructs leading to hostility

* These definitions are Kelly's original conceptualisation. The others involve permutations of the dimensions which he used.

References

Arnold, M. B. (1960). "Emotion and Personality, Vol. 1, Psychological Aspects", Columbia University Press, New York.

Boucher, J. D. (1969). Facial displays of fear, sadness, and pain. *Perceptual and Motor Skills* **28**, 239–242.

Boucher, J. D. (1974). Faces, feelings and culture. *Culture and Language Learning Newsletter*, March 1, 1974, East–West Center Culture Learning Institute.

Bruner, J. S. (1956). A cognitive theory of personality. *Contemporary Psychology I*, 355–358, *reprinted in* "Personality, Readings in Theory and Research", (Southwell, E. A. and Merbaum, M., eds.), 1964, Brooks/Cole Pub. Co., Belmont, Calif.

Darwin, C. (1872). "The Expression of Emotion in Man and Animals", 2nd edition, 1901, (Darwin, Francis, ed.), John Murray, London.

Davitz, J. R. (1969). "The Language of Emotion", Academic Press, New York and London.

Dollard, J. and Miller, N. E. (1950). "Personality and Psychotherapy, an Analysis in Terms of Learning, Thinking and Culture", McGraw-Hill Book Co., New York.

Ekman, P. (1971). Universals and cultural differences in facial expressions of emotion. *In* "Nebraska Symposium on Motivation, 1971" (J. Cole, ed.), University of Nebraska Press, Lincoln, Nebraska.

Ekman, P. (1973). Cross-cultural studies of facial expression. *In* "Darwin and Facial Expression" (Ekman, P., ed.), Academic Press Inc., New York and London, 169–222.

Ekman, P. and Friesen, W. V. (1967). Head and body cues in judgement of emotion: a reformulation. *Perceptual and Motor Skills* **24**, 711–724.

Ekman, P. and Friesen, W. V. (1969). The repertoire of nonverbal behaviour, categories, origins, usage and coding. *Semiotica*, **I**, 49–98.

Ekman, P., Friesen, W. V. and Ellsworth, P. (1972). "Emotion in the Human Face", Pergamon Press, New York.

Ekman, P., Sorenson, E. R. and Friesen, W. (1969). Pan-cultural elements in facial displays of emotions. *Science* **164**(3875), 86–88.

Erikson, E. (1950). "Childhood and Society", Penguin Books Ltd., Harmondsworth, Middlesex, England.

Foulds, G. A. (1973). Has anybody here seen Kelly? *British J. Med. Psychol.* **46**, 221–225.

Fransella, F. (1972). "Personal Change and Reconstruction", Academic Press, New York and London.

Fromm, E. (1956). "The Art of Loving", Bantam Books, Inc., New York.

Izard, C. E. (1971). "The Face of Emotion", Appleton-Century-Crofts, New York.

Izard, C. E. (1972). "Patterns of Emotion", Academic Press, New York and London.

Izard, C. E. with Wehmer, G. M., Livsey, W. and Jennings, J. R. (1966). Affect, awareness and performance. *In* "Affect, Cognition and Personality" (Tomkins S. A. and Izard, C. E., eds.), Tavistock Publications, London.

Kagen, J. (1969). Personality development. *In* "Personality, Dynamics, Development and Assessment" (Janis, I. C., Mahl, G. F., Kagen, J. and Holt, R. R., eds.) Harcourt, Brace & World, Inc. New York.

Kelly, G. A. (1955). "The Psychology of Personal Constructs", Vols. I & II, W. W Norton & Co., New York.

Kelly, G. A. (1961). Suicide, the personal construct point of view. *In* "The Cry For Help" (Farberow, N. L. and Shneidman, E. S., eds.), McGraw-Hill Book Co New York.

Kelly, G. A. (1963). The autobiography of a theory. *In* "Clinical Psychology and Personality", (Maher, B., ed.), 1969, John Wiley and Sons, Inc., New York.

Kelly, G. A. (1965). The role of classification in personality theory. *In* "Clinical Psychology and Personality" (Maher, B., ed.), 1969, John Wiley & Sons, Inc., New York.

Kelly, G. A. (1966). Humanistic methodology in psychological research. *In* "Clinical Psychology and Personality" (Maher, B., ed.), 1969, John Wiley & Sons, Inc., New York.

Landfield, A. W. (1954). A movement interpretation of threat. *J. Abnorm. Soc. Psychol.* **49,** 529–532.

Leeper, R. W. (1970). The motivational and perceptual properties of emotions as indicating their fundamental character and role. *In* "Feelings and Emotions" (Arnold, M. B., ed.), Academic Press, New York and London.

Maddi, S. R. (1972). "Personality Theories: A Comparative Analysis" (revised edition), The Dorsey Press, Homewood, Ill.

May, R. (1950). "The Meaning of Anxiety", Ronald Press, New York.

McCoy, M. M. (1975). Foulds' phenomenological windmill: a reply to criticisms of personal construct psychology. *Br. J. Med. Psychol.* **48,** 139–146.

McDougall, W. (1923). "An Introduction to Social Psychology" (first edition, 1908), Methuen, London.

Mehrabian, A. (1968). "An Analysis of Personality Theories", Prentice-Hall Inc., Englewood Cliffs, N.J.

Montagu, M. F. A., ed. (1968). "Man and Aggression", Oxford University Press, Oxford.

Patterson, C. H. (1973). "Theories of Counselling and Psychotherapy", 2nd ed., Harper & Row, New York.

Peters, R. S. (1969). Motivation, emotion, and schemes of common sense. *In* "Human Action" (Mischel, T., ed.), Academic Press, New York and London.

Pribram, K. J. (1967). Emotion: steps toward a neuropsychological theory. *In* "Neurophysiology and Emotion" (Glass, D. C., ed.), Rockefeller University Press, New York.

Rogers, C. R. (1956). "Intellectualized Psychotherapy", *Contemporary Psychol.*, *I*, 355–358, *reprinted in* "Personality, Readings in Theory and Research", (Southwell, E. A. and Merbaum, M., eds.), 1964, Brooks/Cole Pub. Co., Belmont, California.

Rogers, C. R. (1959). A theory of therapy, personality and interpersonal relationships, as developed in the client-centered framework. *In* "Psychology: A Study of Science", Vol. 3, (Koch, S., ed.), McGraw-Hill Book Co., New York.

Rotter, J. B. (1954). "Social Learning and Clinical Psychology", Prentice-Hall, New York.

Schacter, S. S. (1964). The interaction of cognitive and physiological determinants of emotional state. *In* "Advances in Experimental Social Psychology", Vol. I, (Berkowitz, L., ed.), Academic Press, New York and London.

Schacter, S. S. and Singer, J. E. (1962). Cognitive, social and physiological determinants of emotional state. *Psychological Rev.* **69,** 379–399.

Schlosberg, H. (1954). Three dimensions of emotion. *Psychological Review* **61,** 81–88.

Sechrest, L. (1963). The psychology of personal constructs: George Kelly. *In* "Concepts of Personality", (Wepman, J. M. and Heine, R. W., eds.), Aldine Publishing Co., Chicago.

Shand, A. F. (1914). "The Foundations of Character", reprinted in 1962, Danish Academic Press, Copenhagen.

South China Morning Post, June 1, 1973, "What is Aggression?"
Southwell, E. A. and Merbaum, M., eds. (1971). "Personality, Readings in Theory and Research", Brooks/Cole Pub. Co., Belmont, California.
Tomkins, S. A. (1962). "Affect, Imagery, Consciousness, The Positive Affects", Vol. I, Springer Publishing Co. Inc., New York.
Tomkins, S. A. (1963). "Affect, Imagery, Consciousness, The Negative Affects", Vol. II, Springer Publishing Co. Inc., New York.
Tomkins, S. A. (1970). Affect as the primary motivational system. *In* "Feelings and Emotions" (Arnold, M. B., ed.), Academic Press, New York and London.
Tomkins, S. A. (1971). Homo Patiens. *In* "Personality Theory and Information Processing" (Schroder, H. M. and Suedfeld, P., eds.), Ronald Press, New York.
Young, P. T. (1961). "Motivation and Emotion: A Survey of the Determinants of Human and Animal Activity", John Wiley, New York.

The Community of Self

J. M. M. Mair

"What is man that we should be mindful of him?"*

Questions concerning the nature of man are presumably as old as thought itself and we don't seem to be much closer to any definitive answers now than at any time in the past. In fact you could even argue that we are getting further away from any widespread agreement as to the kinds of account that should be given of ourselves and others. At least since Darwin shattered the myth, so dear to so many, of man being but a little lower than the angels by rudely re-allocating him to a position only a little higher than the apes, there has been an increasing ferment of reappraisal. In this, science has become the new certainty for many, the new protector behind which we can shelter from the chill air of confusion. It was, after all, through scientific modes of thought and inquiry that the old theologically supported view of our world as the centre of everything was replaced by the seemingly humiliating alternative of it being merely an outlying speck in an outlying galaxy. It has been through scientific modes of inquiry that the atom has been split and many of the biological mysteries of life amazingly unravelled. Through science also we have developed means of destruction of diverse kinds and vast proportions.

Such have been the advances of science in our time, and the sophisticated power made possible through the specialised techniques of scientists, that many people seem to have lost any firm psychological or personal footing in the world. So many older beliefs have been shattered by advances in science that many seem to have given up the very business of asking any serious questions about the nature of man in a *personal* rather than in a narrower biological or physical sense. Not only is it now difficult to know what sorts of questions it would be considered meaningful to ask, but we are also frightened of hazarding answers to questions we do ask, in case, as so often in the course of this century, both our assumptions and conclusions are swept out from under our feet yet again by developments in science.

* This is a variant of the question, "What is man that thou art mindful of him?" which is asked in Psalm 8.4.

Psychology and the Reduction of Man

Within psychology, as in the other social sciences, versions of the old questions are asked but many of the answers proffered have done little to increase our self respect. Blows to long standing beliefs about the rationality and freedom of man have come within psychology as well as from the discoveries in the physical sciences.

Freud shocked his generation by insisting that sexuality permeated childhood as well as adult life and by implying the shallowness of much human knowledge and the limitations of personal control in creatures driven largely by unconscious desires. The early Behaviourists opposed much of what Freud said and seemed to challenge further man's sense of his own worth. Within the strict Behaviourist view all talk of self or experiences, hopes or purposes, thoughts or reasons disappeared at a stroke. He became little more than a mindless slot-machine proffering his limited responses as the appropriate tokens were inserted. Since then this simple view has been stretched and modified as thinking, imagination, plans and purposes, meanings and even hidden meanings have been "re-discovered". Yet even now on the powerful, fundamentalist wing of the Behaviourist Reformation, Skinner and his followers seem to project a compelling picture of man as an empty organism jerkily dancing to the barrel-organ of his circumstances.

It is not just science, but also technology which has flourished during this century. Perhaps it is because of this growth in the power, complexity, variety and availability of machines that so many of our models for explaining man have been "mechanical" or "electronic". Thus man, who only recently found himself relegated to a more lowly position of one among the animals, has increasingly found that both animal and human behaviour are being accounted for by analogy with machines. For many practical purposes this, no doubt, makes a lot of sense. Machines have the great appeal in a mysterious and shifting world of being gratifyingly concrete, finite, comprehensible and largely explicable in terms of mathematics, mechanics, physics, electronics or other branches of the more developed sciences. But by reducing man to manageable proportions in this way we should not be too surprised to find many people who then treat themselves and others as machine-like and who believe that issues of human change and development can best be solved by "behavioural engineering".

Another important area of psychological concern has been the study of "Individual Differences", but here too, we have often unwittingly impoverished our understanding of personal functioning. We have repeatedly used large groups of anonymous subjects as a basis for making supposedly *general* statements about human personality and abilities. The persistent use of this kind of approach resulted in any individual person seeming virtually meaningless when considered in their own right. The individual person became little more than a "deviant" of greater or lesser degree on "dimensions" he or she had no part in creating and whose relationship to each other in the original standardisation sample might well have little or no bearing on his or her own ways of ordering events.

A further way in which methods of psychological inquiry have incidentally blunted our appreciation of the nature of human functioning is found in the widespread reliance on questionnaires as a means of communication. Almost unnoticed we have come to accept a crude, over-simplification of what constitutes communication between people. So often we seem to have studied individual differences by weeding out first of all whatever questions seem clearly to be understood in different ways by different people. Thereafter, the hope seems to be that the simplified remaining questions will be understood similarly by pretty well everyone. Here it seems to be assumed that if we perform some equivalent of shouting loudly and clearly enough our one-syllable questions, then subjects, like foreigners, will almost certainly get the message. Firm limits are usually placed on the kinds of answers which are allowed and some investigators go further still in diminishing the personal significance of whatever the subject may squeeze into his YES, NO or SOMETIMES replies. This final reduction comes when the experimenter claims that, after all, it doesn't necessarily matter if people do interpret the questions, the context, or the task differently. It doesn't matter if they do mean different things in choosing any particular answer. The psychologist after all needn't care what meaning or sense the subject may have been trying to convey. All that is important after all is the behavioural outcome, the pattern of marks on the paper.

Alternatives

All this should not be interpreted as an attack on science in general or psychology in particular. My concern has been only to suggest that in both direct and indirect ways our conceptions of ourselves and others have been, for many at least, radically changed. Older beliefs about the dignity of man and his special status in the order of things have been unsettled in many different ways. Former certainties and securities have repeatedly been discredited or become untenable. Many people, both within psychology and more widely in society, are searching for alternatives with considerable confusion and anxiety.

One of the confusing things within psychology, as many people question the adequacy of the "old giants" of psychoanalysis and behaviourism, is that *so many* alternatives seem to be forthcoming. In what has been called "humanistic" psychology there has been in recent years a vigorous out-pouring of alternative approaches, theories and methods in education and therapy. The unhappiness felt by many in being constrained by too few choices seems to have given way to an unhappiness about having too many choices. As we thus try to escape from the old frameworks which formed and directed our understanding, we may well find it is not particular, concrete courses of alternative action that we need, but alternative "ground" on which to stand, different assumptions about the "reality" of ourselves and others, a different kind of basis for belief and action. All this is obviously a very tall order, but indicates none-the-less the area of my present concerns in this essay. In undertaking such a task as this, my only hope is that by attempting to spell out a few possibilities, it will become clearer to both myself and others what their limitations as well as their uses may be. In this way it may be just a little easier to do better next time round.

Metaphors of Man

When we attempt to understand anything unfamiliar, unknown, mysterious or beyond our present comprehension we seem to resort to the use of metaphor. Consider for instance how we make some sense of our "feelings". We talk of feeling "heavy" or "light", "high" or "low", "stirred up" or "settled", "soft" or "hard", "warm" or "cold", "bright" or "dull", "stretched" or "cramped", "falling apart" or "coming together", and so on. Here as elsewhere, we reach for some understanding through the use of more familiar structures laid over the less familiar events which are engaging us. The process of metaphor seems to be an activity whereby we carry across some frame of reference, which is usually applied in some other sphere of action, to view, grasp, explore, open up, structure or otherwise re-appraise the events to which we are now attending. In the process of this act of transference we act *as if* the events which we more conventionally ascribe to one set of categories really belonged to another. Through this procedure we see things afresh and make it possible to approach events differently. We entertain alternative possibilities for further action by willingly making a kind of "mistake" whereby something, or some set of events, are treated as if, in various important respects, they were something quite different. This kind of willing self-deception seems to pervade all our experiencing and can be considered quite fundamental in our understanding of and action in relation to the events through which we constitute our experience of the world.

In this light we can readily begin to see all claims about man as a "developed ape" or a "fallen angel", a "mechanical toy" or a "super-computer", as a "mindless epiphenomenon of a mindless universe" or an "open system in the larger ecosystem of the planet earth", not as bald assertions about what man *is*, but as metaphors for exploring the endless mystery of what man may yet make of himself. The danger here comes, as it has so often come in the past, when we mistake metaphor—our own acts of intentional mistake making in the service of understanding—for truth itself, or when we confuse, as we so often do, an invitation to inquiry with dogmatic assertion. Not one of the metaphors of man which has yet been elaborated can provide a full account of his nature nor need we assume that any single or any combination of metaphors in the future will do so either. This does not mean that any particular "pretence" about the nature of man is trivial or futile. Quite the reverse seems true. It is only as we construct and involve ourselves in further imaginative possibilities concerning what man may be that we will both reveal and create for ourselves further visions of what we may become. By the same token if we commit ourselves to and lose ourselves in metaphors of man which are too small for us we may gain a certain kind of security for a time, but are likely to pay a high price in despair, confusion and the denial of freedom we can ill afford to lose.

I have introduced elsewhere the notion that metaphor, or rather the activities which we label as "metaphor", should be considered a far more important process than has been customarily recognised within psychology (Mair, 1976) and will not repeat that argument here. It is, however, within this frame of reference, viewing man as one who explores and creates the realities of his existence in the world, realising new possibilities through the process of metaphor, that I want to turn now to consider a particular metaphor of man. My intention is not primarily to advocate the use of this particular metaphor, or to suggest that it is universally relevant, but to open up one alternative viewpoint which may offer us glimpses of others beyond it. More especially, I want to elaborate this particular metaphor as a perspective from which to approach aspects of George Kelly's psychology of personal constructs (Kelly, 1955) since I believe that many of the possibilities in Kelly's writings are almost unreachable at present because of the route by which we approach them.

Self as Community

The cultures of the West have been for hundreds of years oriented towards *individuality*. Within Christianity generally, and especially since the Reformation, there has been great emphasis on individual responsibility and the notion that each man must work out his own salvation. One of the corollaries of this general belief in the essential separateness and privacy of individuals seems, however, to be that if you are not *somebody* then you are a *nobody*. People who are not recognised or known by us as "individuals" tend to be clustered as undifferentiated "units" called "workers" or "foreigners" or "subjects". Thus we find ourselves with a psychology which stresses "individual differences", but paradoxically it might seem, loses sight of individuals almost completely in amorphous clumps of interchangeable "subjects". We find ourselves also with an "external" bias in our psychology whereby our "respect" for the individual has all but ruled out the possibility of developing public understanding of our "private" concerns. Whatever John Donne may have said, we do seem often to experience ourselves and others as "islands", even though we also know that we are not complete unto ourselves. Though there are many arguments to the contrary, we do often experience and construct our *selves* as fortresses set up against a hostile world and assume, on the psychological plane, that the Englishman's home is still his castle.

Instead of viewing any particular person as an individual unit, I would like you to entertain, for the time being, the "mistaken" view of any person *as if* he or she were a "community of selves". This is clearly an invitation to make-believe, though I hope it will become apparent that we seem to construct our "realities" through just this kind of willing self-deception. It is not, though, a specially novel idea since this kind of notion has repeatedly been used through the centuries and, as I'll indicate later, is still widely used in many forms. My belief, however, is that by exploring the notion of "self as community" explicity as an exercise in metaphor we may be able to grasp some of its possibilities for understanding and action.

Perhaps it is easiest to introduce the idea of "self as if a community of selves" by referring to the smallest form of community, namely a community of two persons. Most of us have probably, at some time, found ourselves talking or acting as if we were two people rather than one. We talk sometimes of being in "two minds" about something, part of you wanting to do one thing and part wanting to do something else. Quite often we hear people talk of having to "battle" with themselves, as if one aspect of themselves was in conflict with another. Often we pass this sort of thing over as only a form of expression. However, the explicit invitation to consider oneself, for the moment, as two people rather than one-self can make it possible for us to pay attention first to one of the "people" and then to the "other" one. In this activity the person can be encouraged to ignore, for the moment, one of the "selves" and "get inside" the other. From this vantage point of being "inside" one side or party in the dispute, rather than only being vaguely aware of the two sides as an "outsider" from some separate vantage point, it is sometimes possible for the person to sense more fully some of the hopes and fears, values and plans, concerns and confusions of this "other person". Thereafter, a similar activity of "entering" and "experiencing from" the other "party" to the dispute can be undertaken. Not infrequently a graphically clear impression is obtained of the incompatibilities and suspicions, assumptions and tactics of the two "selves" in relation to each other and between them and "members" of other "communities".

The notion of oneself as a "community of selves" can readily be elaborated further by some people to incorporate three, four or any other number of "selves". Some of these "selves" will be found to persist and others may be more transitory, some will be "isolates" and others will work in "teams", some will "appear" in many circumstances and others only on a few special kinds of occasions, some will be "more powerful" and others will give way to them. Sometimes, people can offer and use quite elaborated accounts of their "community of selves". Since these show up some of the features of this metaphor in action, a few examples may be helpful.

Examples of "Personal Communities"

1. JOHN

John had difficulties in stopping smoking and asked for some help. As we discussed the matter, it seemed that many personal issues were tied up in this presenting problem. He seemed to be trying to cope with confusing aspects of his own experience in his dealings with both himself and others. I outlined the idea of the person as a "community of selves" and suggested that he might find it useful to think of his various and contradictory feelings, desires and concerns as if these were different "selves" constituting his personal "community". He responded immediately by saying that in these terms he could see exactly what he had been doing all these years. In terms of the way in which he interpreted the idea of "community" (and different people make different and sometimes changing interpretations here), he now "recognised" that the most powerful person in his "community" was his "Foreign Secretary". His "Foreign Secretary" spent very little time "at home". Instead of this, he was continuously travelling around amongst other "communities" —other people—trying to help and impress them at all times and at almost any cost to his own "community". He was for ever trying to make friendly overtures to other "communities", dispensing "foreign aid" frequently and whenever he was asked for it. The "Foreign Secretary" couldn't and wouldn't refuse any request for aid since he was very concerned to be needed by every other "community". Over the years he had therefore repeatedly committed his "home producers" to all sorts of gigantic production tasks to meet this endless foreign demand and had never checked that they were able or willing to undertake all these "export" orders. Thus while the "Foreign Secretary" went around impressing other communities with his "mother bountiful" act his whole "home community" lived in a state of strain, uneasiness and often down-trodden resentment. The pervasive feeling within the "home community" was of living with the persistent uncertainty as to whether they would be able to meet the heavy demands being made on them to provide goods and services for all kinds of other "communities".

When he returned a week later, he brought with him a typewritten page on which he had outlined what he felt to be some important aspects of his personal "community" which he called "The Home Team". Britain, at the time, was engaging in discussions as to whether or not it should enter the "Common Market" of the E.E.C. and John had used some of the concerns involved in this in giving an account of his own "community". In this document he identified a number of groups and individuals within his "community" and discussed their various activities, concerns, limitations and strengths. He considered major "groups" which he called "the Wise Ones" and "the Common Marketeers", and discussed the positions and policies of both "the Home Secretary" and "The Chancellor of the Exchequer".

Along with this written summary he indicated verbally that he had sacked the old "Home Secretary" who had been too weak and replaced him by a more powerful one. Simultaneously, the powers of the "Foreign Secretary" had been curbed. Now all requests for help from other "communities" had to be discussed "in Cabinet" rather than being passed on automatically because the "Foreign Secretary" wanted it so. John went on to say that for the first time that week he had been able to say "No" to some request for help because he was already heavily committed. Previously, he was sure, he would have felt obliged to agree without delay to add this to his list of obligations. This time he had felt able to pause and think about it "in Cabinet", and then to act with that authority. He had also found himself, for the first time that he could remember, feeling happy to be by himself for periods of time. In the past he had always felt too lonely if he were alone for even a few minutes but now he had experienced something quite different. How, he asked, could he possibly feel lonely with so many different "people" and "groups" emerging in his "community"? Along with all this, John also found that, almost without effort, his cigarette smoking had been reduced to only a few each day and in a number of other ways, his time and his life felt more "his own" to do with as he chose.

John's immediate use of the metaphor of "community" was striking, although it provided only a means for tackling some of his problems over time and was certainly not a ready-made solution to them. It seemed to provide him with the beginning of a personal "language" within which to conceive and begin to control aspects of his ways of dealing with himself, others and the world.

2. DAVID

David also found the metaphor useful as he worked, over a long period of time, to resolve his problems which centred around his stammer. He came from Northern Ireland and had stammered since childhood. Like John, he also chose a "political" interpretation of the metaphor of "community". Within these terms, David talked as if he were composed of a number of political factions. Initially he outlined three groups. The "Hard Liners" were aggressive, right wing, impatient, bigoted, unforgiving and cynical. The

"Soft Liners" were concerned in all circumstances and on all occasions with finding and taking the easy way out, with appeasement, maintaining the status quo and with letting sleeping dogs lie. Between these was the "Middle Group" who were less clear cut but were more or less reasonable, without very definite opinions, liable to be swayed in one direction or another and fairly down to earth.

These groups, and especially the "Hard" and "Soft Liners", were engaged with each other and the world in continuing confusions of warfare and resistance, appeasement and withdrawal. The "Speaker" in David's "Parliament" had the unenviable job of trying to present to the world the common view of the whole "community", but in most circumstances any single viewpoint was simply non-existent. Instead of this, strident claims and counter-claims were the order of things. Indeed, as David explored his "community" further, he became aware of even more extreme "Hard" and extreme "Soft-Line" groups. Between these extremes there were huge and persistent differences in policy and values. The "Hard liners" didn't seem to know the meaning of ideas like withdrawal or surrender and would fight anywhere and on the slightest provocation. The "Soft Liners", on the other hand, had an equally intense policy of appeasement, of peace at any price. If presented with the choice of speaking or not, the "Hard Liners" would absolutely insist on being heard, while the "Soft Liners" would prefer not to speak at all. David felt that the "Soft Liners" sometimes retreated into total silence to keep out of trouble, but were then dragged out by the even greater anger of the "Hard Liners" to assert some kind of position.

On being asked if these two sides ever "spoke" to each other, David instantly replied that neither of these extremist wings ever wanted to admit that the other side was there at all. The "Hard Liners" took the attitude that they should never acknowledge that there was a stammer, while the "Soft Line" side was ready and eager to admit to stammering in everything and anything if necessary. David indicated that the side which had been strongest for most of his life was the one which claimed that he didn't stammer at all, but he pointed out that a friend of his, who also stammered, had chosen the opposite position and become a virtual recluse.

As David made further use of the metaphor of "community" he became aware that the "Middle Group" was more like a loose collection of various individuals and groups. He began to talk about occasional experiences of feeling and acting in a more relaxed manner. By personifying this experience he began to work out some of the implications and possible involvements of this more "Relaxed and Uncommitted Group", which occupied some of the "middle ground" in his political spectrum. This group did not seem to be aligned with any other factions and because of this, David thought they might well frighten a number of other "people" in his "community" who were alarmed by anyone who did not take definite stands in relation to conventional issues.

About this time, David noticed that occasionally he was speaking without any forethought and in a relaxed manner. He found, though, that when he did this, he stammered a lot. He felt that as soon as this more "relaxed" and "spontaneous" involvement in speech occurred there was a reactive feeling of shock and anxiety within the different "factions" in his "community", with the various committed "groups" taking over firm control again. David indicated that he wanted this "Relaxed Group" to wander all through his "community" to spread their kind of feeling and concerns.

In pursuing this, David felt that this "Relaxed Group" might belong to a further grouping which he referred to as a "Friendly Group". These seemed as though they might be "people" who would be friendly with everyone, who cared about others and were charitable in their actions and thoughts. David felt that this, gradually emerging, "Friendly Group", might well be more prepared than any one else to tolerate the views of others in the "community", even though they might personally disagree with them. It was through this clustering of "relaxed", "friendly" and "free-and-easy" groups that he began to develop more active involvement with people at work and home. The "battle" for control within his "community" was however, fought long and hard. The "Hard" and "Soft Liners" only very gradually released control as other more "flexible" and "warm" groupings spread their influence and tested their competence in coping with the everyday affairs of living.

3. PETER

A third example may be useful to indicate that other interpretations of "community" than a "political" one can be used and also to indicate that people with the "same" problem (in this instance, stammering), are not necessarily very similar to each other. Rather than talk in terms of "Parliament" or "Political Factions", Peter developed his interpretation of the "community" theme in terms of a "Troupe of Players". His "Troupe" seemed to him to be guided by a "Council" whose general task was to keep performers and performances "in balance". The main controller of day to day activities was the "Producer". His job was to take responsibility for what was happening "on stage" at any moment in time. This "Producer" seemed to like being fairly easy going and often tried to be on equal terms with the various players. This could be pleasant when no great demands were being made on the "Troupe", but even then, in the lax atmosphere, there was a recurrent tendency for arrangements to come unstuck. Jobs didn't seem to get done and gradually the "players" would find themselves in dire trouble, without sufficient cohesion or organisation to pull things together. At times like this the "Producer" would be forced out of his laissez-faire approach and become an autocratic martinet. He would clamp down and impose a kind of marshall law or a state of emergency. Sharp orders would then be given and would have to be obeyed and many of the "Actors" who couldn't or wouldn't respond to this kind of treatment were banned or suspended till the emergency was over.

Among the "Actors", Peter initially outlined the following. There was the "Conversationalist" who loved company for company's sake, enjoyed the give and take of ideas and was never happier than when relaxing in good company. Balancing and off-setting him was the "Businessman" who was very practical and down to earth. He always had to be getting somewhere and had to be engaged in organised, constructive action or he became very frustrated. For long periods of time these two could occupy complementary "parts", each taking spells of being advanced "on stage". However, there seemed always to be a tendency for one or the other to "hog the limelight" when he got into it. This repeatedly led to trouble and "performances" would get completely out of "balance" in one direction or the other.

Then there was the "Country Bumkin" who had no mind for detail or finesse, but saw the main points in anything and would just batter on with things in crude but purposeful ways around these main issues. He was balanced by the "Metropolitan Smooth Man" who, in contrast, was very much concerned with detail. He considered that refusal to master detail was just a sign of laziness and stupidity. Then there was the "Adventurer" who used to love going on long cycle runs, enjoyed climbing mountains, and relished any challenge like facing a new job or a new situation in which the slate was wiped clear and he was free to act in accordance with circumstances. He had been a prominent "character" for a long time in the past, but was less apparent in the present. Along with him came the "Sentimental Lover" who was all emotion, sentiment and tenderness. He expressed himself a lot in the context of Peter's marriage. It seemed he could come "onstage" there but otherwise was kept in the background. He was very fond of animals and children and was generally very loving and warm.

Another very important "Actor" in the "Troupe" was the "Dreamer". He hated being tied down to anything but was quite a pleasant character. He always wanted to take part in things which were relaxing and pleasing. He loved the countryside and nature generally. If he ever found himself being tied-down or having to be highly organised, he just withdrew. Thus, when the "Producer" was in an autocratic phase, the "Dreamer" was always one of the first to disappear. However, it was apparently the "Dreamer" who was especially fertile in the production of ideas and quite a number of the things which the other characters made capital from came from him in the first place.

Peter commented on how "real" these characters seemed to him, but as we continued it became apparent that very little communication existed between the different "players". The "Producer" seemed to be the only "person" with continuing responsibility for getting the show on the road. The other "players" seemed to be very much "character-part actors" whose only concern was with their particular "act". There seemed to be little sense of shared responsibility or interlacing of interests among the various "actors". When the "Producer" let things slide into a free-and-easy democracy, "everyone" seemed just to slouch around and let things slide further. This in turn would lead to yet another dictatorial purge by the "Producer" who would precipitately grab back all the authority he had recently relinquished in order to get some sort of adequate "show" underway. Peter found that his stammering became particularly bad during the phase of laissez-faire disintegration when the "Producer" relinquished firm control and the "players" lost any sense of direction and order. As part of a programme of "community development", Peter undertook a number of tasks to try to increase understanding and communication between important "members" of his "Troupe". The intention here was to try to foster more sharing of responsibility among the different "members" so that less crisis-control action would be necessary by the "Producer". In the course of this, Peter tried to "get inside" each of the main "characters" in turn to experience "from the inside" something of their personal concerns. On the basis of this we then tried to develop more exchange between the "players" and more participation in each other's actions. The long term concern here was to find and extend common areas of interest and ways of working together so that they need not continue in virtual isolation, relying on the peremptory commands of an often harrassed "Producer".

Some Comments on "Community"

Hopefully, these examples will convey something of the ways in which the metaphor of "self as community" can be interpreted in use. This metaphor is, of course, not a solution in itself for most problems in living, nor does it make sense or seem useable by everyone. Even those who do make something out of it for their personal purposes find that they can sometimes make more use of it than at other times and seem to be more "ready" for it in some circumstances than others. However, my intention here is neither to expound its practical application nor to provide a full critique of its limitations. I want only to indicate some of the more obvious features of this "image of man".

The "community" metaphor seems to provide, for some people at least, a flexible framework within which to represent and express many aspects of their experience in relation to themselves and others. It seems to provide a means by which they can "open out" and sense something of the pattern of their engagements which were previously "cramped-in" and "hard to grasp". One of the advantages of this framework is that it offers to everyone something already familiar in many different forms. Everyone has experiences of ving in some sorts of communities whether they be family, neighbourhood roups, recreational clubs or teams, work situations in shops or factories or ffices, larger scale communities like nations with governments and modes of dministration, or even constellations of nations in the world at large. There s endless variety and complexity available here and also simplicity, since seful elaboration can be done within a "community" composed of two people". No esoteric language need be involved, although any amount of it s available should it be desired. For some people there is also something very obvious" about the metaphor when used in relation to their own experience, vhile at the same time its use often provides surprisingly fresh perspectives.

Clearly the metaphor lays emphasis not on isolation but on inter-relation- hips. Within this view, methods and kinds of communication and ways in vhich communication is restricted or prevented are clearly important. 'ossibilities for increased communication can here be made available between ur "selves" and between "members" in different "communities". In all this, ne potentially fertile feature of the metaphor is that it invites us to explore ur personal experiencing in the world in the same sorts of terms which we ormally reserve for social events. Of course, any metaphor can be pressed oo far and result in meaningless confusion, but it may be that some useful levelopments could arise from understanding personal functioning in terms ormally restricted to social psychology and sociology and conversely from xploring aspects of social structure, group pressures and prejudices, organ- sational control and change and such like within the "laboratory" of ersonal "communities of selves". We can see, indeed, that in using this netaphor, a psychological understanding of any person could make use of oncepts and methods drawn from many other disciplines. In giving some ccount of personal "communities" we may find useful concepts already vailable in the fields of politics, group processes, diplomacy, debate, propa- anda, industrial organisation, labour relations, international trade, law, heatre, literature, arts and sciences. Anything which we find useful in onceiving or guiding action in the various communities within which we live, ncluding the "ecological community" wherein man lives in relation to other spects of his environment, may be *potentially* relevant by metaphoric ransference in giving some account of our "communities of selves".

"Community" in Context

As I have already indicated, the general idea of viewing a person as some kind of organised grouping of sub-selves or tendencies is by no means new. From ancient times man has been understood in these sorts of terms and certainly since Plato outlined a conception of the individual by analogy with a city state, such notions have been widely used in Western Civilisation. Within psychology there have been and still are many examples of this mode of interpretation. Ellenberger (1970) indicates that McDougall talked of th rivalry and struggle which could develop between different "tendencies" i the person and that Janet and Binet both viewed hysteria as a form of dua personality. Freud's writing is also relevant here and among many othe similarities, his notions of "ego", "id" and "superego" could be viewed as kind of "community" structure. Post Freudian theorists, as Brown (1961 points out, developed even more clearly various conceptions of the "social nature of self. Sullivan, for instance, regarded the self as being made up fro the reflected appraisals of others and the roles which are prescribed b society. He recognised not only what goes on between two or more rea people as constituents of self-functioning, but also suggested that there ma be "fantastic personifications" or ideal figures with whom an individua "interacts". Somewhat similar ideas on the social nature of self are expresse also by Erich Fromm and David Riesman. These writers suggested that basi types of character and personality change as society develops, and in particu lar, as the methods of social control change in society so also the kinds c rules, government and controls operating within individuals change t reflect these.

Jung (see Storr, 1973) made a great deal of use of personification an postulated various kinds of enduring archetype "figures" in persona experience. Others like Melanie Kline, Fairbairn, and Guntrip (See Guntrip 1971) structure some of their concerns in Object Relations Theory in term which can readily be interpreted in relation to the metaphor of "community" Many in the "humanistic" movement show similar concerns. Eric Bern (1961) outlines a kind of "community" in talking in terms of the thre figure-roles of "parent", "adult" and "child" as well as many elaborations o this. R. D. Laing talks of the "divided self" (1961) and the "politics c experience" (1967) and thus indicates both a concern with the multiplicity o self-structure and the relevance of "politics" to the understanding of selves i relationship. Perls *et al.* (1951) can also be seen as functioning in terms of idea about "community" in our experiencing. Many of their techniques could we be viewed as ways in which we can identify and bring together aspects of ou experience whose intimate relationship with the remainder we have denied lost sight of or as yet failed to realise.

This list of instances could be almost endlessly extended and any reader will I'm sure, be able to add many important names which have been omitted here And, of course, it is not only in psychology that this sort of idea finds expres sion. The "social behaviourist" and philosopher, G. H. Mead (1964) discusse consciousness as an interiorisation of the actions of others and thought o reasoning as a form of symbolic "discussion" between several individuals He outlined a form of "minimal community" when discussing his concepts o "I", "me" and "self". More recently, the sociologist, Peter Berger (1963) ha ıade considerable use of the metaphor of life as "drama" and talked of the ıany "roles" or "parts" which we play within ourselves and with each other. Iore recently still, the social psychologist, John Rowan (1976) provided a ıost explicit account of the range of "sub-selves" which may constitute our xperience. He outlines a viewpoint which is very close to the notion of community" in his explorations of what he calls "internal societies".

Such ideas have also played a large part in religious and philosophic ıinking through the centuries. Herbert Fingarette, for instance, discusses oth the pervasive issue of self-deception (1969) and also relationships between sycho-analytic ideas and Eastern religious/philosophic views (1963) in terms 'hich are entirely compatible with the notion of "community". In *The Self in 'ransformation* he provides an engrossing discussion of what could be ·garded as our "other lives" or "other selves". "We become responsible gents", he suggests, "when we can face the moral continuity of the familiar, onscious self with other strange, 'alien' psychic entities—our "other selves". Ie goes on to say that "we should perhaps speak of an 'identity' with other :lves rather than a 'continuity'. For we must accept responsibility for the ıcts' of these other selves; we must see these acts as *ours*". He proceeds ırther to claim that we here deal with, "a special, startling kind of intimacy". It calls me", he argues, "to recognise that I suffer, whether I will or no, for ıe deeds of those other selves. It is an intimacy which, when encountered, ıakes it self-evident that I must assume responsibility for the acts and ıoughts of those other persons as if they were I".

Many of the writers mentioned here have made implicit rather than explicit se of some version of "self as community". Before passing on, therefore, ɔ consider aspects of Personal Construct Theory in relation to "community" may be worth mentioning three ways in which my use of "community" iffers, to some degree, from other and earlier uses.

First of all, my intention is to keep persistently in mind that in considering urselves or others in terms of "community" or "mechanism" or anything lse, we are engaging in the use of *metaphor*. We are thereby exploring and iving form to the mysteries of our existence through the pretence that we are ıdeed formed in ways made familiar in other contexts. So often we become he victims of our own fictions and begin to treat them as conclusions rather han inventions. The metaphor of the "community of selves" is not an asser- ion but *one* invitation among many possible invitations, to personal inquiry.

Second, in the *explicit* use of the notion of "community of selves" there eems to be greater flexibility and less pre-judgment and Procrustean fitting f angular people into the round holes of psychological theory. Within the cope of the "community" metaphor comes a wide variety of kinds of ommunities which may be interpreted in many different ways and used in lifferent manners by different people.

Finally, I have chosen the metaphor of "community" rather than notion like "society", "state" or "groups" because it seems to be a more discrim inating idea. As John Macmurray (1961) points out, every community i a society, but not every society is a community. The notion of "community" points to the possible importance of further notions like "communication" and "communion" and "fellowship". Although these extra dimensions wil not be followed further here, they reflect guiding principles, assumptions o values which I wish to incorporate in developing an alternative "image o man".

Personal Construct Theory and "Community"

I want now to turn to Personal Construct Theory and consider it for a littl in the context of the metaphor of "self as community". George Kelly wrot his "Psychology of Personal Constructs" when academic psychology i America was dominated by Behaviourism and an emphasis on operationa definitions, objectivity and a conviction that psychological as well as physica phenomena were amenable to precise measurement. People, like myself brought up in the weary maturity of this tradition sensed something liberating ly different in Kelly's writing. Yet somehow only a little of what we in articulately sensed as specially innovative in Kelly's perspective has yet bee realised in practical or theoretical developments.

Why should this be? Obviously many different kinds of explanations coul be given, but I want to pursue only one possibility here. It seems likely tha Personal Construct Theory, like anything else, may look somewhat differen depending on the direction from which you approach it. Many people, suspect, while recognising something importantly different in Kelly's writing have tended to lose sight of these differences as they have assimilated onl those of his concerns which fitted most comfortably into the perspective *from which* they approached his work. Thus in Britain, for instance, with it scientific traditions of objective measurement and statistics, it is not surprisin that it was the appealing formality of the repertory grid which caught ou interest and channelled much of our attention. Much of Kelly's philosophy and theory remained hard to handle and difficult to integrate with issue already familiar in our empirical scientific approach.

My hope is that by laying aside, for the moment, whatever implicit directio of approach we may be employing, and explicitly viewing Personal Construc Theory from the perspective of metaphor, and in particular the metaphor o the "self as community", some aspects of Kelly's concerns may stand ou more clearly and some possibilities appear which might otherwise remai obscure. Any adequate understanding of Kelly's theory would require muc more than the use of one metaphor, but I hope that this narrowing o attention will be of some use none the less.

Persons and Constructs

George Kelly emphasised the central importance of "the person" in his psychological approach and stressed this in the wording of his Fundamental Postulate and in nearly all the corollaries. He indicated also that for his purposes "the person is not an object which is temporarily in a moving state but is himself a form of motion". Yet, in spite of this stress on the central, living, and perpetually moving mystery of the person in Kelly's writing, it seems to me that most writers have focused primarily on "constructs". The active "person" seems often to disappear and all attention is directed to the "constructs", or at least the verbal labels which may indicate to some degree some of the persons "network of pathways". In this process we may readily fall into the "from–to" trap indicated by Polanyi (1958). We are likely to attend "from" the person "to" his constructs and thereby to hold on to what we have thus objectified as being the central substance of our concerns. In this way the mystery of the person may once again drop out of our awareness because it is hard to fit into the objectifying traditions in much of psychology.

Approaching Personal Construct Theory *from* the direction of the metaphor of "self as community" there seems little likelihood of overlooking the centrality of "the person". Here the person is considered to live through many "selves", with each or any construct, so to speak, being potentially the centre of an alternative "self". Thus, instead of seeing "constructs" as objectified and impersonal dimensions which the person looks at and manipulates from some central "control-tower", they become guises and forms through and in which the person can participate actively in experiencing and exploring his world from numerous perspectives. He may also gain some sense of the "motion" Kelly talks of as he "enters" and "leaves" and moves between the different "persons" in his "community". He may thus, to some degree at least, gain a sense of experiencing and having available to him different lives, different feelings, different choices, different ways of being in the world.

Thinking and Action

Kelly also explicitly claimed he was offering a theory of human *action*, rather than a "behavioural" or a "cognitive" theory. Even so, most of the research work concerning Personal Constructs seems to be more concerned with thinking and thought disorder, conceptual structure and conceptual change. The charge which is often brought against Construct Theory of being too "intellectual" or "cerebral" seems not without foundation if much of the available research in the area is considered. But again we may have been unwittingly induced into assimilating Kelly's ideas into familiar and respectable categories related to "concept formation" and "thinking" already sanctioned within scientific psychology.

If we approach Kelly's ideas through the metaphor of "self as a community of selves" or the person as if composed of a number of different "persons", we can, in considering any individual, gain a remarkably powerful sense of actions, interactions, transactions and counteractions. The weaving complexities of action by different "people" in relation to many and differing commitments and concerns can convey a sense of patterned movement like that between players and teams on a football field as the ball is passed and held, gained and lost. The metaphor, indeed, tends not to convey a sense of people "thinking" but rather persistently engaged in purposive action in pursuit of agreed or unacknowledged ends. Thus, when Kelly talks of a person's processes being psychologically channellised by the *ways* in which he anticipates events, these "ways" would here be considered as "procedures" for going about the business of living rather than as "conceptual labels" to pin on to passing events.

Within "personal communities" it is also, sometimes, easy to note how different "members" can oppose each other, interfere with, alarm, invalidate and otherwise hinder the acting out and bringing to fruition of any particular line of concern. If, as Kelly suggests in his Experience Corollary, a person's system changes as he successively construes the replication of events, he is likely not to change if he persists in the same aborted, inappropriate or hostile procedures of acting in relation to events. Within the "community of self" metaphor it is easy to see that "experiencing in full cycle" for any "person" may be important at times, if the person as a whole is to be able to change. So also it is possible to recognise how the same person may be both hostile and aggressive, in Kelly's sense, at the same time, as different "members" of his "community" pursue different policies in living. Indeed, each of Kelly's constructions concerning transition—fear, anxiety, guilt, as well as aggression, hostility, the c-p-c cycle and the creativity cycle—can be illuminatingly explored as interpersonal strategies in action between "members" of "personal communities".

Organisation and Control

Within communities of different kinds it is easy to see that organisation will be needed if any coherent action is to be undertaken. Procedures have to be developed whereby choices can be made in adopting any particular lines of action among the many possibilities which may be canvassed. Very many different ways have been developed by different kinds of communities for arriving at choices and for putting such choices into practice. This whole complex of activities is normally considered under headings like, government, administration, policy making, planning, politics, law, business and such like.

In considering "personal communities" ideas from any or all of these areas of discourse may also be appropriate at times and any individual may be considered as one complex, organised, self-controlling "community" in complex relations with many other interdependent, but uniquely organised "communities" in the wider world.

Then we come to Personal Construct Theory and find that Kelly clearly recognises in his Organisation Corollary that for any person it is important to order his constructions in ways which will establish priorities in action. In the Range Corollary he recognises that any particular construct has limits to its range of usefulness, just as any person in a community has limited functions. His Choice Corollary could be seen as highlighting an important principle in government action, and his whole range of "diagnostic constructs" deal with different aspects of power, (e.g. super-ordination, subordination, regnancy) government strategy, (e.g. constriction, dilation, suspension, submergence) levels of administration, (e.g. core, peripheral, incidental, comprehensive) and such like. In the Modulation Corollary, too, he seems to be dealing with the ways in which a person, like a community, maintains his identity by regulating the entry of new ideas or actions. In all this it may seem clear that Kelly is inviting us to explore not only the politics, but also the government and administration, diplomacy and negotiation, production and destruction of experience.

But in fact almost nothing of this can be found in the research literature. Perhaps again, it may be because we have tended to approach Personal Construct Theory from a "scientific", measurement-dominated and individual unit-counting conception of man, that we have tended to make only minimal use of many of Kelly's constructions concerning organisation and control. Generally we have rested content with studies of "structure" depending almost entirely on the intercorrelations of repertory grid choices. Even though this has resulted in some interesting work, I suspect we are generally tied to "static" and "group averaged" notions of "conceptual structure", "organisational consistency" or "construct change". We have, as yet, found no ways of representing the intricate, interweaving, *functioning* of individuals in action as they struggle to organise their affairs and control their destinies. The metaphor of "self as community" may at least indicate some of the sorts of issues which seem to have concerned Kelly and which we have, so far, found elusive to the point of meaninglessness.

One and Many

The metaphor of self as a "community of selves" seems somewhat paradoxical though, in that it incorporates the ideas of unity and multiplicity at the same time, where each "part" of a person needs a "whole" person to account for it and not something less than the whole, as is so often assumed. This kind of idea may seem strange in relation to many of our approaches in psychology, but in fact the notion of many lives constituting one life, many selves constituting one self and at the same time, one self being constituted by many selves is a very ancient one. It is certainly not the kind of view likely to be entertained though, by those who believe that only reductionist and analytic approaches to explanation are scientifically respectable.

However, it does seem to be the kind of view which is all but explicit in Personal Construct Theory. Within the Repertory Grid procedure it is not unusual to find that different "self" constructs are used. Thus a person may distinguish between people in terms of them being like or unlike "Myself as I am" or "Myself as I'd like to be" or "Myself as I used to be" or "Myself as I will be after treatment". In this way it is recognised that any person may subsume different "selves". So also in using any other particular construct dimension, since all constructs necessarily gain their meaning for a person in relationship with other constructions, then any construct could be taken as the central and defining feature of another "self" or version of "self". In this way any one person can be considered as being constituted of numerous alternative, but potential, "selves". Not only is this so, but Kelly recognises in his Fragmentation Corollary that a person can successively employ a variety of subsystems which seem unrelated and even incompatible. In this way one person can function as if he were more than one person, with each being out of touch and working at cross purposes with the others. In all of these ways Kelly seems to be recognising many "persons" or "potential persons" within the one person, and one self as if permeating all a person's "selves".

As we will see in the next section, this general view has important possible implications for our understanding of ourselves and others. It is also interesting to note that Kelly, while inviting us to view any man as if he were in some important respects like a scientist, at the same time outlines a view of personal functioning which shares much with forms of thought and explanation long recognised within different traditions of religious thought in both East and West. In this, as so often, Kelly entices us to break down old and often futile boundaries between disciplines in a search for more adequate understanding of human functioning.

Self and Other

Often, I think, when we consider the notion of "self" or "our selves" we vaguely suppose something contained more or less within the envelope of our skin, or spreading out around our bodies a little to lay claim to a patch of "personal space". What we consider as "self" tends to be identified as "inner", while that which is distinguished as "other" is likely to be classed as "outer". In addition we tend to accept very often, that what is "external" or "out there" is more "real", "substantial", and "objective" than what is "internal" and therefore "subjective". In using the metaphor of "self as community", however, we find that distinctions between "inner" and "outer" are blurred. We find ourselves oriented to a person's relatively more private concerns in terms which are similar to those used in dealing with his more public activities in the world at large. So also, Kelly (1955) laid little emphasis on the necessity of distinguishing between "internal" and "external" in this context.

> "Persons", he suggested, "anticipate both public events and private events. Some writers have considered it advisable to try to distinguish between 'external' events and 'internal' events. In our system there is no particular need for making this kind of distinction. Nor do we have to distinguish so sharply between stimulus and response, between the organism and his environment, or between the self and the not-self".

What Kelly seems to be suggesting here is that while some people may find it useful to use distinctions like "inner" and "outer", there is nothing "given" or fundamental about them. Particular people may use this kind of discrimination "for their convenience in anticipating events" and in giving viable form to their own experiencing. Kelly invites us instead to consider "self" as a personal construction rather than a geographic location.

> "The *self* is", he suggests, "when considered in the appropriate context, a proper concept or construct. It refers to a group of events which are alike in a certain way and, in that same way, necessarily different from other events. The way in which the events are alike is the self".

So Kelly is suggesting that "self" can be considered as a personal construction and a means by which any person may order, sort or otherwise determine his undertakings in the world. Like any other construction it will be used differently by different people.

He is also saying that "self" is, by itself, not the construction but only one side of an integral discrimination and grouping. Thus the personal construction is always something like "self/other" or "self/not-self". Events are sorted, approached, disposed of, undertaken or rejected as "self and not other" or as "other and not self" or as being outside the range of convenience of either. Although we often talk in a casual way as if "self" and "other" were separate things, this is not the view taken by Kelly. What he suggests is that "other" defines "self" just as much as "self" necessarily defines "other". As the bounds of whatever we take to be "other" or "not self" are moved, so also and by the same token, our understanding of "self" is changed. In trying to cope with events, therefore, one person may group only a few events as "self" set over against, and highlighted by, a vast range of "otherness" in the world. Another person may construct himself as being at one with a great many of the events he experiences and distinguish himself from little. Both of these are considered by Kelly as personal undertakings in coping with events and not "givens" in nature. In all this Kelly is emphasising that "self" is not something quite distinct from "other", but that they rely on each other, they are intimately united in their separation, they symbiotically depend on and define each other.

This kind of view, however, whereby "otherness" is conceived as an integral and necessary side of "self" is not easily grasped if we approach Kelly's writing with our common, Western assumptions about the separateness, sovereign integrity and independence of individuals. But if we approach instead through the metaphor of "self as community" we find we are immediately considering ourselves as composed of many *other* and possible "selves". Each person may, from this perspective, be seen as being constituted of many "others". Indeed within this context, "self/other" distinctions are easily recognised as relative to the particular perspective from which personal events are experienced at any moment. Thus within his "community" a person may be able, to some extent anyway, successively to "dwell in" and then "break out" from the various *other* "persons" he has identified. From the perspective of each alternative "person", what is taken as "self" and what as "other" will be different. In this way it would seem that even for one person, the construction of "self/other" is not settled and done with, but may be hugely variable depending on the vantage points chosen within his "community" or "system" *from which* he constructs his experience and experiences his actions.

In considering "self/other" constructions in this way we are led on to a somewhat different understanding of "self" and "other". We can consider "self" not as an "object" of our attention, but as a "base" *from which* we experience "other" events. Thus each person can be considered as having as many "selves" as he has vantage points *from which* to act. Each of these possible "selves" can be identified only as the person somehow steps away from that base of experiencing and makes it "other" in relation to yet a further vantage point. So we can see that "self" in this sense, is realised *only* as "other", never as "self". "Other" is our vital means by which the mysterious possibilities of "self" can be materialised. We can see then, that when we talk of "self" as an object of our attention and assign characteristics of various kinds to it, in terms of "self/other" perspectives we are more correctly referring to an "other" *to which* we attend *from* some different and still undisclosed "self" position. All this can be seen clearly in using the metaphor of "self as community". While a person is acting from some particular "self" base, that base cannot be identified and characterised by him since he is, by definition, acting *from* that base *to* other events or objects. He can, of course, move to another base and take another perspective on events, and from here the first base may, to some degree, be understood as "other". Thus when a person tries to turn around, so to speak, and experience himself, he is engaged in treating himself as "other" in relation to, and from the perspective of, a further and now unidentified base. When we talk about "self" or "ourselves" we are more correctly referring to something "other" than the experiencing centre from which our "other selves" are being characterised.

So it seems that the "self structure" through and from which a person acts in relation to events at any time is *realised only in others* or *as otherness*. In this sense, we necessarily need "others" to realise or become our "selves". The construction of "other" is the very means by which "self" may be expressed, explored or otherwise grasped, and Kelly recognised this kind of essential "bipolarity" in his basic definition of a construct. In the Repertory Grid type of procedure he also offers a means by which we can gain some sense of the experiencing, living person through the constructions he places on others. In Fixed Role therapy, he suggests that a person may make something more of him-"self" by becoming an-"other" for a time. In the extremely fertile, and still largely unexplored, notions of "role" and "core role", Kelly succinctly recognised that we elaborate ourselves in relation to others by making more of others in relation to ourselves. Yet, though all this is true, perhaps Kelly did not explicitly specify some of the important implications of "self/other" perspectives which can more clearly be recognised in considering "self" as a community of "other selves".

All this can be generalised further and all our constructions might usefully be considered as "from-to" perspectives through which we live and give form to our worlds rather than as flat, bipolar "dum-bells" floating about somewhere in psychological space. Thus any of us can be considered as having many different bases from which we continuously construct and reconstruct our acts of experiencing. Often, therefore, we may be acting from many different positions without recognising this and so perhaps confuse ourselves by the apparent inconsistencies in our deeds and words. Often, too, we may insistently limit the number of perspectives from which we are prepared to experience events and so constrict ourselves within familiar, manageable, even if painfully narrow limits. The possibility of involving ourselves in others of the perspectives we may already have within our systems may indeed prove an unsettling experience. Our initial, routine and perhaps limited ways of defining ourselves over against the world may be shaken. We may sense a terrible loss of the familiar ground that we have long taken to be *ourselves* and experience the sense of dreaded change or loss of self which Kelly described as threat and guilt. But if we persist in entering and living through other perspectives we may also find ourselves involved with and understanding many "others". Having "lost" some of the narrow restrictions of our old selves we may begin to "find" hints of a larger sense of self in others. As Kelly indicates, what we take to be "self" and what "other" is a "convenience" constructed by us for our purposes and limited only by the ventures we are able or willing to undertake.

Making-up and Scaling-down Ourselves

I started by suggesting that over the last century our sense of ourselves has been scaled down considerably till, in important areas of psychology, the individual human being seemed insignificant and all but meaningless. I then suggested that we view all our attempts at giving an account of man's nature as exercises in the elaboration of metaphors, rather than direct descriptions of how things really are. From this base I then offered and explored some of the implications of a particular metaphor. Through this perspective on oneself *as if* a "community of selves", a flexible and elaborate sense can sometimes be gained of any person in action. This metaphor, in its turn, provided us with an alternative base from which to approach Personal Construct Theory afresh so that different aspects of Kelly's concerns may, perhaps, become more available to us. At the same time, through both the metaphor of "self as community" and Kelly's notion of the bipolarity of construing a view which is different from our conventional understanding of "self" and "other" was developed. It seemed reasonable to suppose that we have available to us many "self/other" perspectives through which to act and elaborate our understanding of ourselves and others. This fundamental concern shown by Kelly with the essential bipolarity of construing is something we have as yet only partially understood. By recognising that all our actions and experiencing is *from some base to something else* we highlight the need to recognise and learn to make something more of these integral and numerous "from-to" perspectives through which we direct our living. We need to learn how to attend, when necessary, *to* the bases *from which* we and others act, as well as *to* the events towards which our or their actions are directed. In doing this, of course, we need also to remind ourselves that we are thereby now attending *from* some further and as yet undisclosed base which structures and limits, highlights and extends our possibilities in experiencing and *making* sense of our worlds.

All this is, of course, to engage in a kind of "make believe". I have tried here to make-up and illustrate an alternative view of ourselves and others. Alternative metaphors involve different "mistakes" and provide us with different "perspectives" on events. I have tried to show how a particular metaphoric view may enlarge, to some degree, our common understanding of self and others, but I am certainly not suggesting that this particular perspective will be universally appropriate or useful. Rather am I suggesting that we may need to acquire a greater "command of metaphor". This in its turn requires that we learn also about the dangers involved therein since metaphor is, "a dangerous game" and "an inviting quagmire". It is easy to reduce any metaphor to meaninglessness by pressing it too far or not far enough. It is very easy, too, to fall into the trap of reifying that which is metaphor, "pretence" or "make believe". Thus, for some people, it seems remarkably easy to slide into believing that a person *really is* composed of many other selves, and that we *really are* communities of various kinds.

In all this, I hope it may be apparent that we have some *choice* in what we take ourselves to be. In talking of "taking" ourselves to be this or that, or of "making" more of ourselves, we are, of course, again engaged in metaphor. We have no means of *really* knowing whether we can "make" anything of ourselves or can "take" ourselves anywhere different from where we now "find" ourselves. We have no ways of knowing *other than by doing it*, by acting in and experiencing through our own "best bets". It is an invitation and not an assertion to say that we are engaged, at all times, in the active process of "making" what we may individually and communally yet become.

References

Berger, P. (1963). "Invitation to Sociology", Penguin Books, London.

Berne, E. (1961). "Transactional Analysis in Psychotherapy", Grove Press, New York.

Brown, J. A. C. (1961). "Freud and the Post-Freudians", Penguin Books, London.

Ellenberger, H. F. (1970). "The Discovery of the Unconscious", Allan Lane Penguin Press, London.

Fingarette, H. (1963). "The Self in Transformation", Harper Torchbooks, Harper & Row, New York.

Fingarette, H. (1969). "Self-Deception", Routledge and Kegan Paul, London.

Guntrip, H. J. S. (1971). "Psychoanalytic Theory, Therapy and the Self", Hogarth Press, London.

Kelly, G. A. (1955). "The Psychology of Personal Constructs", Vol. 1, Norton and Co., New York.

Laing, R. D. (1961). "The Divided Self", Penguin Books, London.

Laing, R. D. (1967). "The Politics of Experience", Penguin Books, London.

Macmurray, J. (1961). "Persons in Relation", Faber and Faber, London.

Mair, M. (1976). Metaphors for Living, *In* "The Nebraska Symposium on Motivation, 1976: Personal Construct Psychology", (Editor: A. W. Landfield), Univ. of Nebraska Press, Lincoln.

Mead, G. H. (1964). "George Herbert Mead on Social Psychology", (Editor A. Strauss), Univ. of Chicago Press, Chicago.

Perls, F. S., Hefferline, R. F. and Goodman, P. (1951). "Gestalt Therapy", Julian Press.

Polanyi, M. (1958). "Personal Knowledge", Routledge and Kegan Paul, London.

Rowan, J. (1976). You're never alone with yourself. *Psychology Today* **2**, 1, 18.

Storr, A. (1973). "Jung", Fontana/Collins, London.

Towards a Personal Science

J. Brenda Morris

"The psychology of the first half of this century was absolutist, outer-directed and intolerant of ambiguity. When a college student carries this unholy trio of traits he is called authoritarian, and such has been the character of the behavioural sciences. But the era of authoritarian psychology may be nearing its dotage and the decades ahead may nurture a discipline that is relativistic, oriented to internal processes and accepting of the idea that behaviour is necessarily ambiguous." (Kagan, 1967, p. 30)

In psychology we often go to great as well as absurd lengths to make objective our pursuit of knowledge. Whilst this goal is a worthy one based, in the era it was conceived, on the highest ideals of the quest for truth; there is a growing body of theoretical work and research which is overtly indicating that for psychology, the study of persons, the methods of science are changing and that a new set of ideals and values are emerging with the desire to understand subjective and personal knowledge. There is in this a healthy dialectical process of thesis and antithesis leading on to a new synthesis which represents progress forward. Some, like Broadbent (1973), fear that by questioning current scientific method in psychology, that we will revert to a dark age of superstition, mythology or even worse unquestioning faith, dogmatism and the tyrannical oppression of dictators.

This however is the wrong antithesis. That is where we came from. Not where we are going. Such fears, however genuine, belong to a certain era in history and a certain context which has changed. The quest for new knowledge brings many things which are now held to be true under question. To question is to quest. The war is now over. Today's problems are different and so will be today's and tomorrow's science.

An Experiment with Science

Bannister (1970) has argued that because of "the peculiar and particular nature of psychology as an endeavour, 'science', must be enlarged if it is to encompass the study of the person" (p. 48). But one thing psychologists don't do very often is experiment with their scientific methods. There seem to be lots of restrictions and lots of do's and don't's. So that by experimenting with our methods in psychology we are bound to break some of the rules.

In my experiment I am experimenting with some of the rules of scientific method as psychologists use them, thus by definition breaking some of them but not all. As I am not breaking all the rules, I can assert that I *am* elaborating the notion of scientific psychology instead of standing it on its head.

To be more explicit, I am in particular elaborating the notion of empiricism. The hypothesis here is that the central notion to empiricism is the "testing out" of one's hypothesis or assumptions and that the form in which it is recorded is secondary. So that empirical tests can be expressed numerically or non-numerically, that is, purely through language. This notion is explored through recorded interviews. (H1)*

I am experimenting with the idea that personal interests and values inevitably influence our experiments (H2) so I shall explicitly state what my interests and values are and how they influence what I do.

In doing this it is no longer appropriate to use the impersonal form in writing up experiments so I shall use the word "I" and change to a more personal form. As such this is an experiment in scientific literary form as well (H3)

I am experimenting with the notion that psychology is one of the few sciences which is truly reflexive (H4), i.e., it is the study of people by people so that whatever you say or do should be applicable to you the researcher.

I experiment with this on several levels:

(1) I am exploring Kelly's notion that if I am a scientist then my subjects are scientists and they experiment with me too. (H5)
(2) I include myself in the sample of subjects so that I can experiment with my experimenter to explore the concept of reflexivity one step further along the line of infinite regress. (H6)
(3) I am exploring the notion that I as scientist and my subjects use a similar model of science which is basically empirical in nature and can be conceptualised as one form of the hypothetico-deductive method. (H7) This notion of Kelly's also seems to have some support from Bronowski (1956) who says that science is a fundamental value in our culture.

* I use the following notation: A = aim, H = hypothesis, NH = null hypothesis, E = experiment, R = result, UR = unpredicted result, C = conclusion.

Where several of any of the above occur they will be numbered, e.g., H1, H2, H3, etc. If these have several interlinked parts they are numbered H1a, H1b, etc. Where possible the numbering will be logically linked so that, e.g., E2, R2 and C2 will relate to H2. Where any of a series is missing or not numbered it means the person has not followed the experiment through.

I suggest you read the interviews through first without taking too much notice of my notation. That way you get to see them through your own eyes first. Then look at the analysis to see if you agree with it.

therefore hypothesise that my subjects will have no particular difficulty talking about aspects of their lives as if they were experiments. (H8)

Experiments with Life

K

"Can you construe aspects of your life as experiments? Pick any experience and let's see what kind of experiment it was."

"I'd like to talk about my experience in a T-group. I'd never been in a T-group before so I wasn't quite sure what it would be like. I thought it would be kind of unstructured and verbal and specifically about relationships and how people related to each other."

"Was there anything specific that you wanted to deal with concerning your relationships?"

"We were asked that. I just said I wanted to learn more about the way I relate to people and get some feedback from them. (A) And my biggest fear was that I would be bored."

"What does being bored mean?"

"That I wouldn't move. That I wouldn't learn anything from it. (NH) I felt as though I had very few close relationships. And thinking about it I wasn't very happy with my relationship with the people I lived with. I wasn't very sure if I had a relationship with Liz any more. All my life I've not been completely happy about the way I relate to people. It's the thing I feel about being invisible. (H2) Very often people don't see me because I presume people know what I'm thinking and feeling. So I think more of me is actually coming out than they actually receive. Sometimes I feel that my facial expression and what I'm feeling don't quite match up. And that's very much to do with realising that I very much didn't trust my instincts. (H3) I tend to use my head rather than believing what seems right at first. I kind of weigh up decisions internally."

"You try and make an intellectual or rational decision?"

"Yes. And also try and predict the consequences. (H4) Then I realised that predicting consequences was something that really wasn't necessary in the group. I found I could do something very positive without actually knowing what was going to happen." (R4)

"What sort of thing?"

"I did it on Thursday. (E1) I had a fantasy about what I wanted to do in the group and what else I wanted to get out of it and I thought: I want to regain my sense of humour; I want to show more of myself to all these people who can't see me very well. So I got everyone to stand up and I changed their seating positions. I placed them in chairs in relation to each other so that it would either change some kind of relationship which had been set up or it would stimulate another relationship or provoke something else. So the idea was to change what was going on and to stir it. (E1a) I was taking a personal risk. I was putting myself in a kind of position of power and yeah, mainly I wanted to see what would happen to me and whether I could tolerate it. So when I'd rearranged everyone I sat down in the middle of the circle just to see whether I could stand being there. I felt very uncomfortable and didn't know what to do." (R2)

"What happened while you were sitting on the floor?"

"Well various people got a bit irritated. I put all the women together and then sat down with my back to them. (E1b) One of them made a comment about feeling like the Supremes."

"Why did you put all the women together?"

"Because there were so few women in the group and I wanted them to support one another. Actually I wanted them to support me because I felt excluded since the day I was nasty to Andy. (H5) That had got sorted out during my lesbian announcement. (E3) I'd already told one woman and she felt better when I told the group because she felt it was quite a responsibility that she knew and no one else did.

Anyway I knew it wasn't right, me just sitting there in the middle doing nothing. So I just walked around to get back my strength or courage or something. (E1c) And then I sat down and it so happened that the empty chair that was left for me was in an interesting position. Then I just watched what was going on. There were several verbal fights in the group that seemed to be to do with leadership and power. (UR) Even Ken the leader said at one point: 'You set this up.'

So we went through a lot of confusion (R4) and the guy on my right kept on having flashes of insight and kept saying 'Aha'. He kept fitting everything in and that was exciting. It took quite a long time but all sorts of things were sorted out and everything fell into place. (UR)

Another thing that came out was about androgyny and various people in the group began recognising parts of themselves that were other than the gender they associated themselves with. One of the men said that after I had got up from sitting in the middle of the circle and I was walking around, sometimes I was very male and sometimes I was very female and it kept changing and this was very apparent." (R2)

"What did you learn from the group?"

"I felt somehow that everyone had caught up with me, that they had recognised my position in themselves. It also somehow became OK to be lost and it's to do with tolerance of uncertainty. I felt other people were with me in that and so it was a validation for me. (R4) Several people gave me quite positive feedback afterwards; they felt that I had understood them. It's a kind of unconscious instinctive understanding. (R3) Oh yes, there was a joke about my unconscious being the sixth trainer in the group and Andy actually thought he and I had been the leaders.

The most important thing I got from the week was the feeling that I belonged in the world. (R2) There was a multi-dimensionality in the group but also a patterning. Everything fitted in and each person had their place in relation to the whole and each person had to be there so, I had to be there and I had to be in the world. So I'd come back to earth and I felt I'd come home."

"Were you still invisible?"

"No, I wasn't." (C1)

"How did you feel after the group?"

"The first few days after I was very positive and effective in stirring things. It didn't take any effort. I felt that I had the power to change people and get them moving. (C2) Generally since then I've felt pretty happy." (C3)

K. is quite a neat experimenter. She has dealt with all the issues she raised for herself in a surprisingly systematic way for the quite complex situation she was in. One of her main hypotheses was that she is often invisible to people, a theory which she elaborates on. By the end of the experiment she had certainly achieved visibility. Her experiment, like her H4 was nicely open-ended allowing all sorts of possible outcomes—a bit reminiscent of Luke Rhinehart's "The Dice Man". One presumes that her null hypothesis was disconfirmed.

GEOFF

"Some years ago I very much wanted to do a Ph.D. in Einstein's theory of relativity. (A) This was after a fairly protracted rejection of the whole idea of any further study and of my own ability."

"Why was that?"

"Well, to go back a bit, I decided to come to England. (E) One of the strong reasons, I think, was that my mother was here, and I suppose I felt that I needed something from her. There had never been a good relationship between me and my mother. When I got to England, I discovered that in fact she couldn't give me what I wanted. (R) I went through a terrific period of self-analysis. I got into a deep analysis as well with my sister, who gave me lots of information about my childhood, and this enabled me to come to a certain understanding of why the relationship between me and my mother was as bad as it was. After several months of being very depressed, I rejected any more study. I felt I wasn't going to be a mathematician, I completely rejected myself because I was so depressed. (H1)

But one day—incidentally, after having started a reasonably successful affair with a girl called Sarah, (E4)—I had a box in which I had put all my books, all my mathematics and all my Einstein and everything, and I had refused to open this box for six months and one day it suddenly struck me that this box was there and I said 'I am going to open it', and that perhaps was, if you like, a conscious experiment. (E5) I opened this box and I took out Einstein's little book and his original papers on relativity—and I started thinking about relativity, and it was glorious, because I came up with a unique way of seeing some of the fundamental concepts in the special theory of relativity. (R3) I immediately, at that stage, went along to the college, enrolled as a part-time student in relativity under Penrose, and was so impressed with this guy, who in fact is the top relativist in this country. (E6) I felt very fortunate that I was able to study under him at all. He encouraged me because I was so enthusiastic and because I seemed to have something. (R6)

I then went back to South Africa. It was during that time that I started my relationship with Lesley. (E7) The studies in relativity went reasonably well; I worked nicely, I worked hard once I'd settled in.

Then something happened. I had friends and I was getting into a reasonably stable state and then one by one everybody left, almost at the same time, and that was a blow which I don't think I appreciated the magnitude of at the time. I became very lonely. (R7) Then, when Philip left—damn his eyes—he left his record player behind and he wrote to me one day and said, "Would you please take this record player to Jean." And I did that. (E8) The only unfortunate thing about it was that afterwards I fell in love with her and she fell in love with me and we got married. (R8, E9)

We went up to Johannesburg, involved with the family, happily married for a year—yes, everything was nice. I was then doing the M.Sc in relativity from Johannesburg by correspondence.

When the M.Sc was finished and when we'd saved up enough money, I came back to England with Jean and I enrolled at Birkbeck as a student with Roger. (E10) I also tried to get a job in England, some post which would pay me something and at the same time leave me enough free time to study, and this didn't happen. Months went by, and all the avenues seemed to be closed, there was no money to be had, and I got very, very depressed about it, and my studies were not going well at all. (H2) Some people argued to me that it was because my studies were not going well that I became depressed. (H3)

Now, to some extent, that was true, but something even more fundamental was happening, because I found, when I was sitting in the seminars with Roger, that I was not able to really follow what he was talking about. (R10)

So eventually two things happened. I went to see my sister in Israel, (E11) and in the meantime also the relationship with Jean had become just diabolical. (R9) It was tense all day long, so when I went away it was like a great big cloud lifted from my head. (R11a) I didn't have any pressures on me and I was able to clearly think that I didn't love her anymore and I'm going to confront her with it. (R11b) It was perhaps one of the most severe experiments I have taken with myself; and at the same time I decided I was going to quit relativity. (R11c) Up to then, whatever I had tried intellectually, I was able to succeed at, and nicely. This time, no, my intellectual limits had been reached. (C1) I was working at my intellectual limit without having anything in reserve. I know what I'm talking about because the work I'm doing now is mathematics. It is stimulating, (UR) but it's at a much lower level, which means that I've got plenty in reserve for the problems that do come up, and I can solve them easily. Penrose could solve his problems in relativity as easily as I can solve mine in the work I'm doing at the moment. That is the difference.

And so, that is what has happened. I am now, I think, far more integrated through that experience than I have ever been because I know what my limits are. (UR) I don't have to get depressed about it any more. (C3) And through that, other things started coming through, I turned to art. (UR) I've got a good sense of colour, I've got a terrific sense of space and visual spatial relations because I've always been good at geometry in mathematics, so sculpture is almost a natural."

Geoff has done a whole series of experiments on two themes; relationships with women and his own ability as a physicist. Just to clarify, in his H1 he links his frustration with his relationship with his mother to a subsequent depression during which he rejects his ability. In his H2 lack of money might have caused depression which then affects his ability to study. In H3 his poor progress in his studies (i.e., lack of ability) might have caused depression. There seems to be a chicken and egg dilemma for Geoff here as to which causes which. There seems to be a lot of evidence that frustration in his relationships with women causes depression which then leads to him rejecting his ability. However, in his conclusion (C3) he firmly opts for H3. Perhaps the question of what limits his ability and why, is not so important as what he actually does with that observation, which, in the final analysis, seems very positive indeed.

LES

"What is doing your painting for you? Can you conceive it as any kind of experiment with yourself, your life?"

"The activity of painting is just to defeat time, for me. (A1) And so any form of experimentation with myself and my material comes after that."

"What's defeating time about?"

"That puts a value on time. When I paint I'm not aware of time passing. I do this activity compulsively to give time a meaning to me. I'm not too sure why time is important. But that's just a starting point. For me its also an organic process.

I think I can say I'll be painting in 50 or 60 years time. I can say that because I don't have much choice at the moment, its not a choice for me." (C1)

"You couldn't not paint?"

"No, it's a muscle, a muscle I have to exercise, I mean, I do go through patches of self doubt . . ."

"You've always talked about it in terms of a struggle."

"I've found in retrospect, that if I persist, and there have been millions of instances when I've given up and given into the distraction. But at the exact instance when I persist, something happens. And that's what I mean by the struggle. If you can hold on long enough—even if its six months of putting something away (E2a) and say I don't know what I'm going to do now—in six months time I'll take it out and have a look at it. (E2b) Even to destroy it, that's alright. (R2) I won't throw it away until it's really obvious that it's trash—it often is. (C2)

I recognise that I make a lot of decisions that are to do with quality. (A2)

I found about three years ago that I could put chance and instinct into the work. That followed a long period of very tight intellectual preconception and of being very rigid. I suddenly realised, you know that chance, options, the unpredictable! And that after many years of not being surprised by myself, but being successful. But the actual producing of the work wasn't exciting. (H5) So I had a sort of a watershed and the mathematical thing fell apart. It wasn't replaced by anything and I felt very frustrated. I began throwing paint on canvas without planning it. So I began throwing more and more paint. (E5) Yeah, I did a lot of work like that. It was very exciting. (R5a) And in that period of painting I found that I could reproduce a stroke of paint to within an inch of what I wanted, to within an eighth of an inch of thickness of a stroke. But I'd actually perfected the technique of making a gesture consistently time after time, (R5b) and that as much as I wanted, you know to be instinctive or free of an intellectual process, it was going on all the time. I was evaluating. (R3) You know, it's not good enough to have good intentions or be sincere. I think the ability to evaluate and for it to be as instinctive as the need to enjoy the painting is the key to things that last, that remain interesting."

"Was leaving Australia something to do with your art work?"

"In the sense that I am what I do and I realise now that that works the other way around, that I do me, I couldn't do me in Australia. I didn't know it at the time, I knew there was something dreadfully wrong in my fourteen stone heart of hearts as I picked up my baggage and emotional pieces of broken friendships and nightmare faces and friendships returning again and again, and ran.

I trust my despair, I'm not frightened of being alone. The worst things that happen to me alone are acceptable, not tolerable, but acceptable because it's mine. It's my pain and my experience and I want to follow it down the line to see where it goes. I didn't use to trust my ability to go down the line with people. (NH) There have been times when, if there has been a question of fulfilling a relationship and good work conditions, I've chosen good work conditions, with all the costs involved to other people and to myself. Recently, I've begun to correlate the changes in my work to produce a good painting, not directly and not always, but sometimes, which is a step from nothing, with my changes as a person. (E) I didn't use to trust my ability to go down the line with people so that's new and that's exciting. (R) I might find it doesn't get anywhere I might go back to the loneliness.

There is another part of my mind that is aware of success and failure and that is often divorced from the process, and they are genuine end products. So yes, in 50 years time I will be painting but I may be a very bitter person, because I'm not successful." (H2a)

"What would you consider success?"

"I want success." (A2)

"What is it?"

"Everything. You name it and I want success in it, money, prices of work, (A4) quality of work, consistency of production, satisfaction. (A5) You know, I just want. What interests me, is the degree to which I have persisted without success. I think I will be very stubborn though. I think I will not accept failure." (H2b)

"How did it feel talking about your artwork?"

"I think that's the degree of ruthlessness in me. It doesn't matter, you don't matter to me and the machine doesn't matter. But if I can use you to string a sequence of words together, which I haven't done up to now and which will enable me to see something which I haven't done up to now, I'm prepared to be generous, even stupid, in the hope that something will come out that I just haven't come across before."

"Did anything come out?"

"Yeah, sure. Just realising how conscious my new figurative work is becoming and it's got to do with the world and with people."

Les' painting is an experiment with his whole life. What is striking is how many aims are included in this activity. It could be that this is not really an experiment at all. Usually being alive is not in itself an experiment for people because they do not have much choice over it, but within the context of life, people do experiment. Les' involvement with art has that same quality of life itself. Yet within the context of his art he is experimenting with techniques, personal satisfaction and excitement, the quality of work and whether certain relationships will help improve his artistic work or not. When it comes to quality he's an extremely honest experimenter. When it comes to success or failure, he is setting up some pretty definite criteria but on the other hand he is virtually saying it will need an enormous amount of evidence to convince him of it, and it sounds as if there will never be enough.

ROGER

I first got into politics in the early 1960's. It was quite an exciting time because it just followed the Campaign for Nuclear Disarmament. This had attracted a lot of young people to come into politics. There were a large number of people at my school who were in some way associated with the C.N.D. and lots of discussions took place. Also one of my favourite subjects was history, especially contemporary European history which was also tied up with contemporary politics. So I became interested in politics then, but I didn't become in any way active until 1966.

I remember watching the election results in 1964 on the television and I thought it was very exciting. (H1) When it came to the 1966 election, I volunteered my services to the Labour Party who I decided most represented my general opinion in politics, which weren't terribly well formulated. After taking part in the 1966 election, I was approached to become a member of the Labour Party and I did. And then I was approached by some people at school to join the Young Socialists which I did in 1966 with a considerable number of reservations and just gradually got more and more drawn into the spectrum of politics. Going on from there, I became active in various ways. I attended more and more meetings. Took part in discussions and debates. (E1) I gradually discovered that my viewpoint was beginning to become more definite. (UR) When you haven't got very clear opinions about where your political beliefs lie what tends to happen is that they are very erratic at the beginning, they fluctuate greatly from one side of the spectrum to the other. (H2) You go through a period where you are trying out all sorts of different angles (E2) and then eventually they tend to crystallise into a more definite form within the context of the party of which you are a member. (R2) And this is what happened to me.

For example, I couldn't see any strong argument in Britain giving up its East of Suez role. It struck me as a rather way out viewpoint that a lot of Britain's economic problems were somehow tied up with Britain's traditional role as a world military power. My views on defence would have been termed very right wing and then over a period of years they have become very left wing. And I think this is what happens over a number of issues and it's really a question of trying to find your own philosophy and your own viewpoint. I think the only way you can find it is by challenging various ideas that you have. Well, for example, you listen to what other people have to say who have different viewpoints on the same particular issue, you ask a lot of questions about it, you read different view points and then also as you become active in politics you begin to see how many things are inter-related and how issues don't just exist in their own right but are a part of an overall context which is political philosophy (UR2) which everybody has, but the difference between somebody who is politically active and somebody who isn't is that somebody who is politically active inter-relates most of these things within a political ideology whereas non-activists tend to regard them as one-off issues. (C2)

On the purely personal level apart from the philosophical reasons, there was also the egotistical reasons. In politics, I get a great kick out of the excitement attached to it, it makes my adrenalin run faster and it makes me feel good and there is a self-perpetuating momentum attached to it which I personally enjoy. (H1) Also running for the Council elections (E3) and winning it, (R3) gave me a sense of achieving something which is confined to a relatively small number of people, it's given me a sense of personal fulfilment." (C3)

Roger's series of experiments are to do with becoming active in politics. The first hypothesis is that it needs to be exciting and enjoyable, the second is that you begin with very unclear and erratic ideas and gradually develop a political philosophy.

His experiments in politics are interesting because many of the early ones are very open-ended and hence, full of unpredicted results. This type of experimentation seems very appropriate for a learning situation. He seems to have benefited from this approach and learned a lot because his E3 is a very precise and not at all an open-ended experiment, and he wins.

J

"I think I tend to assume that if things have changed, they change for the worse (NH1) so therefore it's difficult for me to make changes in my life. I tend to look at the world I've got now and think well, it's a pretty good world, whatever else could I get out of it if I changed it? So I resist change very much."

"So what brings you to the point where you are prepared to contemplate it?"

"It usually seems to be forced on me from the outside in some way. Well, let's say the opportunity comes from the outside and I then have to decide whether to accept or reject the opportunity. My immediate impulse is to say no. I would never volunteer for anything like that. Like that T.V. programme I did. (E3) in fact my first impulse is to be angry with the person who disrupts my style of life. But for some reason I don't refuse it in fact!"

"What do you think draws you on?"

"I think its partly a question of pride. I'm afraid of people saying, you know, J's so narrowminded, (NH2) she won't take any chances. (NH3)

Something like that. I saw it as something I had to do to oblige other people. Don approached me to do it and I accepted, not to let Don down. I also saw it as being something that would be good for other people, like it would be good for my patients. (H4) I was prepared to be adventurous on other people's behalf.

What happens is, I'm frightened about these things before they happen and when it actually came to the time, I actually coped with it quite well, because I had worked through all my fears about what would happen beforehand. They were fears of being shown to be inadequate in front of an enormous television audience. (NH5) I think one of the fears is 'who does she think she is?' You know, sitting there on television pretending she is a well known group therapist. (NH6) So again that was very much concerned with other people's opinions. I've always been afraid of teaching situations or seeming to know. Not wanting to put it over that I was a good pyschologist or a good therapist. So really it was a real kind of testing situation for me because, you can't get people watching you any more obviously, than on a television, can you?

I felt in charge of the situation and able to cope with it in a very surprising way. (R5a) We had to use some group techniques, like feedback, role playing and family sculptures. They interviewed each of us and asked us what was the point of the various techniques and what were we trying to do, which I seemed to be able to explain with great lucidity which surprised me, because, I had thought I would stutter and bumble and not know what I was talking about."

"Then there was quite another phase, actually seeing yourself on television."

"Yes that was quite different. By the time the film was finished, I was already quite pleased with what I'd managed to do. But still didn't know how I'd look from the outside, so it was still another obstacle to overcome, actually sitting down to see what I looked like. I didn't actually stand on my head or look particularly opinionated or anything. (R6) I looked like a normal human being. I also thought I looked very nice actually, and sounded super. (UR)

By the time that was finished, I felt very confident for some time, (R5b) I felt well, from the outside, I look as if I know what I'm doing—I may not believe it from the inside. I remember when I was doing the interviews, by the time I'd finished talking about what role playing is all about I was by then believing in it. I was doing a sort of role play myself. I was role playing a person who believes in role playing. I wonder, now that I look back on it, whether I should have, before it happened, been able to guess what good things I could have got out of it too. (C1) Because I think in fact, I got more out of it than the patients did (R4) and yet if it had been left to me, I would never have done it.

I think I've got some strange idea about change being for the worse

Something will be lost if something's gained. (H1) I still think its only acceptable to change if it's forced on me by others." (C4)

All her hypotheses were confirmed and her null hypothesis disconfirmed and she begins to question her overriding fear that change is for the worse. But she does not entirely give it up and refuses to be an active experimenter. Change still has to be forced upon her. By saying she was only role playing she is beginning to invalidate evidence about her ability to cope with work. T. likes the evidence but is not very good at following through the implications to change her original hypothesis, however gloomy. She seems to have a great need to stick to NH1. One would have to talk to her about her history to find out what brought about her fear of change.

STEVE

"When we were first married, we knew that we would have children, (E1) perhaps 2, perhaps 3. The main decision was when. (E2) In order to decide when, various assumptions, hypotheses and predictions have to be marshalled, some short-term, others long-term. We wanted to have them sooner, rather than later, so that we weren't too old to remember our adolescence when they were going through theirs. (H2a) We also wanted our children to be 'chronologically' close. We imagined these preferences would be beneficial, of course. (H2b)

In practice (and in retrospect), I think it was far less important deciding *when* rather than *whether or not* to have children at all. (C2) It is easy for parents to underestimate the problems and disadvantages of childbearing, especially the early ones. (UR1a) It seems to be that cultural and family practices and attitudes have emphasised the joys and delights of parenthood. (H3) The expectations and preparations for parenthood are often totally unrealistic. (R3) I was struck for example by the total irrelevance of ante-natal classes which we went to. (E3)

We were just unaware of the amount of time involved in looking after a young baby; of the physical and psychological strains imposed by interrupted sleep; and of the number of decisions that have to be made and agreed to regarding feeding, sleeping, toilet training, babysitting, and so on. (UR1b)

It seems to me that there is a conspiracy to avoid unpleasantness and not to face difficult situations in traditional roles. (C3a) That's unfortunate I think. It does get through, but the lesson is usually a harsh practical one!"

"How does all this alter you personally?"

"There are some things at the moment that I can't get back to. Janet's given up work so I'm the sole provider and I'm occupied quite a bit, 5 days a week and 3 nights a week. (R6a) The only thing I indulge myself in at the moment for physical exercise, is squash, (E6) which I enjoy immensely, (R6) it's so different from the work I do. We don't get out very much anymore, (R4b) time is at a premium now, but I can see my way clear of that, I can put up with it for a couple of years. Certainly, in the first 12–18 months of the child's life, I foresee a predominance of problems and restrictions, (H4) overshadowing the pleasurable landmarks (smiles, teeth, crawling, etc). I personally will look forward to my daughter's company more when she can walk and talk, when the relationship is more reciprocal; when she is far less dependent." (H5)

"Have you any particular hopes as to what sort of daughter you would like her to be?"

"I haven't really looked that far ahead. I don't think I have any long term expectations that I would be very disappointed if she didn't fulfil them.

I think I can see having children as an experiment, (E1) but one for which there can be no adequate preparation, except having done it before." (C3b)

"What do you think you got out of it?"

"I think the biggest thing I've got out of it so far is not quite so definable and effable, it's just the kick, thrill, or boost or knowledge, of having your own offspring, whereas before you didn't. It's the creation from nothing, it's a very vague thing which you don't think about very often, but that's probably underneath it all and that's what keeps you going I suppose, from one landmark to the next and keeps you going over the first 18 months. The actual pleasure of having someone there, who is now your daughter." (C1)

I put H3 as his hypothesis because I think that was his expectation too. This is a heavily loaded experiment with a vast number of results, a couple vaguely positive and a lot heavily negative. Conclusion 1 and 2 seem contradictory, conclusions 3a and 3b do too, i.e., there should be adequate preparation in antenatal classes, but even so it may not help. The problem is that despite his awareness that it is an experiment that went badly wrong he cannot act on that. He cannot opt out of this particular experiment, it is irreversible, hence the emergence of contradictions as he looks for something positive to cling to. He is coping by projecting his positive hopes into the distant future and trying not to have anymore definite expectations.

GRAHAM

"As those who have read it know, incorporated in Shakespeak Vol. 1 is such a thing as a dictionary, a play, and a tabulation of locations of spaceships, a Sufi story, a mock satirical emulation of the Pope, and then again there was a bit of philosophy which was written in a crypto-eighteenth century style and even a totally random index at the end. I was looking at all these literary forms and trying to see what I could express through them. (E1 to 7) I can't really disentangle in what sense the book was a literary experiment (H1) and in what sense it was a write up of an experiment I was doing on myself." (H2)

"What was the idea of having lots of literary forms instead of just one?"

"Well, I would have to go into the background. Having been through a period of self therapy to sort out where I was at, I then found myself in a space where I was seeing the world as being completely ambiguous. You know, if I wasn't a pyschologist, I would probably in some senses be classified as suffering from schizophrenia. It wasn't at all clear to me what level of meaning one should attribute to any given signal in the environment, because it seemed possible to attribute to them meanings at a whole variety of levels. In a way that's why the whole meaning of Shakespeak hit me. First of all it hit me as an opposition to all the world's newspeak. But then also it seemed to me that all things at one level were very funny and liberating and at the next minute they seemed terribly kind of oppressive.

Initially, I had this idea that part of the problem with science fiction is that you are committed to a plot and what I wanted to do was get away from the notion of a plot because what I really want to do is just depict a situation. The way round this seems to be that you can build up in a pointillist fashion by dictionary, an image of situation. That was the first idea. (E1) I then had the idea that . . . I was then at the time getting very stoned. I mean it would be intellectually dishonest to try and pretend one wasn't getting stoned a lot of the time. But then so was Baudelaire, so was Rabelais, so was Van Gogh. I don't think that invalidates anything. I was in a sort of trip frame where it was a question of actually reporting what was going on as it was going on. (E2) In some parts, I was walking around like Chaucer with this notebook and every time a new kind of word hit me I would jot it down, and I get back in the evening and I'd think, 'what could that mean?' And I'd think up a jolly definition for it. In fact the sequence of words and sequence of definitions seem to constitute a unity. It was almost as if I was monitoring what was going on in my unconscious. In fact this first section, the dictionary, was actually done overnight, when I first hit this idea. Each time during the night that some idea hit me, I would write it down. And so I went through a whole single night in effect reporting on my dream sequence and so in some places it's quite bizarre. I think I'm the first person to have done that.

But then, getting back to the question of why I had chosen a lot of different forms. I think in a way, I was refusing to be limited. I was saying to myself, 'don't set any limits on it, just let it happen, don't make any judgments, don't make any discriminations. Play it like it's happening.'"

"That sounds a bit like a commitment to ambiguity or a multiplicity of meaning."

"Yes, very much. But then I feel in some senses artists are working on the edge of chaos. (H3) That the job of both the artist and the scientist is to put some order into chaos before it hits everybody else. Your average man imagines that everything is really under control, that there are experts somewhere that understand it. But we are moving through times where changes are just so bloody fast, so what defined reality for the average man in say 1975/6 are the kinds of films and books published in that year and that's the kind of thing that keeps it together for everybody.

As far as the identity problem is concerned, I think at the time I started writing the book, I don't know how it looked to other people, but to me it looked like my identity was of some academic in his 30's, who's just bust up from his wife and he is sorting himself out, which I didn't particularly dig as a really interesting identity. Well, the next section was as author of Shakespeak. It was like that would be my identity for the period of writing that book. I also felt that it wasn't sufficient, that it wasn't a good basis for living. There's the problem of getting labelled.

After I finished, (R1) I felt relaxed, I felt as if I had succeeded in doing something. (C1) I still don't quite know . . . obviously once you have done something it's out of your hands and it's up to posterity to decide whether it's good or bad, significant, mad, sane or whatever. I do not know which it is. I haven't got a clue how it's going to be received. I genuinely still oscillate between thinking that it's really incredibly good and the most important thing that's been written in English Literature since Ulysses (C1a) and thinking it's the most incredible load of drivel ever unleashed on the public under the guise of creativity." (C1b)

Graham's is an amazing experiment with his identity. He is toying with the notion that either he is a recently divorced academic who is going mad or a creative writer. He succeeds in becoming a writer while integrating part of the former. But the fact that he has to use dope to get into touch with the madness part of the first identity, suggests that the distinction if becoming clearer. The result in my opinion is a profound, imaginative and excruciatingly funny book.

PAUL

"The two areas I'm working in are both very animal and very experimental. Both should have something to say about people. (H1a) The areas are memory and depression. I would like to be in a position at some point to start trying to bridge what gaps there are. Memory, I was working on for a Ph.D. (E1) and I'm dropping it now. I started with the problem, what is memory physiologically? (A) There are many ways of studying memory. I chose a rather esoteric approach which was working on the simplest animal I could find, which would learn something, which turned out to be a headless cockroach. (E2) I did get a Ph.d out of it, (R1) but the actual work didn't get anywhere. (R2) I'm taking it as given that the physiological understanding of memory would have interesting implications—I'm not sure what they are. The results might tell us something about development, about ageing." (H1b)

"You mean relevance to people?"

"Yes."

"I'm curious about your choice of the cockroach, why is a decorticate cockroach a simple system?"

"There were at the time two such systems on which it looked sensible to work. One was the Californian Sea Snail, but that's very nice only if you are in California, the other was the isolated ganglion of the cockroach."

"Did you find any learning taking place after decortication?"

"It wasn't actually decortication because they don't actually have a cortex. They have effectively a series of small brains which control different parts of their body. They have 3 pairs of brains. You then cut off their heads which means they've got no antennae, and they've got no eyes so they can't sense very much and their mouth has gone so they can't feed, so you are a bit limited. So what you do is suspend them in mid air and they have a leg waving around and you put a bowl of water under their leg and give them an electric shock whenever they touch it. They then learn to keep the leg up in the air. In the refinement of this technique, which I developed, you can do the opposite and give them electric shocks for not putting their legs in the water. (E3) You can use yoked controls which are shocked whenever your animals which are learning are shocked. So that the change in their behaviour can be explained on the basis of shock alone, it's an association of shock and position, and they learn these things—with a little difficulty. But unfortunately for a student of memory, a few minutes later they have forgotten it all, (R3) which is a terrible drag, (R4) which means that this system is totally useless for working on memory, (C2) which is why I'm not working on it anymore."

"It seems an enormous leap from that to human memory?"

"Ah, but you wouldn't do that, what you would do is use any such result as a source of hypotheses for testing on more complex animals. (H1c) But as I said the problem with the more complex animals is that you don't know what to look for."

"I'm interested in why you are particularly interested in animal experimentation".

"I think it was very largely accidental, at least there is a big accidental component. I did my degree in Oxford and in those days you studied psychology in combination, either with physiology or with philosophy, I had no time at all for philosophy."

"What didn't you like about it?"

"Lack of achievement in the field. It seemed to me then a total waste of time and never has since seemed much less than a total waste of time." (C)

"Is there anything about the particular area of memory which poses interesting questions to you?"

"Not as such, no, I think the area of memory is one of the most important and interesting questions we have left and I'm fascinated by the fact we have absolutely nothing worth knowing about it. I suppose there were reasons over and above why I was interested in animal work. I have a bias towards experimental work. Now doing experiments with animals is really a lot more fun than doing experiments with people." (H4)

"Why is that?"

"I don't know, I haven't the first idea why one enjoys the things one enjoys, I don't know, I haven't tried to analyse it. It might be a feeling of control over the world. (H5) I don't know. The other thing is you have very much more experimental control over what you are doing than with people, so you can plan much less ambiguous experiments, that's quite agreeable. (H6)

I've moved on now to work on anti-depressants. I want to know what effects they have on the brain. You could give them to animals over long periods and look for unusual effects . . ."

Paul's experiment is very precise and its results clear. There seem however, to be contradictions between his hypotheses 1a and 1c which do rather undercut his rationale for his experiment. Another possible conclusion which he doesn't draw is that decorticate humans with no eyes, no sensation and being unable to feed wouldn't be able to learn either. But as there is no direct evidence available for this we would have to formulate it as an hypothesis.

MYSELF INTERVIEWED BY K.—ONE OF MY SUBJECTS

"What did you want me to ask you?"

"Just generally why I did the experiment which is a series of interviews on the topic, 'Experiments with Life' and what I think I learned from it. Also whatever you want to ask me about."

"What was your initial idea?"

"I felt that a lot of so-called scientific experiments in psychology were kind of missing a point, perhaps missing the main point. I was taking the notion that man is a scientist whether he is in a laboratory or not; that there are a lot of people out there struggling with life, making sense of it, experimenting, finding out, checking their hypotheses and speculations and changing them, and that this is basically a part of living." (H1)

"Do you see people as far more complex scientists than other researchers tend to do?"

"On the whole, yes. I think that people are conducting multiple inter-related experiments all the time (C1) and I wanted to explore that and whether it wouldn't inform and enrich the particular way we tackle experiments in psychology."

"How did you choose the people you did choose?"

"First of all, I thought I would ask anyone picked at random if they could construe part of their life as an experiment and you were the first and you picked the experience of going to a T-group. And then it occurred to me that there were in fact specific questions I wanted to ask of people and hear the answers of. So instead of trying to keep myself out of the interviews I spent some time asking myself—if I was going to interview so and so what would I really like to ask them to talk about. Out of those I picked eight that interested me most. It turned out that mostly the questions I asked of people had some relevance to me personally, so I really learned a lot. (C2)

I think it also made a difference in that I was obviously interested in what people had to say and they responded very well to that and at some point the interveiw would take off and the person would really tell me what it was like being in that situation. You could tell by the way they began to flow and become quite lyrical at times. (C8) Unfortunately, I had to cut the interviews down a lot—they were originally an hour long each.

I must point out that Paul's interview was quite different from the rest. I didn't ask him about his experiments with life as such, but about his experimental work. You see, after doing five interviews it suddenly struck me that I wasn't really sure that experimental psychology was still like I remembered it and whether it was such a contrast to what I was doing as I thought it was. So I was going back to source as it were. (E) He confirmed for me that there was quite a contrast. (R) His work fits into all the descriptions of scientific method I say that I'm no longer using." (C)

Then your interview with me was of course last because I was talking about the experiment so I had to have done it first, or at least most of it because it's still unfinished as I'm doing it now."

"Yes, it's a real dialectical thing isn't it? You are still in it now."

"And yet I'm talking about it."

"How does that feel?"

"I'm enjoying it. I think though that it would be more appropriate to describe the situation as reflexive. You are experimenting with my having experimented with your experiments with life. (C6) It seems to me that the idea of reflexivity is really very interesting, it has lots of echoes in my life. Even the word echo is a reflexive one. Just for an example, I often see people doing things that I do and I have all sorts of reactions to them which I don't have to myself. Just at a certain moment they have stepped out of me and yet as such are still part of me and it's like seeing yourself on a screen and suddenly realising what you are doing from a somewhat different perspective." (H3)

"Could you see that happening in the interview?"

"Oh yes, there were all sorts of echoes for me. The one that springs to mind most is Les and his artwork. Perhaps what he said is closest to me at the present, but they were all important to me in different ways. Roger's one felt very familiar. I've been quite involved in politics too. (R3)

You know what was also fascinating? It was unusual to give people the opportunity to talk about themselves for a whole hour and it was even more unusual to give the dialogue the legitimacy of science. I said to people that the interview was part of an experiment I was doing related to the philosophy of science and in particular I wanted them to see if they could make use of the idea that they actually did experiments with their own lives. (H4) Like I was doing with you a few moments ago, I was putting a pretty definite framework on my interviews with people and out of that came something quite fresh and new. In a sense *they* made the experiment work, they and you used these rules and found them useful. (C5) I particularly included Les' comments at the end of his interview as an example. It gave him the opportunity of possibly learning more about himself. I felt in a very definite way people were experimenting with my idea about experiments. That's pretty reflexive." (C4)

"What model of science did you take?"

"Well, to begin with I was taking the notion that science, its methods and philosophy are in the process of change and development and that no method we have yet, strictly defines science. In psychology we have the logical positivist tradition as expounded by Hobbes, Ayer, Bertrand Russell, Popper, Medawer and Broadbent for example. Then there is the Bridgeman school where concepts are strictly defined by the operations one can perform. And then there is a not terribly unified group sometimes called, by their opponents, humanists. Scientists like Bronowski, Kagan, Polanyi, Liam Hudson and George Kelly to name a few. I would align myself with the latter group from whom I feel my ideas have gained a great deal of nourishment. Amongst the community of scientists there are often hot debates as to what can be included as scientific or not and I am asserting that what I have done is a scientific experiment, although many would disagree.

I feel that in the quest for rigour and in the attempt to make sense of the complexity of human beings, experimental psychologists have had a very restrictive notion of what they can legitimately do and have consequently wandered off into what Polanyi so aptly calls a "desert of trivialities". This is largely due to the fact that the rules for the choice of a topic for research are based on the hypothetico-deductive method where each proposition or hypothesis should be logically derived from previous research findings. This is part of a reductionist model of science which more precisely, assumes that you can study people by looking more closely at smaller parts which supposedly make up the whole, e.g., neurones, physiology, the elusive behavioural atom and even animal behaviour. Each new finding is supposed to be another building block adding to the edifice of knowledge already obtained. There have been many arguments against this view and I don't want to go into them all, but the general theme is that human behaviour cannot be reduced to anything else otherwise it loses its meaning and its sense of value. Zangwill (1956) has also pointed out that in any case psychology has not by any means achieved a coherent body of scientific laws by using this model.

Despite all these arguments, the rules of research topic choice based on reductionist form of hypothetico-deductive method are still followed. I suggest you look at the titles of all Ph.Ds in psychology done last year. If you can understand them at all, my guess is that you would find yourself yawning pretty soon. They are not interesting partly because they are not related to the personal interests of the researcher.

Closely related to this formalism in research is the rule that you do not use the word 'I' in research. This is not just linguistic modesty; it's supposed to be in keeping with the idea that the research being done is the next logical step and that the research worker as the faceless servant of science, is interchangeable with any other research worker or preferably a computer, and that any notion of personal interest or bias is thereby done away with. Of course it is some time since worried psychologists with a modicum of integrity have furrowed their brows over the work of Orne (1961) and others who found that experiments tend to work out the way experimenters want them to. Some have since taken to sententious finger-wagging to safeguard against this evil, whilst others have ventured more timid suggestions that perhaps research workers should state their values and intentions beforehand. But not much is ever done about it and the ban on 'I' remains.

We could well start experimenting with putting the "I" back in psychology. That's what I have done in this experiment. Related to this I am also using the approach of clearly stating my values and intentions. (C2)

I hope this won't be confused with the old hoary issue of introspection. It's still in some academic circles considered a dirty word. But actually the wolf has crept in in respectable sheep's clothing—garbed somewhat dowdily I might add, as attitude questionnaires, opinionnaires and the enigmatic interview, structured or not. In fact a large portion of psychological research using human subjects relies on their subjective reports of what they are perceiving, hearing, thinking or emoting at the time. So the argument as to whether it is acceptable or not is becoming ridiculous. We cannot proceed without it." (C3)

"What about the hypothetico-deductive method. Have you used it?"

"I think I and my subjects were to some extent using it. (C7) But I must qualify this. It only forms a small part of it, just as the hypothetico-deductive method is only a small part of science. In fact Medawer (1969) says there is nothing particularly scientific about this method either. I found that an interesting statement as he uses it a lot. Anyway, I want to point out that there are several formulations of this method. One I have described already which embeds the hypothetico-deductive method in a larger model, i.e., the reductionist one. This model also has in it the assumption that what we are after is universal laws concerning behaviour. Medawer limits the claims of this model. He says that the hypothetico-deductive method is a very valid way of proceeding once you have your hypotheses, but can make no statement as to how you arrive at them. He says we arrive at our hypotheses intuitively and by inductive logic which is extra-scientific. Scientists like Bronowski and Polanyi are vehemently opposed to splitting off the creative process of searching for hypotheses from the rest of the scientific endeavour and I agree with them. However, Medawer is really side stepping that by saying that you can ask whatever question you like, the hypothetico-deductive method is only applied afterwards. The idea of the edifice of accumulative facts falls away and instead of deriving universal laws, you get their poorer cousins, high probability statements.

We only have to turn to Asimov's brilliant science fiction triology, the Foundation series, to see clearly illustrated how statements of high probability can tell you nothing about any one individual's behaviour. It only enables us to predict the behaviour of large groups or populations.

However, understanding any one person's behaviour is not totally outside our grasp. It is more true to say that our methodology to date is very inadequate for this. Some may argue that it is not the proper subject matter for psychology to be studying one person's behaviour and that it is just general predictions about human behaviour we are after. I think it depends. If you are a politician or a town planner or an actuary you would be interested in the general predictions. I am a clinical psychologist with a waiting room full of troubled people each with a different problem in their lives and I know I have to understand each one of them separately before I can help them. Some say that psychotherapy is an art. I think this is a meaningless and less than useful distinction. I formulate hypotheses about my client and how to help him and I test them out and whether he feels better or he doesn't, I then check back on my hypotheses and reformulate or confirm them. And the hypotheses are linked to my theory of human behaviour which is partly grounded in current psychological knowledge and partly grounded in my own experience which is both personal and to do with checking that knowledge in reality with people.

So getting back to your question, I think I and my subjects were using the hypothetico-deductive method, but it's a third formulation of it. (C7)

There is another aspect of the model of science I am using which I think is very important and that relates to the notion of empiricism. Many psychologists seem to believe that the empirical method is to do with making observations which are then quantifiable and subject to mathematical and statistical analysis. In fact I discovered much to my surprise during my reading for this article that quantification is not an essential part of empiricism at all. Bronowski (1956) quotes Macbeth as using the empirical method in the following:

> 'Is this a Dagger, which I see before me
> The Handle toward my Hand? Come, let me clutch thee—'

The thing is to be tested by its behaviour. I really like that example. Nowhere does a single number appear. What he is also saying is that the data from the senses is the way we come to know. The implication is also that there is nothing particularly scientific about using mathematics or statistics. In fact, most psychologists can tell you, you can lie just as easily with statistics as you can with words, yet it has lent psychology an air of scientific respectability for decades and no M.Sc or Ph.D thesis in psychology would be accepted without it.

Also it seems to me that the use of numerical empiricism in psychological experiments has the disadvantage of producing a condensation or loss of meaning. For example, Broadbent (1973) in one of his experiments on the perception of emotive words found that unpleasant words are harder for his subjects to hear at the 0·01 level of significance. If I was a psychotherapist I would regard it essential to know which words, and in what context and why .The kind of empiricism I am using in this experiment may be able to tell you these things. (C1)

I am also aware that the trouble is that because clinical psychologists are gaining non-numerical empirical knowledge, they don't communicate it because in current psychology journals that just would not be accepted. And moreover they would probably be afraid of ruining their reputations and careers if they did. The tragic aspect of this is that each clinical psychologist is becoming his own scientist and each has to start again from the beginning and struggle through similar problems, issues and clinical experience and each is building up his knowledge and skills in relative isolation. Going through the experiences oneself is necessary but there is a great deal which is communicable and which could immeasurably add to our psychological knowledge, but the channels for communication are incredibly restricted and rule bound. Perhaps the old forms are not appropriate for this and like Graham, we need to experiment with new literary forms such as I have done." (C6)

"Are you sure that that knowledge *is* communicable?"

"I think the answer is that in no way have we tested out the limits yet. (H7) We are essentially articulate beings. We have at our disposal the most unbelievably subtle and flexible way of coming to know, understand and act on the world, and that is language. It's so beautiful. Within language we can come to know all kinds of different realities, whole universes are embedded within it. This is illustrated in a most fascinating way by Watson (1973) in his book 'The Embedding'. We are not just limited by language as Whorf has pointed out, although we certainly can be. Language also gives us enormous possibilities for exploring and expanding our knowledge through it. So we are essentially articulate beings and it is through the articulation of our experience that we can make objective, we can make real for others and ourselves (Polanyi, 1958). I think this a valid definition of objectivity. A lot remains uncommunicable and subjective but there is also a large grey area which I believe is not out of reach of our articulate powers. It requires a tremendous self-discipline you know, trying to put into words a lot of our experience. Try next time you feel really good, to write down how it feels and why perhaps you feel like that. It's difficult and we have all sorts of inner resistances and people will even put you down for it and tell you to stop trying to rationalise or intellectualise. But it is because we lack the habit of articulating certain areas of experience in our culture, rather than other areas. For example, you would never be told that if you were giving a blow by blow account of an exciting football match.

I think that some clinical psychologists are developing methods and languages for articulating experience which are very complex and meaningful and I certainly feel it is an area which scientific psychology could pay a lot of fruitful attention. (C1) It's also linked to what Graham said about artists and scientists making sense out of chaos for our society. They do indeed only I'd like to take that further because I don't believe it's all chaos. It can also be quite simple and beautiful."

"Did you learn anything that you really didn't expect, that surprised you?"

"Oh yes, lots. It was a pretty open-ended experiment. I particularly want to relate three things. One was that the notation analysis which I began after transcribing some of the interviews, gave me a quite different perspective on what people were doing in their experiments from when I first listened to them. I began to see what constituted a good or bad experiment and perhaps why people sometimes didn't test all their hypotheses or refused to admit contradictory evidence or drew certain conclusions. And yet other people tested things out for themselves very systematically and with great subtlety and were incredibly honest about their successes *and* failures even at great cost to themselves. Bronowski said that one of the few general statements you could make about science was that it was about the habit of truth. I think in some respects this is a study in honesty too.

Another thing was that I *was* convinced that people did not behave or think according to the null hypothesis and felt that Popper's notion was an artefact which scientists, like most other people couldn't cope with conceptually. And the fact is people can and do and I am amazed and I have proof of it.

The other discovery I made, possibly only a minute ago or less. I was going to write a separate epilogue on this experiment after my interview was finished and I've just realised that I've said most of what I want to say within the same form as my subjects, which makes the experiment into an integrated whole and that is really exciting and rather unexpectedly validates what I'm trying to do. That's really fun and I'm chuckling to myself no end about it."

"This study certainly seems to integrate psychology with the person. How has doing it affected the way you see people in general, and the way you see life?"

"That would involve a more personal statement. What I set out with was the idea that people experiment with their lives and I experiment with my life and I am in the process of experimenting with doing sculpture and I feel that there are links between that and my doing psychology. I am trying to integrate these two things to some degree. It seems to me that I am also experimenting with myself when doing sculpture, perhaps in a more articulate way than I would have, had I not done psychology and learned a lot about science and its creative basis, and also learned a lot about myself and how to express that. So I'm taking all that with me into my artwork even though one day I may get to the stage of not working as a professional psychologist. So yes, this experiment is a statement about me and my way of bridging things." (C9)

"Has this affected the way you see people that you interviewed?"

"Yes, that occured to me too. Having asked the questions I did, I got to know them that much better. It confirms what I sometimes feel, that there are great depths which for some reason people choose not to reveal and that's strange and I'm still puzzled about it. But I do appreciate all those people that much better. I was very moved by what many of them said. I hope I do them justice."

"It sounds like a great responsibility."

"I think psychologists do have a great responsibility and I think they often avoid that. I get very angry at what I see as the systematic trivialisation of people through traditional experimental methods. I don't think all psychologists are like that but there is a substantial body of them that are.

Oh yes, I just wish that psychologists would get out of their laboratories and mental cages, restrictive methodology and rituals of pointless question asking and get out there and experiment with life. I would like to say 'yes', to all the work that's gone before. We have been there and it's time to move on and one of the ways to do that is to elaborate the concept of scientific psychology."

K. was in a sense experimenting with my idea but I don't think my part of the interview can be regarded as an experiment in its own right. It consists mainly of hypotheses elaborated theoretically and of conclusions. The experiment and the results are in the preceding interviews. Nevertheless, conceptually and in form, it integrates with the whole. It is also an interesting reflection on reflexivity in its third stage of reflection.

References

Asimov, I. (1962). "The 'Foundation' Triology", Panther Books, London.

Bannister, D. (1970). Science through the looking glass, *In* "Perspectives in Personal Construct Theory" (D. Bannister, ed.), Academic Press, London, pp. 47–62.

Broadbent, D. E. (1973). "In Defence of Empirical Psychology", Methuen & Co., London.

Bronowski, J. (1956). "Science and Human Values", Pelican, London.

Hudson, L. (1972). "The Cult of the Fact", Johnathen Cape Ltd., London.

Ingleby, J. D. (1970). Ideology and the human sciences. *The Human Context* **2,** 159–187.

Kagan, J. (1967). On the need for relativism. *The American Psychologist* **22,** 131–142.

Kelly, G. A. (1955). "The Psychology of Personal Constructs", Vols I and II, Norton, New York.

Koestler, A. (1964). "The Act of Creation", Danube Edition, Hutchinson & Co., London and New York.

Mair, J. M. M. (1970). Experimenting with individuals. *British Journal of Medical Psychology* **43,** 245.

Mair, J. M. M. (1970). Psychologists are human too, *In* "Perspectives in Personal Construct Theory" (D. Bannister, ed.), Academic Press, London, pp. 157–184.

Maslow, A. (1969). "The Psychology of Science", Gateway Edition, Harper & Row, New York.

Medawer, P. B. (1969). "Induction and Intuition in Scientific Thought", Methuen & Co., London.

Orme, M. T. (1961). On the social psychology of the psychological experiment, *In* "Problems in Social Psychology" (C. W. Backman, P. S. Secord, eds), McGraw-Hill, New York, pp. 14–21.

Polanyi, M. (1958). "Personal Knowledge: Towards a Post-Critical Philosophy", Routledge & Kegan Paul, London.

Popper, K. (1959). "The Logic of Scientific Discovery", Hutchinson & Co., London.

Popper, K. (1972). "Objective Knowledge: an Evolutionary Approach", Clarendon Press, Oxford.

Watson, I. (1973). "The Embedding", Quartet Books, London.

Zangwill, O. L. and Pryce-Jones, A. eds (1956). "The New Outline of Modern Knowledge", Gollancz, London. p. 168.

Construing in a Detention Centre

Margaret Norris*

A Detention Centre, as Don Bannister commented when this chapter was in an embryo stage, forms a "natural laboratory", and the findings reported here describe some of the effects of custody in such an environment upon a group of young men. The research project was designed to investigate the usefulness of a repertory grid test for an evaluative study concerned with a population which included some "deviants".

"Follow-up" enquiries in that study were not only, as with this sample, difficult, but in some instances quite impossible. Without such enquiries evaluation of the effects of different kinds of environmental circumstances by measures such as recidivism, even when this might otherwise be appropriate, are unsatisfactory because of the numbers of unknown intervening variables.

The theoretical focus of the main project was concerned with changes in self-esteem and in the self perceptions of those involved about rule-breaking and independence, which were also relevant in studying change in a Detention Centre.

A series of studies including Tannenbaum (1938), Lemert (1951), Hewitt (1971), Reckless (1957, 1960, 1961), Reckless and Dinitz (1967), Schwartz and Tangri (1965), Jensen (1972) and Quinney (1970), argue that people with high self-esteem are less likely to indulge in deviant behaviour and that a self percept as not deviant or not delinquent assists in insulating against deviant behaviour. The acceptance of a self percept as deviant is an important stage in the adoption of a deviant career, according to Lemert (1961), Becker (1963) and Wilkins (1964), amongst others. Self percept as independent may be related to self-esteem and was also of interest because of suggestions by Goffman (1968), Clemmer (1962), Sykes (1958), Morris (1963) and Cohen (1972) that custodial treatment creates dependencies upon sub-cultural peers or institutional authority.

It was assumed that the explicit goals of the Detention Centre would be to establish "The will to lead a good and useful life", as the 1949 Prison Rules

* Acknowledgments for assistance are made to the Leverhulme Trust, Surrey Probation and After-Care Service, The Home Office, and Surrey Community Development Trust.

put it. On the basis of the theoretical approach briefly summarised above, it could be argued that these goals would have been achieved if the trainee left (1) with high self-esteem, (2) regarding himself as less rule-breaking than when he arrived and aspiring to continue to be so and (3) seeing himself as reasonably independent and aspiring to remain so, rather than being dependent and "institutionalised".

The methodology and analysis employed permitted investigation of changes in relationship with family and peers. Both sets of relationships have been regarded by theorists as important factors related to delinquent behaviour, and therefore (4) improved relationships with parents and possibly (5) more distinct relationships with peers might indicate changes towards a less delinquent life style.

Most readers of this book will be familiar with the theory and techniques involved in the use of repertory grid. The main sources of reference for the methodology adopted included Kelly (1955), Bonarious (1965), Bannister and Mair (1968), Bannister (1970) and Bannister and Fransella (1971). Grids were analysed by Dr. Patrick Slater's programme at the M.R.C. Unit and sources of reference for the interpretation of this analysis included Slater (1964), (1965) and Hope (1966, 1969).

For readers unfamiliar with this analysis, Slater and Hope's useful visual images of the way in which the construct worlds of individuals are composed may be helpful. Slater suggests that a geographer's globe may be used as a practical visual aid; Hope offers the concept of a wire model in which each wire represents a construct and the wires all cross at a common origin. Hope's image is perhaps more helpful in enabling visualisation of varying "lengths" or proportions of variance of constructs from one pole to another; Slater's helps the reader to visualise the way in which elements may be plotted in relation to constructs, in the same way that stars in the sky have a spatial relationship to map references on the globe below. Slater's globe makes it easier to understand the problems of comparing measurable changes of variance between individuals. It is easy to see that one person may have a component "globe" the size of a pea, and another a globe the size of a planet.

Absolute measurements of change in these circumstances are only useful in comparing grids for the same person. However, shifts from one hemisphere to another, in Slater's terms, *are* comparable between individuals, and so is the relative positioning of elements in relation to supplied constructs. It may be easier for readers to think in terms of North and South Poles for negative and positive poles of constructs and of the shifts of self concepts of groups from negative to positive poles of similar constructs as migrations from one hemisphere to another. It is then possible to say exactly how many migrations occurred from the northern to the southern hemisphere, or the reverse, and it is also possible to chart accurately how many individuals moved towards the opposite hemisphere though still remaining in the same hemisphere in which they began their trek. The direction of movement can be seen to be of importance, and the effect of "crossing the line" of perception of self as rule-breaking or not rule-breaking may be extremely significant in terms of theories concerning deviant careers, to which reference was made earlier.

The Environment

Detention Centres receive young men immediately after sentence, usually for a period of three months of which one month is normally remitted for good conduct. In the sample in this study only a few entrants lost a few days of their remission. Those few who had longer sentences were amongst the eight transfers described in the section below which gives fuller details of the sample.

The Detention Centre where the research was carried out provided a superficial impression similar to that given by a modern prison. It was surrounded by high wired fences and uniformed officers unlocked the entrance gates and also the doors of buildings where trainees lived and worked. Trainees wore denims for most of the day, grey flannel trousers and white shirts when not "working". They were accommodated in single cells for some days after arrival and were then gradually promoted to dormitory accommodation and other privileges. By the end of the two months they were allowed to move about the grounds unescorted and some worked under escort in parties outside the Centre.

The regime included a strictly supervised programme of work, education and physical exercise. Outdoor parades were frequent. The goals of the Centre, which is a "total institution" (see Goffman, 1968) were explicitly reformative. Inmates were called "trainees" (a subtle shift from the more obvious description of "detainees") signifying that they were to be resocialised as more useful and well integrated members of the community outside. The regime was described by some members of staff and some trainees as being intended to make the boys fitter, better disciplined and to have more self-respect. However, comments from probation officers recommending detention (for example, "needs to be shocked by authority", "needs a sharp reminder of the unacceptability of his behaviour") supported the impression that the regime was still regarded as punitive by others as well as the trainees.

Entrance procedures appeared to conform to those described by Goffman (1968) as "mortification of self". The obligatory short hair cut was, for instance, generally perceived as a most humiliating experience. Comments were volunteered by a number of trainees about the probable stigmatic effects, when they were discharged, of this personal defacement. Their views seemed reasonable considering the appearance of most young people of this age group at the time and the procedure is difficult to defend on rational grounds in view of the attitude towards women prisoners with long hair and, recently, towards male service personnel in some armed forces. "Deference patterns" (Goffman, 1968) were enforced. Three days after arrival almost all the sample addressed the (female) research worker as "sir" at the end of every sentence at the beginning of the research session. "Contaminative exposure", also mentioned by Goffman (1968), brought officers into contact with any visiting relatives and such letter writing which was allowed was censored. Aims might therefore be summarised as explicitly reformative but implicitly punitive.

The Sample

This consists of 50 entrants to the Detention Centre who all served a two months' sentence and who all completed one grid on arrival and one just before leaving. In addition eight other entrants were transferred after serving about a month of their sentence. They also completed a second grid and, where appropriate, findings relating to these eight are shown separately.

All 58 were young men between the ages of 17 and 20 with one exception, (who was four months short of 17 years on arrival). Social Enquiry reports were available for all except eight and trainees also supplied information about occupation, father's occupation and some family details during the research sessions. Where it was possible to crosscheck, these tallied with the social histories.

The sample was as homogeneous as could be expected in any comparable fieldwork situation. Investigation of social histories showed the following distribution of age.

Age	16	:	17	:	18	:	19	:	20
No.	1	:	20	:	19	:	9	:	9

Almost all were unskilled manual workers, with the exceptions of one driver and two who claimed to be skilled tradesmen at an age when this seemed more likely to be an ambition than an accomplishment.

Missing information made it impossible to calculate a social handicap score on the lines of that devised by West (1969) but only four trainees would not have scored on at least one item. Of those for whom information was available, 36 had some unusual factor in their family background such as being adopted, fostered, having half- or step- or other quasi-parents; 15 came from families living on social security payments or at a similar level of income, and 18 were from families with six or more children.

Where official information was available, IQ when assessed seemed to be normally distributed. Subjective impressions of trainees when interviewed were similar. One young man had at one time been assessed as ESN. Two or three were unable, or almost unable, to read or write, although at least one of these gave the impression of being well above average intelligence, sufficiently to manage to conceal his disability in most circumstances.

Probation officers had recommended 26 of the 58 as suitable for custodial detention and 22 for alternative penalties. A number of the latter recommendations gave specific reasons why the young man seemed unsuitable for a custodial sentence.

Three trainees were married, two or three others were living with common law wives and some of these relationships had produced children, or children of previous relationships formed part of their present family.

Offences for which trainees were sentenced, where known, fell into three main categories: 31% were driving offences (mostly taking and driving away) and allied crimes; 26% were brawls and assaults; 30% were offences against property (theft, breaking-in etc.). The remainder were mostly breaches of probation etc, plus a couple of drug offences. (A few trainees were charged with more than one offence.) Elizabeth Field (1969) describes similar proportions of offences in her résumé of research on detention centres.

The Procedures

Every entrant was seen as soon as possible after arrival, usually on the third day and again the day before departure. Bank Holidays and Sundays sometimes intervened and, on occasions, a large number of arrivals and departures on the same day necessitated minor variations in this schedule for a handful of the sample. Trainees usually arrived for the research session under escort but were interviewed in private.

It might have been difficult for subjects to exercise their right to refuse to see the research worker in this environment but co-operation was subsequently obtained by briefly outlining the objectives of the project, explaining that it was hoped to obtain a great deal of information in a short period from the Detention Centre to help the worker understand how best to deal with some other fieldwork data. Trainees were told that it would take too long to explain every part of the procedures, but any questions about these would be answered. Anonymity was guaranteed and at the end of the second session differences between the two sets of data were discussed with those trainees who showed any interest. A statement about professional independence from authority figures in the environment was made, primarily to emphasise the guarantee of anonymity.

It was stressed that there were no "pass or fail" standards and that the results were mainly concerned with changes between sessions and with group patterns of change, not with any individual's performance.

Criticisms of the techniques were accepted and answered where possible, a procedure which made the establishment of rapport progressively easier as more of the resident population of trainees became aware of the project and apparently satisfied themselves about the research worker's professional credentials.

The form of repertory grid used was a ten by ten matrix composed of ten constructs and ten elements, in this instance roles supplied by the research worker. These included "As I am", "As I would like to be", "Father", "Mother", "Brother", "Sister", "Best Friend", "Girl Friend", "Happy Person", and "Someone I feel sorry for". These were prepared as cardboard slips about the size of visiting cards. Each trainee supplied his own name for the occupant of each role where this was appropriate. Substitutes were accepted (e.g., for relatives) if necessary and this was noted. Eight constructs were elicited by the usual process of using elements in triads. Two were supplied—"Breaks rules—doesn't break rules". "Stands on their own feet—depends on others". Trainees ranked all the elements for each construct in the usual manner. The use of cards enables people with little reading or writing skill to rank elements quite easily. For the very few illiterate or nearly illiterate young men, sketches were drawn on the cards according to the individual's description of the person concerned until he could identify each card without assistance.

Findings

It will be recalled that the main interest was in changes in percepts of self and ideal self, and in the interrelationship between those elements and the supplied constructs which were concerned with rule-breaking and dependency. Relationships with significant other persons were also investigated and so was the relationship between the two supplied constructs.

The mean of the general degrees of correlation (i.e., the accumulated covariations for all constructs in each grid) between first and second grids for the 50 trainees who completed two months' sentence was 0·69 and for the other eight 0·76. The difference is one which would be anticipated if changes due to the environmental situation increased as a factor of the length of time spent in the environment.

Analysis of the percentage of variance of the first three principal components extracted for all these grids confirmed that these accounted for most of the variations for all trainees, 40% of whom had 50% or more variance accounted for by the first component, another 43% being within the 40–49% range on this component. Sixty-seven per cent had a further 20–29% accounted for by the second component and 77% had 10–19% in addition accounted for by the third component.

The elements and constructs of most interest were next examined for importance on these components in first and second interview sessions. A construct or element was regarded as "important" for the trainee concerned if it fell within the first quartile of the weightings for this item on any of his

TABLE I. *Percentage of total variation accounted for by components I, II and III for all grids for 50 trainees*

	10–19%	20–29%	30–39%	40–49%	50–59%	60–69%	70+ %
Component I		2	15	43	26	9	5
Component II	23	67	10				
Component III	77	3					

first three components and not important if it fell within the fourth quartile, regardless of negative or positive affect. The findings are shown in Table II.

All these items at the first session were salient on one component in the majority of the sample, but it was interesting to note that the importance of

TABLE II. 58 *Trainees: first session*

	Not important on first 3 components	Important on component 1	2	3
Rulebreaking/nonrulebreaking	9 (15%)	36 (62%)	9	7
Dependence/independence	5 (9%)	25 (43%)	18	15
Self	16 (28%)	25 (43%)	19	10
Ideal self	15 (26%)	15 (26%)	14	25

Second session

	Not important on first 3 components	Important on component 1	2	3
Rulebreaking/nonrulebreaking	10 (17%)	35 (60%)	15	2
Dependence/independence	5 (9%)	32 (55%)	18	15
Self	33 (57%)	11 (19%)	11	9
Ideal Self	20 (34%)	20 (34%)	18	13

(Some elements and constructs are "important" on more than one component and, of course, many would fall midway.)

the rule-breaking construct changed least and the importance of the dependency construct increased only a little, as did that of the element of "ideal self", but the salience of the element of "self" was considerably reduced. However, the elements and constructs appeared to be of sufficient importance in the construct worlds of all the individuals involved to make the analysis of changes in the group a worthwhile undertaking. Findings are grouped in the order of relevance to goals listed in the first paragraph of this section.

Self Esteem

At first glance this might be thought to have increased, since for 32 of the 50 "two-monther" subjects self and ideal self were more closely correlated at the second session. However, in order to ascertain the proper interpretation of this finding, which is a percentage unlikely to occur more than once in a hundred by chance (Bruning and Kintz, 1968), it is necessary to consider how stable each element has remained in the perception of the individual concerned.

In order to do this, grids for each subject were revised so that the pole of each construct which the trainee regarded with positive affect for "ideal self" was rated positively throughout the analysis. This does not, of course, affect the main factor analysis, but allows a more accurate assessment of the differential rating of himself and other people in the trainee's construct world in terms of the attributes which he considers desirable, "better" in terms of his value system.

Bearing this in mind, self esteem was only held to have increased when "ideal self" perception remained stable, or rose, in the analysis of the grid of differential changes for elements in the construct system and "self" was upgraded amongst other elements to converge with "ideal self", and in three cases where "self" rose substantially more than "ideal self" fell. In all other cases the down grading of "ideal self" concepts was always very large compared to the small rises in self concept.

It then became apparent that in these terms, self and ideal self were converging only because for 82% of these trainees ideal self had been considerably downgraded by the second session and, indeed, for 42% both self and ideal self were perceived much less favourably in terms of all desirable attributes compared to other significant persons in the trainee's construct system. (The alternative explanation, i.e., that so many significant persons in so many different universes should have coincidentally been viewed far more favourably by 82% of trainees seems highly unlikely.) Self was viewed more and less favourably in roughly equal proportions. (A similar phenomenon was observed for the eight "one-monthers", seven of whom perceived ideal self less favourably and for one no change was registered.)

This is a finding of considerable interest, suggesting that, far from providing trainees with higher standards and improved ideals, the detention period considerably reduced the aspirations of more than four-fifths of the sample and about half of these boys also perceived themselves less favourably, emerging with both diminished self-respect and lowered aspirations.

The closer correlations may be interpreted as increased insight into society's negative view of themselves, plus an internalised acceptance of this view as correct and with little possibility of being changed. There are implications here of confirmation and reinforcement of a deviant life style.

Rule-Breaking

On entry 88% of the 50 "two monther" trainees saw themselves as rule-breaking, but 66% aspired not to be so. On departure 90% saw themselves as rule-breaking and only 48% aspired not to be so. No significant change occurred in the relationship between self and rule-breaking concepts, but the changes in relationship between the percepts of ideal self and rule-breaking were significant at the 1% level and in the reverse direction to that which would be intended by the administration.

Independence

On entry 88% of the 50 trainees had aspirations to be independent; on departure 84% aspired to be so. On arrival 38% saw themselves as "standing on their own feet" and 58% viewed themselves in this light on departure. The change in perception of self is significant at the 5% level. Twelve trainees who changed their outlook from dependence to independence also achieved the goal of being less rule-breaking (of these twelve, five also increased their self-esteem.) Two of the eight "one-monthers" also achieved these two goals, so that the total success rate for these two factors for the 58 was 25%. However, those achieving more independence but also seeing themselves as more rule-breaking numbered 13 (of these 13, 5 also increased their self-esteem) and an additional two "one-monthers" achieved both goals, thus producing a very similar percentage. 32% saw themselves as less rule-breaking, but also as less independent and only four of these increased in self-esteem. 18% saw themselves as more rule-breaking, more dependent and lost self-esteem.

In considering the above findings, the distinction should be borne in mind between the numbers of trainees whose self percepts, for example, are positively and negatively related to the constructs and the numbers of trainees whose self percepts and these constructs become more (or less) closely related whether these were positively or negatively associated in the first place.

This is an instance where the visual image of migrations may help. Self percepts may have been placed in either hemisphere and crossed the equator. Or they may have moved towards the equator from either hemisphere without crossing it. The total number of movements towards either pole will be greater than the numbers who cross the equator.

The complex pattern of the findings are shown in Tables III, IV and V. Changes in self concept are given separately in Table III, changes in ideal self concept are shown in Table IV. Finally both sets of changes are combined in Table Va. Those who by the second session saw themselves as breaking less rules and standing on their own feet more are described in Table III as "CONFORMISTS"; those who saw themselves as breaking more rules but also more independent as "REBELS"; those who saw themselves as less rule-breaking but also more dependent as "INSTITUTIONALISED" and those who saw themselves as both more rule-breaking and more dependent as "PROBLEMS". In Table IV, the same typology applies but refers to the aims of the trainee, his aspirations or ideal self. In Table Va the numbers in each cell show the percepts for self and ideal self combined, and Tables Vb and c show the distribution of loss or gain in self-esteem in each category in

Key to Tables III, IV, Va, b and c.

CONFORMIST	breaks less rules: stands on own feet more
REBEL	breaks more rules: stands on own feet more
INSTITUTIONALISED	breaks less rules: depends on others more
PROBLEM	breaks more rules: depends on others more

TABLE III. *Changes on self percept during two months/one month*

Sees self as more	Self esteem	Two months (50 trainees)	total	One month (8 trainees)	total	Total (58)	total
Conformist	lost	7	12	2	2	9	14
	gained	5		0		5	
Rebel	lost	8	13	4	4	12	17
	gained	5		0		5	
Institutionalised	lost	13	16	0	1	13	17
	gained	3		1		4	
Problem	lost	9	9	1	1	10	10
	gained	0		0		0	

TABLE IV. *Changes in ideal self percept during two months/one month*

Aspires to be more	Self esteem	Two months 50 (trainees)		One month (8 trainees)		Total (58)	
			total		total		total
Conformist	lost	8	10	2	3	10	13
	gained	2		1		3	
Rebel	lost	10	18	3	3	13	21
	gained	8		0		8	
Institutionalised	lost	7	8	1	1	8	9
	gained	1		0		1	
Problem	lost	12	14	1	1	13	15
	gained	2		0		2	

Table Va. It should be borne in mind that the figures represent *changes* achieved over one or two months.

The more "conformist" self and ideal-self concept probably represents the best outcome, and the more "problem" self and ideal-self concepts the worst outcome from the administrative point of view. Only 2 of 58 trainees appear in the first category and of these only one increased self-esteem which would,

TABLE Va. *Changes in self and ideal self*

	Sees self as more				
Aspires to be more	Conformist	Rebel	Institutionalised	Problem	Total
Conformist	2	3	5	3	13
Rebel	4	11	6	0	21
Institutionalised	4	2	2	1	9
Problem	4	1	4	6	15
Total	14	17	17	10	58

TABLE Vb. *Gains in self esteem in categories in Table Va*

	Sees self as more				
Aspires to be more	Conformist	Rebel	Institutionalised	Problem	Total
Conformist	1	0	2[a]	0	3
Rebel	3	5	0	0	8
Institutionalised	0	0	1	0	1
Problem	1	0	1	0	2
Total	5	5	4	0	14

[a] 1 is "one-monther".

theoretically, reinforce a good future prognosis. Six of the 58 are in the worst category, all losing self-esteem, reinforcing a poor prognosis.

Attribution of significance to the distribution in the typology presents difficulties because of the small numbers involved. However, almost three times as many trainees appear in the cell for those who see themselves as more rebellious and increase their intention to remain so, as might be expected by an even distribution* and the highest number of gains in self-esteem also appear in this cell. These are flanked by people whom Goffman

TABLE Vc. *Losses in self esteem in categories in Table Va*

	Sees self as more				
Aspires to be more	Conformist	Rebel	Institutionalised	Problem	Total
Conformist	1	3[a]	3	3[a]	10
Rebel	1	6[b]	6	0	13
Institutionalised	4[a]	2	1	1	8
Problem	3[a]	1	3	6	13
Total	9	12	13	10	44

[a] 1 is a "one-monther".
[b] 3 are "one-monthers".

* Applying chi square test to Table Va results in a figure of 16·05. (16·92 is required to achieve 5% degree of significance and the result is of course significant at the 10% level.)

might describe as "playing it cool", seeing themselves as more conforming or institutionalised at present, but aspiring to be more rebellious. These together total almost as many (10) as the pure rebels (11) and these three categories account for 36% of the trainees and more than half the small total number of gains in self-esteem. From the point of view of psychological survival alone (Cohen and Taylor, 1972) these categories may be regarded as successes, though they are the reverse from the administrative point of view.

Various other characteristics of trainees were examined for some kind of relationship with the typology of change, with little success, except that of those sentenced for offences roughly lumped together under the general heading of brawls, the majority preceived themselves as more institutionalised and aspired to be more rebellious.

Relationships with Family and Peers

Elements were ranked and those which accounted for the three greatest variations on all constructs for all subjects were analysed in greater detail.

(The occasional substitution of a friend (peer) for a relative, or the reverse when for example trainees had no sister and substituted a girl friend, or had no girl friend and substituted a sister, was taken into consideration in the analysis. Nineteen trainees had surrogate fathers but an extensive analysis revealed no significant difference between findings for these or other trainees.)

As might be expected from the earlier findings, ideal self featured more often than any other element amongst the first three ranks and ranked first for about a quarter of the trainees, usually, as already mentioned, downgraded by comparison with other elements on all constructs.

The elements in second place in the first three ranks were girl friends (for 36% of trainees) and they were perceived more favourably twice as often as they were downgraded on constructs compared to other elements. Considering the difficulties involved in visiting and letter writing, this relationship survived rather better than might have been expected.

Self was almost as frequently amongst the highest ranking varying elements (for 34% of trainees) but moved equally both up and down compared to other elements and all constructs.

Best friends featured next frequently, also varying equally in both directions.

Other elements appeared about equally in the higher ranks, sisters being distinguished by being perceived more favourably three times as often as less favourably, other family members being perceived equally more or less favourably.

Overall, sisters and girl friends were perceived more favourably more often than any other elements.

Changes in similarity between self and ideal self and each of the elements when analysed show a significant percentage of trainees who saw themselves as having become more like their parents (and like the person they pitied) during the two months' detention, but aspiring not to be so, but to be more like their brothers and sisters. The possible interpretation here is that the conformity imposed in the centre made trainees feel that they were at present more like their parents, but they had no wish to remain so. However, the trend was to increase perceived similarity of trainees to family members rather than to increase identification with peers. Despite the improved perceptions of girl friends, no significant similarity was perceived or desired.

Relationship Between Supplied Constructs

Over the two month period, 62% of trainees came to regard rule-breaking as a construct more closely related to independence than when they arrived (a percentage unlikely to occur more often than five times in a hundred by chance). This would be inconsistent with the goals of the administration, whose custodial regime would apparently stress that the consequences of rule-breaking are a loss of independence. However informal norms equating successful rule-breaking with demonstrable independence of coercion in the environment may be more powerful factors in the socialisation process.

There is a trend towards a relationship between changes in percepts and the typology, but the numbers are too small for further analysis in this study.

Other Analyses

A small clutch of additional findings are reported below.

BLACK TRAINEES

Because of some reported differences (Jensen, 1972) concerning self-esteem in black and white subjects in American research projects, all data for the nine out of the 58 trainees who were black was compared with the data for all the others. No significant difference was found, and the only noticeable trend was for black trainees to relate to supplied constructs less closely at both sessions than the white trainees, though changes over the period were similar for both groups.

NEUROTICISM

On the basis of some evidence by Ryle and Breen (1972) that above average distances between self and ideal self compared to other elements may be equated with some measures of neuroticism, this data was examined. More (29) trainees were "normal" than not (21) at the first session, despite the fact that this occurred at a period of considerable stress. Changes during the two months shifted eight more into the "deviant", or neurotic level, and two into the opposite direction.

GATE-HAPPINESS

Some doubts had been expressed to the research worker about the validity of analysing grids composed by entrants in the state of despondency often apparent when they arrived or in the "gate-happy" euphoria just preceding departure. Theoretically the analysis of grids should not be affected by ephemeral emotional states, although constructs which express these may be elicited in stressful circumstances.

For 56% of the sample of 50 "two monthers" the construct "happy–unhappy" was elicited. It was not only frequently volunteered but also accounted for a great deal of variance in the construct systems—ranking in the first five of the ten constructs when ranked for contribution for variance in 78% of these cases.

However, contrary to what was presumably predicted by the dubious, this variance was not significantly related to "self", but was spread over many elements, family and friends of trainees. The fact that the whole construct system takes on a rosier glow does not necessarily mean that the individual perceives himself as any happier than before compared to other significant persons in his construct system, which depends on a rational process of ranking. Eight trainees moved from the "unhappy" to the "happy" pole, but four moved in the opposite direction.

The considerable variation in "happiness" is so dispersed that it will not unduly affect the overall analysis of the relationship between elements nor, of course, of the relationship between the two supplied constructs.

However, in the course of the administration of other procedures, trainees were asked during the second research session if they thought they had changed at all and, if so, in what way. Subjective insight into change was not correlated with the grid results but *was* probably much affected by the euphoria referred to as "gate happiness" which is quite evident in the general behaviour of trainees on the day or so before release. Twenty per cent thought they had changed for the better in some socially approved way ("swear less", "more disciplined", etc.) but a number of these said that they would "adapt back" to their normal life style after release. Twenty-two per cent thought they were fitter, an impression which was reinforced by the fact that many of these appeared to have put on weight (and said that their clothes no longer fitted them, when they tried these on in preparation for leaving), which may be attributed to the diet and exercise.

Only four thought they had changed for the worse. In view of the 82% downgrading in aspirations and 42% downgrading in self perceptions in the grids, where it is very difficult to mislead, it is suggested that reliance on such statements by trainees on discharge would be unwise.

Constructs

An attempt to rate constructs using Landfield's (1971) scales was unsuccessful, because of lack of interjudge agreement. A rough summary of the most frequently occurring constructs was made for comparison with other groups.

"Happy" and similar constructs such as "jolly", "a laugh", "joker", with "serious" or "miserable" at the opposite poles, occurred 35 times (in 58 grids). Expressions of liking/dislike for other people, or constructs denoting perceived similarity, or friendly contact or affection appeared 52 times. "Drinking a lot" was mentioned 22 times, "goes out a lot"/"stays at home" 16 times. Expressions of affection, love, understanding, support and help received from others account for another 40 constructs. Liking the same kind of clothes, music etc. "as I do", or "as me" appeared 13 times and getting on well with people, or a significant liked person, appeared 12 times. Another construct which recurred concerned "trouble" or "crime" and appeared six times; money was mentioned nine times and sport eleven times. "Speaking their mind", "not shy", "noisy", appeared twenty times with a further seven "quiet"/"less so" in the same continuum. Critical comments such as "bitchy", "snidey", "creepy", "weird", "silly", "nasty", "sloppy git", "hits people", "mad", "swines", "tearaways", "eats like a pig", "lazy", "thieves", appeared about 25 times.

The stress throughout is noticeably on affect and social relationships and a system for more precise comparison with other groups is under consideration.

Discussion

The usefulness of the grid technique seems to have been demonstrated by the findings of lowered aspirations and self-esteem and increased self perceptions as rebellious in a substantial proportion of trainees after two months' exposure to the Detention Centre environment. These findings are in accordance with the considerable body of evidence that custodial sentences are detrimental to individuals and fail to achieve the intended goals. (See, for example, Mays, 1970.) That evidence, however, is mainly based on reconviction rates ranging from 30% upwards, which must be contaminated by many other factors affecting the individual and his environment in the intervening follow-up period. Recidivism rates certainly fail to take into account those members of the sample who are still engaged in a deviant life style, but whose activities are undetected.

The advantages of using a measure which gives immediate and specific information about changes which result from treatment intended to be reformative are obvious. Grids can, of course, be devised to investigate other aspects of change. Their use in measuring change in clinical work has sometimes proved disappointing, as Caine and Smail (1969) pointed out, possibly because the groups concerned have been less homogeneous, or the treatment, if effective, less continuously administered than that inherent in the environment studied here. The results in this study, at any rate, suggest that inconclusive findings are not due to the use of grid techniques as measuring instruments.

References

Bannister, D. (ed.) (1970). "Perspectives in Personal Construct Theory", Academic Press, London and New York.

Bannister, D. and Fransella, F. (1971). "Inquiring Man", Penguin Books.

Bannister, D. and Mair, J. M. M. (1968). "The Evaluation of Personal Constructs", Academic Press, London and New York.

Becker, H. S. (1963). "Outsiders: Studies in the Sociology of Deviance", Free Press, New York.

Bonarius, J. C. J. (1965). Progress in the personal construct theory of George A. Kelly: Role construct repertory test and basic theory. *Progress in Experimental Personality Research*, Vol 2.

Bruning, J. L. and Kintz, B. L. (1968). "Computational Handbook of Statistics", Scott, Foresman and Co, Illinois, pp 197–204.

Caine, T. M. and Smail, D. J. (1969). A study of the reliability and validity of the repertory grid technique as a measure of the hysteroid/obsessoid component of personality. *British Journal of Psychiatry*, **115,** 1305–1308.

Clemmer, D. (1962). Prisonization, *In* "The Sociology of Punishment and Correction" (Johnston, Savitz and Wolfgang, eds), Wiley, pp. 148–151.

Cohen, S. and Taylor, L. (1972). "Psychological Survival", Penguin Books Ltd.

Field, Elizabeth (1969). Research into detention centres. *Br. J. Crim.* **9,** 62–71.

Goffman, E. (1968). "Asylums", Penguin Books, pp. 15–22, 24–28, 31–38.

Hewitt, J. P. (1971). "Social Stratification and Deviant Behaviour", Random House, pp. 33–48.

Hope, K., (1966) Cos and Cosmos. Considerations on Patrick Slater's monograph—the principal components of a repertory grid. *Brit. J. Psychiatry* **112,** 1155–1163.

Hope, K. (1969). The complete analysis of the data matrix. *Br. J. Psychiatry*, **5,** 1069–1079.

Jensen, G. F. (1972). Delinquency and adolescent self-conceptions: A study of the personal relevance of infraction. *Social Problems* Vol 20, No 1, 84–102.

Kelly, G. A. (1955). "The Psychology of Personal Constructs", Vols. 1 and 2, Norton.

Landfield, A. W. (1971). "Personal Construct Systems in Psychotherapy", Rand, McNally and Co., USA.

Lemert, E. M. (1951). "Social Pathology", McGraw Hill, New York.

Mays, J. B. (1970). "Crime and Its Treatment", Longman.

Morris, T. P. and P. (1963). "Pentonville", Routledge and Kegan Paul.

Quinney, R. (1970). "The Social Reality of Crime", Little, Brown and Co., pp. 13–15, 234–248, 277–280, Boston.

Reckless W. C. (1957). The self component in potential delinquency and potential non delinquency. *Am. Sociological Review* **22,** 566–570.

Reckless, W. C. (1960). The good boy in a high delinquency area. *J. Crim. Law, Criminology and Police Science* **48,** 18–26.
Reckless W. C. (1961). A new theory of delinquency and crime. *J. Federal Probation* **25,** 42–46.
Reckless, W. C. and Dinitz S. (1967). Pioneering with the self-concept as a vulnerability factor in delinquency. *J. Crim. Law, Criminology and Police Science* **58,** 515–523.
Reckless, W. C., Dinitz, S. and Murray E. (1956). Self concept as an insulator against delinquency. *Am. Sociological Review* **21,** 744–746.
Ryle, A. and Breen, D. (1972). Some differences in the personal constructs of neurotic and normal subjects. *Brit. J. Psychology* **120,** 483–489.
Schwartz, M. and Tangri, S. S. (1965). A note on self concept as an insulator against delinquency. *Am. Sociological Review* **20.**
Slater, P. (1964). "The Principal Components of a Repertory Grid", Vincent Andrews and Co., London, p. 55. (Obtainable only from author.)
Slater, P. (1965). The use of repertory grid technique in the individual case. *Br. J. Psychiatry* **111,** 965–975.
Sykes, G. (1958). "The Society of Captives", Princeton.
Tannenbaum, F. (1938). "Crime and Community", Columbia University Press, New York.
West, D. J. (1969). "Present Conduct and Future Delinquency", Heinemann.
Wilkins, L. T. (1964). "Social Deviance", Tavistock, London.

The Interplay Between Mothers and Their Children: A Construct Theory Viewpoint

Joy O'Reilly

Introduction

ADULT–CHILD RELATIONSHIPS

There seems to me to be a problem for a worker in the field of child psychology which, at first glance anyway, does not arise in the adult area. This concerns what I see to be the difficulty of assuming the same degree of validity for one's interpretation of what a child means by what he says, as one does with an adult. In the therapeutic situation, which is the one in which I was working, there were obstacles to overcome which appeared to result specifically from my way of responding, as an adult, to the child client. On closer investigation, I discovered that I had accumulated certain adult response sets to children which, instead of enabling me to understand what a child was about, set us both on a familiar question and answer routine which ended nowhere. The most obvious adult behaviours, like patting a child on the head and exchanging certain adult type comments with mother in the waiting-room, for example, could set the session off in a particular vein, with the child behaving in a coy manner or else in sullen incommunication. Either way, the ensuing interaction was unsatisfactory.

Having unlearned such obvious grossness of approach there remained more subtle pitfalls. What does a tried and true adult do with long tension-ridden silences in a session with a child? It was so easy to ignore these by "chatting" oneself or by "interpreting" them away by talking about shyness etc. It was a constant effort not to finish the child's sentences and not to assume I knew what was meant by vague or ambiguous responses. I seemed to be having to unlearn most of my repertoire of adult-to-child behaviours and it was not always easy to cope with what I was beginning to recognise as the anxiety which arose as a result.

I was confused by my own reactions. After all, why should I be apprehensive about having to get to know a child all on my own. And then it occurred to me that I had never before sat with a child in a room for close on an hour with no testing equipment between us or no task to be learned. We were totally on our own and had to get to know each other, a situation unusual in most adult–child interactions.

As a result of having to learn new ways of handling my interactions with children, ways more akin to general principles of therapy, I found myself up against another problem. And this had to do precisely with my attempt to understand what the child was saying in his own terms. I had ceased to "interpret" what was being said in my own frames of reference, because I had discovered that these prevented elaboration by the child of what he was about. As I would do with an adult client, therefore, I was first going to provide a structure within which the child was free to develop his own meaning. It was in understanding what this meaning was that I found myself having difficulties because, unlike an adult client, the child and I did not appear to share what I might call basic consensus about the meaning of interpersonal interaction. I came to this conclusion after many experiences of discussing something which seemed straightforward to me but which, it became clear, had a meaning for the child other than the one I had been preoccupied with.

I would like to illustrate this confusion of meaning by quoting an incident told to me by a mother I interviewed some time ago. Her three-year old son, Johnny, was in the kitchen, more or less getting in her way and paying no attention to her reproofs. Finally exasperated, she shouted at him "Oh, for goodness sake, Johnny, get out, get out". A few minutes later the mother was horrified to see her son come downstairs with his coat on, ready to do what he had been told, to leave home: for this mother there was a clear circumscribed context within which she told her child to leave the room. Her response is to a specific event, misbehaving, and refers to this event having taken place at a particular point in time, the present. She has seen such misbehaviour before, expects to see it again, and believes that such incidents happen between all little boys and their mothers.

If we can hypothesise what the child thinks about the event, we might say that he seems not to share such assumptions about time, place and other people. In fact he seems to perceive the issue as shatteringly unique, and furthermore the issue seems to be his mother's love for him, which apparently can disappear as a result of his mis-behaviour. Whether or not this is a completely accurate reconstruction of how Johnny sees the situation, it does seem reasonable in the light of his behaviour that he and his mother (and probably most adults would be in a similar position to the mother) do not share a consensus about what is going on between them. There are at least two ways the mother might have responded to the child's behaviour, depending on how she herself construed this behaviour. On the one hand, she could have stuck with her original construction of the situation, outlined before, in which case she might construe her child's recent actions as deliberately provocative and attention-seeking. Her response to the child would therefore be consistent with such a construal. On the other hand had she, as the mother in the example, been so shocked at the effects of her words on the child she might have had to make some attempt to construe Johnny's construction processes. It is clear that these two construals among many possible ones would influence subsequent construals by the child of his mother, himself and of the way they relate to each other.

Considerations such as these demand some kind of theoretical framework within which both the therapist may see his work with his child clients and the general worker, in the area of child psychology, may make sense of the child's relationships with significant adults. For the construct theorist, there are some very straight-forward conditions that must be met at the outset of such a venture. We must bear in mind that the subject of our inquiry, a child, is to be construed as a "scientist", as having theories about his world and himself. We must further build into our approach the assumption that these theories are formed so as to provide our subject with ways of acting towards events in his life. We, therefore, right from the start, assume meaningfulness in the child's actions, and cannot content ourselves with analyses based solely on the child's *lack* of physical, emotional or cognitive skills. Obviously this is not to say that we do not accept as integral to our investigation the fact of the child's immaturity in certain clearly discernible ways as compared with adult members of the species. This seems self-evident. What is not acceptable if one adheres to a construct theory approach however, is that an investigation into aspects of childhood functioning should start from the premise that the only way children can be viewed scientifically is as immature members of human communities.

I would like to take the mother–child relationship in particular, as my unit of inquiry, not only because this has already been suggested by construct theorists in the field (e.g., Salmon, 1970) but also because it provides the sort of context most likely to illustrate the difficulties peculiar to an investigation of childhood functioning. That is to say it would seem that the mother in relating with her child will encounter the same sort of difficulties and confusions of interpretation as a therapist or any professional investigator of childhood functioning. More important however, is the fact that a central assumption of construct theory holds that it is in and through his relationship with others that a person elaborates his own particular set of construct systems. Furthermore, it is in his relationship with a caring adult that the child first learns to use constructs. Particular issues that will have to be dealt with in our discussion of the mother-child relationship, spring from two main sources. The first, which can be broadly called the problem of the child's immature cognitive skills, is most dramatically reflected in Piaget's description of egocentrism. Whether or not we accept such a description we have to attempt at least to make sense of the generally agreed upon observation, that a child does not possess mature cognitive skills.

At the same time we have to understand egocentrism in our own terms, i.e., in the light of the contention that the child is cognitively incapable of differentiating his own from another's point of view, what can we say about the possibility of a role relationship between mother and child. The second issue with which we shall have to concern ourselves eventually, refers to what I might call the "problem of the emotional context" of the mother–child relationship. In this we will have to take into account to some extent the extensive work of psychoanalytic workers in this area.

These two issues will be discussed in the context of construct theory's approach to the mother–child relationship which we shall find is an area in need of some elaboration. We shall take what already exists within the theory pertaining to our subject and try to draw out some of the implications. In addition we will make use of general theoretical notions, such as "emotion", in the particular context of mother–child relationships.

I should also say that what will be discussed in certain important respects concerns what might be the possible stance of the investigator in relation to this subject. We have said that there has been a trend among some influential writers to take as their starting point the immature state of the child, and to shape their inquiry round the processes by which he learns various skills which will be recognised as those of a mature member of the human race. Another aspect of this same approach is to be found in many "socialisation" studies. Thus while these tend to take the child in his relationships with others as their focus, as opposed to the straight development of skills approach, the emphasis mostly tends to focus on the socialising effect that others have on a child, rather than on the relationship going on between the child and whoever these others are. These therefore tend to minimise the child's contribution to the relationship or else to see this in an essentially negative light.

Little attention seems to be paid to the mother's construction of what her child is doing, for example, when he persists in unsocialised behaviour. It seems to be assumed that she, like the investigator, simply perceives it as something to be discouraged, as having no other meaning than that it is undesirable. We shall therefore be trying to see how the construct theorist might decide how he or she stands in relation to the socialisation aspect of the mother–child relationship.

All in all, the discussion which follows will be trying to outline aspects of childhood, within a context of the mother–child relationship, and will be attempting to formulate useful descriptions of these which will be consistent with general construct theory principles. Certain difficulties in maintaining the central element of our approach, that the child is a "scientist" in Kelly's terms, will be discussed in the light of general findings about the child's cognitive and emotional immaturities.

I would like to begin by drawing on the work of two writers who have each from their individual points of view discussed the subject in hand. The first of these is John Shotter, a philosophical psychologist, who is concerned with issues pertaining to the development of what we think of as "human". Thus, for example, he would distinguish between people as "individual personalities possessing personal powers and as natural agents possessing natural powers" (Shotter, 1973). He has formulated some of his concerns in construct theory terms and it is to these we will refer in the following section. In particular, we will be interested in his efforts to work through some of the implications of construct-theory with regard to the possible source of "personal constructs".

The second writer whose work I shall be drawing on is Phillida Salmon. She, at a more empirical level, is concerned with the meaning of personal growth in construct-theory terms. She has done most of her work with child subjects and has discussed the difficulties of using repertory grids with children. In what follows, I would like to make use of some of her formulations of the mother–child relationship. These two writers together, I believe, provide us with a useful framework within which we may begin to discuss the subject.

A Construct Theory Context

THE ORIGINS OF PERSONAL CONSTRUCTS: THE WORK OF JOHN SHOTTER

When Kelly first set out his theory of personal constructs he was at pains to point out how his notion of "construct" differed from the traditional notion of a "concept". It is basically the implications of this difference, and Kelly's emphasis on the primacy of bodily encounters, that Shotter uses as starting points of his mythology of the origins of construct systems. "Constructs", says Shotter, "are involved in a form-producing process, not a knowledge acquisition one". Kelly's model of man is of a creature whose "nature" is to "impose upon the chaos around him forms which he finds intelligible". These forms are the schemes of ordering we call constructs and construct systems. Constructs are "invented when a man notices a difference, draws the attention of a fellow to it and they agree upon a way of representing it." "Constructs are used", on the other hand, "when a man makes (expresses) a distinction using the agreed ways of representing it". Construct systems are thus invented and used in human communities and it is through this process that a child is born into the world which has devised for itself a distinctive means of making sense of experience. It is in and through his relationships with others that he gains access to this "system" and in turn through use of this "system", that he devises for himself constructs and construct systems which are distinctly his own.

By ways of illustrating what an inheritance this is, Shotter uses the examples of a wild boy, divorced from human communities, having to rely on his own resources to get to know his world and himself. Originally the boy would be unaware of himself or his world. Gradually, through actual physical contact with the world, in order to relieve states of restlessness, he would get to know his needs in terms of what he had done to alleviate these. Such activity would provide him with knowledge of his world in terms of the actions within which he had encountered it. Eventually he would be able to differentiate between states of discomfort and so would be able to alleviate each with the minimum of searching around. He would have arrived at a stage where he would be able to perform an action that he had done before, *in order to* relieve a certain need. In this way the wild boy would "learn about himself and his world".

But such learning would provide only one construction, one that is tied to the organism's needs. The child born into a human community, on the other hand, not only possesses the capacity to know the world through his bodily encounters, but also gains access to a "special way of seeing the world" which "represents and incorporates knowledge gained from the many different standpoints of his fellows". Furthermore, the child can escape the bondage of a single construal of himself and his world through language which "enables him to re-arrange the component parts of his (construal) into new patterns".

It is through being involved in an organised pattern of social interaction that a child learns to attribute significance to the world in the way that his fellows do. And it is also through learning to request organised sequences of behaviour from another, and have such requests made of himself, in such a social interaction, that the child learns, at a later date, to organise his own behaviour. (In this latter thesis, Shotter is drawing on the work of Vygotsky, 1962). Shotter (1970), in summarising the inheritance of a child born into a human community, says that a child inherits a

> "special way of seeing the world, of attributing significance to it, and interwoven with it, a skilful way of co-ordinating his social interaction in its terms".

It is worth noting that Shotter, in his outline, gives a particular meaning to the term *personal* construct systems. We do not mean that each individual constructs for himself a way of seeing the world, which is completely separate from that of his fellow-men. On the contrary, it is by sharing this "special" view of the world, that an individual finds open to him the many different standpoints of his fellows, from which he can make a *personal* selection; furthermore, he is able to subject his own highly individual knowledge gained in bodily encounters to construal in different ways, as a result of possessing forms of ordering which incorporate the knowledge gained, not only by himself in different situations, but also knowledge transmitted to him by others. The child born into a human community can therefore "construct for himself many different construct systems for alternative construals of a situation—something inaccessible to the wild boy", as well as make a personal choice of constructs and construct systems.

It would seem from this that when we talk of socialisation in construct theory terms we are talking in a general sense of the child's inheritance of this special way of seeing the world. To quote Shotter once more ". . . babies born to us need not grow up to be what we understand as human; their humanity is transmitted to them after birth. It is inherited but like houses and cities not like blue eyes . . ." Basically this inheritance is transmitted to the child within the dynamic exchanges of the mother–child relationship (which is how we shall describe the relationship between child and primary caring adult). Specifically what is involved is that the child learns to attribute significance to the world in the way his mother does and this necessitates, in the development of language, as one example, that he share judgments about the world with his mother prior to developing vocal speech at all. For example, he learns to value certain sounds more than others. Also, and in the same way, the child learns to co-ordinate his social interaction in terms of the person-to-person system represented to him by his mother's interaction with him. This too involves the sharing of judgments as to usefulness, or goodness of certain patterns of activity.

The child thus learns to participate in, and at some stage presumably contribute to, the schemes of ordering by which men attempt to make sense of the chaos around them. What makes it possible for him to contribute at some stage, is the fact that, as Shotter points out, right from the very start, a child is capable of gaining knowledge about the world which is unique to him.

Despite the overwhelming similarities in the day-to-day routine of young infants and the apparent universal nature of their needs, they can all probably be assumed to be gaining marginally different kinds of "basic knowledge" as a result of their physical encounters with the world. As we have seen, this knowledge is originally written in bodily terms and will remain inexpressible and incapable of alternative construal unless a way can be found to provide it with meaning, which is to say, unless it can be expressed in terms of some system of constructs available to the person.

It is worth while pausing here, because this question of "basic knowledge" is an important, though a highly difficult issue. We may consider it in terms of our unit of inquiry, the mother–child relationship. Shotter points out that by getting the child to *select* among his actions the mother is responsible for some of the very fundamental constructs available to the child. It is through this instruction that some of the child's spontaneous actions become deliberate, and thus under intellectual and volitional control, which in turn is the process whereby such basic knowledge can become structured into its components parts and thus organised and re-organised. That is to say it is now capable of alternative construal and thus has "meaning". The origins of a child's personal construct-systems, which are the means whereby he can give expression to what he discovers about the world through his own activity, are to a very significant extent influenced by his mother's "instruction" (Vygotsky, 1962). The effect of this instruction is "not to make the child do what he cannot, but to restructure what he can do already into new forms". This proceeds firstly without language and when linguistic communication is underway it provides the instrument through which the mother "cumulatively restructures the child's actions" (Shotter 1970, p. 241) and later enables the person the child becomes to restructure his own actions.

This selecting of activities by the mother is obviously of fundamental importance, because it is the construing of construct systems that provides the child with the means of behaving in an anticipatory manner towards events in his life. The relationship between a child's developing construct systems and his "spontaneous actions", or non-instrumental activity as Shotter calls it, presumably continues throughout life in some form or other. How much of what a person, child or adult, discovers through such activity finds expression and utilisation within his construct systems, and how such a process might come about, is a question requiring more intensive analysis than is possible in this paper. In this regard, the work of Newson (1969) into the preverbal communication between mothers and their 10–20 months' old infants, should prove very interesting, in that it focuses specifically on the instruction of a young child by its mother. What has emerged so far from this project is the suggestion that the patterns of interaction are, to some extent, idiosyncratic, in that both members of the dyad seem to influence the response of the other. It is not a question of passive receptivity on the part of even these very young infants, but rather a reading of cues by each of the other's response. Shotter has called this characteristic pattern of interaction between mother and child the "creative exchange, par excellence, of people making people".

If we consider Shotter's extension of construct theory to be useful and consistent with the theory as a whole, then we find ourselves with the following context within which to view the development of "persons": men have created for themselves a means of making sense of their world, and, in the process, a means of becoming what we recognise as human. A child, born into a human community, gains access to this system through his relationships with others, and primarily, through his relationship with one caring adult. This relationship involves the instruction of the child by his mother whereby spontaneous actions of the former become deliberate, i.e., under intellectual and volitional control (Vygotsky, 1962). Since the child has the capacity for knowledge about himself and the world through his bodily activity, this interaction between mother and child involves something more than the guiding of the child by the mother. We might say that what seems to be involved is a role-relationship in Kelly's terms, whereby each member of the dyad attempts to construe the construction processes of the other. We will have reason in another section to look at this notion more critically. However, suffice it to say at this point, that the construal of the mother–child relationship in terms of an active interchange between two people who are both seeking to find out what the other is about, represents a position consistent with the formal postulate of construct theory.

In terms of the considerations discussed in the previous section regarding what appeared to be difficulties in any adult–child relationship, it might be said here that such a perception would seem to have at least some basis in fact. That is to say if we can assume that "humanness" is a construction which is only gradually incorporated by the developing person in his relationships with others, then it is likely that in the therapeutic situation we may experience difficulties initially in sharing notions of what we both are. In the light of this kind of proposition also we can understand the apparently very strong set we adults have in relation to children, i.e., to fill a gap by telling the child what he is thinking, feeling etc. In a very real way this would seem to be an essential aspect of all such relationships. However, as Shotter's account has made clear, this is not nearly so one-sided as it appears superficially. At the same time, this does seem to lead us on to a further issue which is what might be described as the susceptibility of adult–child relationships as such to domination by one member, i.e., the adult.

There are very obvious reasons why this might be so and in fact the question as to whether the relationship had the capacity for any other kind of exchange is of relatively recent origins. It is worthwhile in any discussion of mother–child relationships to bear in mind the fact that many workers have referred to patterns of interaction within families as of vital significance in understanding emotional disturbance of all kinds in persons. The fact that this may be so should come as no great surprise when we consider what was said about the child learning to become human within the context of family relationships in particular. I believe that it is useful to regard such work as providing us with descriptions of processes at the extreme of some kind of continuum, a continuum which includes the kind of mother–child relationships usually referred to as "normal". That is to say, in even the most academic discussion, cognisance must be paid to the indisputable fact that the development of persons takes place through processes of relationships which are open to what we might call error, distortion or breakdown.

Shotter has extended the basic theory of Kelly in order to provide us with a construction of how men themselves devise notions of their humanness. Formulations as to how individual children may develop as persons are put forward by Salmon by drawing on aspects of the formal theory itself.

"PERSONAL GROWTH": THE WORK OF P. SALMON

In her article "A psychology of personal growth", Salmon (1970) sets out the broad dimensions of an approach to personality development in construct theory terms. She acknowledges that as Kelly himself has said so little about children and as he was so opposed to the textbook type of classification, she is taking something of a risk in attempting to apply the theory to an area in which children are so essentially concerned. It might also be mentioned in this regard that some construct theorists have questioned the necessity of developing a specific theoretical framework to deal with children since Kelly's model of man contains an implicit notion of change or development whereby a person is seen as constantly developing. While I find myself totally in sympathy with the commitment to meaningfulness in a person's behaviour, irrespective of their age or cognitive status, I would not accept that we have thereby provided ourselves with a construction which will further our understanding of a subject who happens to be a child. In her account, Salmon too seems to implicitly accept the need for some elaboration of the basic theory which will cover that kind of development which is usually associated with childhood.

For our purposes here, it is sufficient to say that Salmon centred her account of personality development in terms of the child's inter-personal relationships in general and the relationship between the child and a significant adult in particular. For a full account of the discussion of what she sees as the distinctive contribution of construct theory to the area of personality development and in what respects it differs from other available notions, the reader is referred to the article mentioned at the beginning of this section. The focus on the mother–child relationship, as opposed to a single study of the child, and the application of two corollaries in particular to the interaction characteristic of this relationship constitute the aspects of Salmon's discussion of most interest to us here.

Since personality is construed as inter-personal in nature within the theory, any investigation of personality development needs to concern itself with more than one individual. As Salmon (1970, p. 216) puts it,

> "The dimensions in terms of which any individual defines his behaviour towards others . . . are derived from the roles which he has played with other individuals, the frames of reference which he has elaborated in common with them and the agreed network of implications which he has shared with others in crucial inter-personal relationships."

One of the most crucial of these is undoubtedly that first relationship with a caring adult, usually his mother. "Her construing gives him the basis of his own, however much he may later elaborate his own view" (Salmon, 1970, p. 216). As we said before Salmon construes this interaction between mother and child in terms of two corollaries in particular. The first of these is commonality which states

> "To the extent that one person employs a construction of experience which is similar to that employed by another his processes are psychologically similar to those of the other person." (Kelly, 1970, p. 20)

The mother–child relationship is one where the mother, in the early stages at least, "helps to define the relevancies of situational contexts for the child, thereby offering him a construction in terms of which he can act towards the situation . . ." (Salmon, 1970, p. 206). Going back to Shotter's "wild boy", we can see that what a mother is doing for her child in contrast to this, is cutting down on the amount of "blind" exploration a child need do in order to achieve his ends. She provides food and physical comfort with the minimum of searching by the child, and by responding to certain cues and not to others, she provides frameworks within which he learns to construe his needs, their satisfaction and thereby a framework within which to construe the relationship between the two of them. Major socialising achievements, such as toilet training, for example, are accomplished through the child learning to construe his own activity the way his mother does. The structuring of vocal activity, prior to actual acquisition of language proper, is another essential intervention into spontaneous child behaviour by mother.

The other way in which the content of the child's developing construct systems is influenced by his relationship with his mother is described in terms of the sociality corollary. "To the extent that one person construes the construction processes of another he may play a role in a social process involving the other person" (Kelly, 1970, p. 22). As Salmon points out, "Kelly lays considerable stress on this aspect of interpersonal relationships". Furthermore, "since relationships with others are central to the concept of personality, and since expanding role playing abilities define the ever-widening limits of the child's interpersonal relationships, the concept of sociality seems a key one in the psychology of personality development" (Salmon, 1970), p. 206).

We will have reason later on to draw out in greater detail the implications of construing a role relationship between mother and child. Suffice it at this stage to define what is meant by role. Bannister (1962) has described the construct as "an ongoing pattern of behaviour that follows from a person's understanding of how others who are associated with him in his task think". To behave towards someone on the basis of what I think he's about, requires that I put myself in his shoes, i.e., what *I* think are his shoes on the basis of my own past experience. The ensuing investment and encounter with him provides information as to the discrepancies in my construction which then is altered and I am ready for a new investment and encounter. The process gradually allows me to differentiate between the two of us, by providing me with ongoing information about myself and the other person. As Salmon (1970) says, "A role, like any other human endeavour, can best be understood by considering the ingenuity of individuals, who launch 'unprecedented behavioural undertakings' ". In terms of the child playing his first role with his mother, we are reminded of Shotter's comment on the Nottingham project referred to earlier, that the patterns of interaction between mother and child are, in some degree, idiosyncratic. "The mother must watch the child, and the child the mother, there are no pre-established rules, only continual improvisation" (Shotter, 1970).

We might say that the individual approaches of Shotter and Salmon provide us with two views of the same situation. Thus while the former is concerned with the macro view of things, the socialisation aspect in the broadest sense of the transformation of "natural agents" into "individual personalities", (Shotter, 1973) the latter provides us with the means of getting a close-up look at the processes in individual mother–child relationships. These represent to all intents and purposes what might be called a construct-theory approach to childhood, socialisation, mother–child relationships etc.

In the following two sections we shall attempt to test out this approach in the light of the work of other major theorists in the field of child development. This is seen as not so much a means of picking out weaknesses in the approach, but as a way of elaborating the basic outline when faced with certain apparently incompatible hypotheses. The first of these is the problem of cognitive egocentrism as described by Piaget. The following section will try to answer the issues which seem to be posed by what Piaget considers to be a fundamental characteristic of childhood functioning. The second issue, with which our final section will be concerned, represents what seems to be the most obvious of all the characteristics of mother–child relationships, i.e., the emotional context of the relationship. I shall be making use of the work of Susan Isaacs in this section.

The Problem of Cognitive Egocentrism

Piaget has used the concept of cognitive egocentrism variously throughout his work, from on the one hand, his account of moral development (Piaget, 1932) to his description of play in the young child (Piaget, 1951) on the other. It is a concept which must be seen within his total theory of development and as such it has not therefore been used in any account of the developing child's interpersonal relationships with others. However, since the concept is used widely throughout the literature of child psychology, though not always in strict accordance with Piaget's definition, it seems important that construct theory should address itself to such a notion in formulating its account of the mother–child relationship. I will therefore present a general description of what is meant by cognitive egocentrism and draw out what seems to be the implications of such a description for our purposes. Piaget (1962, p. 3) defined egocentrism thus: "I have used the term egocentrism to designate the initial inability to decentre, to shift the given cognitive perspective (manque de décentration)." And later in the same section: "Cognitive egocentrism, as I have tried to make clear, stems from a lack of differentiation between one's own point of view and the other possible ones . . .". He makes clear further that until a child reaches a certain age he is developmentally egocentric. "At about the age of seven the child becomes capable of cooperation because he no longer confuses his own point of view with that of others" (Piaget, 1968, p. 39). The behaviour of the child who is described as egocentric is "intermediate between purely individual and socialised behaviour" (Piaget, 1932, 1932, p. 26).

The central issue in these extracts seems to be the child's initial inability to "shift the given cognitive perspective" (Piaget, 1962, p. 3). When we talked earlier of Shotter's wild boy we said that while he did develop a way of seeing the world, it was one "tied to the organism's needs" and that alternative construals were impossible. It was suggested that to a lesser extent this was also the case with the "basic knowledge" acquired by the child in bodily encounters with his world unless a way could be found to express (use) it in terms of the special way of "seeing" the world devised by men. Through language our child becomes able to "re-arrange the component parts of his 'view' into new patterns" and Shotter adds, . . . "it is a view that represents and incorporates knowledge gained from the many different standpoints of his fellows" (1970, p. 240). If this is what is meant by a child's inability to shift the given cognitive perspective, then we have constructions in construct theory to handle the notion. However, one suspects that this is not fully what is implied in Piaget's description. First of all, he is talking about behaviour resulting from cognitive immaturity, which until a certain stage of development has been reached, is incapable of becoming socialised. Whereas in construct theory terms no such distinction exists.

It is clear that the child only *gradually* learns to see the world fully in the special way of his fellows but there is no way within construct theory terms to consider the child as incapable of construing the world, i.e., make sense of it within schemes of ordering devised by men. Thus while these schemes may be few and poorly organised and not very stable, a child, from a very young age, is seen as capable of sharing constructions with another person and thus is behaving socially with respect to that behaviour at least, in construct theory terms. We may recall, as one illustration, the observation by Newson (1969) on pre-verbal children and their mothers and how they develop a mutual construct-system. Newson has described how the mother gets the child to select among his activities and he says

> "out of all this mutual activity, mother and child develop a shared frame of reference. Each knows the happenings to which the other is responsive, and they are thus in a position to communicate their intentions to one another".

Which brings us to the issue of sociality and the question of role-relationships in the light of egocentrism. Construct theorists make certain assumptions about the status of the two people in a mother–child relationship. Apart from the overall hypothesis that *both* persons are forming theories and acting on these with each other, we have from Salmon a description of the type of relating between the two. Using Kelly's illustration of the client–therapist relationship, Salmon says that while the mother's attitude is one of providing "an enabling structure" for the child's "progressively shifting referents of behaviour", the relationship could certainly not be viewed as one-way and in this the mother's position is much like that of the psychotherapist, who "does not know the final answer either—so they face the problem together" (Kelly, 1969). This comparison will be questioned in the next section, but for now it provides us with material for discussion.

If mother and child "face the problem together" then we have on our hands a qualitatively different kind of relationship than that seemingly implied in Piaget's description of the child's egocentrism. The meaning of sociality was made very clear by Kelly when he said

> ". . . my construing of your construction processes need not be accurate in order for me to play a role in a social process that involves you. I have seen a person play a role and do it most effectively—even in a manner quite acceptable to his colleagues when he grossly misperceived their outlooks and they knew it. But because he did what he did on the basis of what he thought they understood, not merely on the basis of their overt acts, he was able to play a collaborative role in a social process whose experiental cycle led them all somewhere." (Kelly, 1970, p. 24)

In this passage we have some of the essential hallmarks of construct theory. First of all what is hypothesised is a dialectical process of relating, rather than a mechanical one and the shape of the encounter is movement along an experimental cycle rather than the comparison of points of view. With regard to this latter point, we are reminded that whereas construct theory considers that

> "reality can never be known in any final absolute way, but only through our constructions which, as a result of the varying validational outcomes of the behavioural experiments we make are subject to continual revision" (Salmon, 1970),

the aim of a Piagetian subject is always the accurate perception of reality, which is subject to universal laws and whose perception can be achieved by mastery of the principles of logic. Piaget made it no less clear than Kelly what he had in mind when he talked about socialised behaviour, which for him characterised the mature adult as does the achievement of abstract thought in the intellectual field.

> ". . . if an individual A mistakenly believes that an individual B thinks the way A does and if he does not manage to understand the difference between the two points of view, this is to be sure, social behaviour in the sense that there is contact between the two, but I call such behaviour unadapted from the point of view of intellectual cooperation." (Piaget, 1962)

It's clear from these extracts that the differences in the two accounts arise from their differing philosophical assumptions. The worker holding one or other theoretical position will consequently not only be interested in different material but will draw different conclusions from the same material. There can be no question therefore of Kelly's theory being seen as an extension of Piaget's in the field of personality as has sometimes been suggested. An issue like egocentrism is one which illustrates these differences.

If we accept a phenomenon of childhood functioning which describes a child's inability to differentiate his own from other points of view, and I think we can in construct theory, then what are the differences in implication following from this for the two theories? For the Piagetian worker with his assumptions about knowledge and reality, egocentrism will be seen within a context of developmental stages. The aim of development has been accepted as the attainment of abstract thought and socialised behaviour which are seen as closely related. Behaviour characterised as egocentric will be seen as inadequate and the adaptive effort therefore will result in systematic errors. Thus the child at the stage of concrete operations for example will be seen as making systematic errors in his adaptation as a result of his perceptual, cognitive and affective distortions arising from his state of egocentrism. From a construct theorist's point of view, on the other hand, such a child may be seen to construe the task situation differently from the adult experimenter to the extent of not sharing those constructions which have a high degree of consensual validation. However, the main focus of inquiry for the construct theorist is: what does the child see himself as doing while failing to do what is expected of him? In other words, the worker in construct theory possesses no ready made criteria of success and thus can have no prior definition of error and is, moreover, obliged to discover the child's meaning in these "errors". In so far as the Piagetian worker attempts to describe the errors in the child's performance it will be in terms of their deviation from formal cognitive structures.

The two theories seem to be in agreement about the issue to the extent that both emphasise the developmental need of the child's gaining the perspective of his fellow men, in order to become what we recognise as human. Both also indicate the importance of social interaction in the achievement of this end. Where they differ fundamentally however is in defining what successful adaptation means. Piaget has linked socialised behaviour and abstract thought quite closely within systems of logic and this seems consistent with his notions of reality and knowledge. The implications of a construct theory position on the other hand have not been so precisely predicted and may not be because "success" includes the transcendence of the rule systems themselves. As Shotter points out "men not only create systematically according to their own rules—they create the rule systems too" (Shotter, 1970).

As has been emphasised before, the interaction patterns of mother and child are somewhat idiosyncratic. The mother and child seem to be involved in trying to find out what each other is about. Though not "equal" in the sense that adult to adult relationships would be they both seem to "launch unprecedented behavioural undertakings" (Salmon, 1970) in order to get to know each other and therefore the risk of error exists on both sides. Once again what is predicted in the situation is the constant creation of new ways of "going on". Despite the differences between the relationships therefore, the criteria of a role relationship, i.e., the mutual construing of construction processes, are present in the mother–child relationship.

It is not essential in construct theory terms that this mutual construing should result in "intellectual cooperation" (Piaget, 1962, p. 8) as is the requirement of socialised behaviour within a Piagetian system. The fact that the child in the relationship is egocentric has different implications in both theories therefore. In Piaget's terms, how ever we describe the interaction between mother and child we may not describe it as representing "socialised behaviour" with all that this entails. On the other hand, to describe a child as egocentric, as a construct theorist, is to describe someone who has relatively few constructs, which are loosely organised and which are subject to more comprehensive change than is the case with adults. In having few constructs the child must consequently make sense of all the events in his life in terms which will be quite generalised and which will still bear the marks of their origins in his own immediate life context. In order to be able to cover all eventualities these constructs will have to be loosely organised. There is not much point, in anticipatory terms, in being able to construe minute events of life at home, if the child is totally at a loss once he steps outside this. The value in having a construct system which is not firmly organised is that the child can more readily respond to the steady stream of new events.

I have been suggesting that cognitive egocentrism as described by Piaget essentially has no meaning in construct theory terms. However, this is not to deny that the phenomenon of egocentrism exists and can be observed in the behaviour of young children. The implications of the notion as described in Piagetian terms are incompatible with a construct theory approach however, as might be expected given the differences between the theories in philosophical assumptions and area of interest. The obligation now rests with construct theorists to devise ways of understanding egocentrism in their own terms.

We have described in construct theory terms the egocentric child as possessing few, loosely organised and relatively unstable constructs. We have much to learn about the implication of such characteristics for mother–child relationships. We also need to know more about these characteristics themselves in addition to other features which may be present in the construct systems of such children. The sort of observation study carried out by Newson (1969) seems a particularly useful model for the kind of investigation needed in this respect. This seems a fruitful area of study not only to throw light on egocentrism as such, but also, more importantly, to increase our knowledge of the ways of relating devised by mothers and children.

In the final section I shall discuss the issues arising from one of the most complex aspects of the mother–child relationship, from the investigator's point of view, that is the distinctive emotional context of the relationship.

The Emotional Context of the Mother–Child Relationship

We would not expect to find a section headed "emotional development" in any account of construct theory. It would be futile therefore to expect to unearth easily what might be called a construct theory approach to the emotional aspects of the mother–child relationship. And somehow one is in sympathy with the determination not to drift into this sectional view of persons all over again. But we must be careful that we are not simply avoiding what is, after all, a very real if complex phenomenon. This is, that the activity that Shotter called "people making people" is one which takes place within a context of strong emotional involvement by two people. This is a fact most obvious to non-professional "scientists" whatever difficulties it may present to those of us who professionally attempt to investigate our fellow-men!

In the literature of child development much has been discussed about the child's emotional state in the early years. Unfortunately, the emphasis in many of these accounts tends to be so one-sided that they prove unacceptable to anyone who has chosen as their unit of inquiry, the reciprocal relationship between mother and child. What seems to be required is an account which will allow us to construe the relationship processes between a mother and her child in terms of the theory-making capacity of both of them. We dealt with this difficulty under another guise in the last section when we talked about the possibility of role-relationships between mothers and children in the light of egocentrism. Here we must tackle it again, as the theoretical problems for a construct theorist in making sense of the emotional aspects of the relationship. In other words, we cannot content ourselves with an account of the emotional development of a child in isolation from the ongoing emotional development of the most significant person in a young child's life, a caring adult. These two aspects can only be viewed separately as an academic convenience.

The beginnings of such an approach has already been made by Salmon (1970, p. 210) when she provides us with a description of maternal attitude. This she sees as consisting in the mother's capacity to "provide an enabling structure" for a child's "progressively shifting referents of behaviour", a capacity that is actualised within a context which Salmon has likened to the client–therapist situation. In other words, the mother's, what we might call, socialising function, takes place within the sort of reciprocal relationship that Kelly described when talking about client–therapist—since neither know the answer ". . . they face the problem together" (Kelly, 1969).

Salmon rightly makes use of this comparison if only to emphasise what a construct theory approach to the subject would *not* be like. The stage has been set, as it were, for an account which must contend with a relationship of two people trying to make sense of each other. However, we must be careful not to rely too heavily on the client–therapist model. We have seen where it can be most useful, but our task at present will consist in trying to show its short-comings in terms of the constructions it offers of the present subject. Thus while we do talk of an emotional involvement between child and therapist, we are aware of the clearly circumscribed nature of this involvement, as well as the *specific* purposes for which the relationship was undertaken in the first place.

We must never lose sight of the fact that the therapist has been trained to be constantly aware of what is "going on" between himself and his client, and he has a professional responsibility to make clear, in self-conscious verbal interchanges, this communication. Aspects of these functions may perhaps exist at times in mother–child relationships. However there seems to be one difference between the two situations which I would suggest is of great importance, particularly for the subject presently being discussed. This concerns the historical circumstances surrounding the beginnings of any mother–child relationship. It is because of the difference in these beginnings that comparison with the client–therapist situation may no longer be of use to us when we come to construe the emotional context of the mother–child relationship.

For the sake of simplicity, I would like to distinguish two aspects of our subject for the purposes of discussion. Firstly, we must consider what is involved for a woman who becomes a mother, in a relationship with a young child. Broadly speaking what we must consider, are the physical and emotional factors involved in giving birth and caring for a young infant in the succeeding months. More precisely in construct theory terms, we will be looking at how a person construes someone who was once part of her own body, who is now separate from that body and who is nevertheless completely dependent and making demands on her. The framework of our approach, therefore, is similar to other investigations using construct theory, in that our subject is a person who also happens to be the mother of a child.

Too often psychological studies in this area seem to construe mothering as an activity carried out by particular subjects called "mothers". One of the implications of thinking of persons, rather than "mothers" is that we can make sense of the fact that such a person has preoccupations and relationships other than with her child. This is a distinctive aspect of human mothering situations, and one which can be expected therefore to have a bearing on the child's perception of his world, as we shall see later on. Thinking of persons also allows us to consider mothering in terms particularly fitting for a construct theorist, that is in terms of change.

As we said before, mother and child meet each other in circumstances which seem unique. Giving birth to a child and caring for it in the succeeding months seem to be events likely to provide a person with much new information about herself and her world. In a very particular way these seem to be opportunities for physical activity of the type Shotter and Kelly had in mind when they talked about the "priority of behavioural encounters" (Shotter, 1970, p. 224), in the gaining of knowledge. Of course unlike her child, the mother is not totally reliant on these for her construing of what is happening, nor does the construal of these bodily encounters take place within a vacuum. Her already existing construct systems probably contain constructions of birth, babies etc. formed in the experiences of her life to date, most likely going back to her own childhood, her construction of mothering from her experiences then right up to present-day encounters with her own peer group. Many of these constructions will be at a preverbal or unconscious level, arising as some of them must from experiences far back in the woman's life.

Another set of constructs which seem likely to be of importance are those about herself as a child in a relationship with a mothering person. As yet we have no way of knowing what the influences are on a person's development of constructs of mothering and hence we can only speculate. However, it seems reasonable to suppose that women do have constructions of the role before the birth of children and that these do influence their early contacts with their new-born infants.

It is into a context such as this that the new knowledge provided by physical encounters with her baby, must eventually be integrated. And this new source of knowledge is likely to provide information about herself as well as her baby. How she first construes her baby therefore, is going to be intimately tied to how she construes herself. In other words a woman's construction of her baby is closely related to her own core structures. We begin to get some notion of what we mean when we say that there is an emotional involvement between a mother and her child.

Construct theory definitions of emotion in terms of "imminent or active construct changes" (Bannister, 1962) seem of particular relevance in the light of the foregoing discussion. We may state the position anew: The emotional involvement of a woman with her child may be seen to represent a coming together of several factors in her life which are likely to demand some changes in her existing construct systems. These factors include prior constructions of self, which may or may not be related to constructions of mothering, which in turn may or may not be related to early constructions of herself as a child in a mothering relationship. There is clearly an infinite number of ways that these constructions may be combined in any one woman's life. However, no matter what way these are related or maybe not related when her child is born a woman is faced with the task of making sense of herself, her body and what has emerged from her body, in the light of new information, which must be integrated in some workable way with what has gone before.

When one sees the relationship in a context such as this, one may more easily accommodate the notion of a woman experiencing feelings of anxiety, guilt etc. towards her child. For example, in a situation where she is having to define herself anew, as it were, what could be more understandable than that she should at some stage be aware of the "dislodgment of the self from (her) core role structure" (Bannister, 1962). One of the implications of considerations such as these is that we would expect any child to be born into a situation which contains elements of which neither he nor his mother may be fully aware. There remains much work to be done on aspects of the mothering relationship such as these. At the moment, we can only rather tentatively mark out the area for inquiry. But it does seem likely that there is much to be gained from a construct theory approach particularly with regard to how a woman forms constructs of herself in the relationship.

We have seen some of the elements which may be at work in the relationship, as far as the adult is concerned. We must now turn to those aspects which seem to arise from circumstances due particularly to the child in the situation. That is to say, can we outline any dimensions of the relationship which come about because the other in the relationship is a child.

I shall now turn to the work of Susan Isaacs, a noted authority on childhood functioning of the psychoanalytic school. Her work seems to be generally representative of a line of thought, which must be acknowledged as the most extensive in the field of the emotional development of the child. To begin with, Isaacs distinguishes the child from an adult in terms of the "intensity of feeling" exhibited by the former, which is often not "accessible to reason"

(Isaacs, 1933). In order to understand such a development Isaacs returns to the very young infant and describes what she considers his characteristic state. In her book "Social Development of Young Children", she quotes M. N. Searl on this:

> "An infant normally has few and simple but urgent and imperative desires connected with strong feelings. He has, however, no power to satisfy these urgent desires without the help of mother or nurse. He has no power to know anything of the whys and wherefores of any lack of satisfaction. When he has what he wants he feels 'good'. When he has not what he wants he feels 'bad'. Similarly, the person or object from whom he gets what he wants feels to him 'good': the person or object from whom he does not get what he wants feels to him 'bad' " (1932, p. 285). Isaacs believes that the young child "brings to his experiences in the real world enormous intensities of feeling and desire, along with an almost equal degree of actual helplessness to effect such changes in the outer world as will lead to satisfaction of his desires."

A further dimension to the child's position in relation to the world concerns the mother's involvement with other people.

> "This central situation, in which, at the time when the child is utterly dependent upon his mother for all his satisfactions, he is nevertheless obliged to realise that she gives herself and her love to his father is almost, if not quite, unique, in the mammalian world."

The fact that the human child spends a much greater length of time than the young of any other species helpless and dependent on others, along with the fact that at no time does the child have the total attention of his mother, but always shares it with others, is considered by Isaacs to constitute a caring environment which is distinctly human.

We are reminded of Shotter's comments on what was involved in babies growing up to be what we recognise as human. It might be interesting to speculate on what appears to be the inherently conflicted context within which human beings are created. For example, it was mentioned before that a woman must be seen as a person with relationships other than that with her child. The child on the other hand is incapable in the beginning of relating to anyone else, since she and he constitute a barely differentiated duality to him. The conflict arises for him because she cannot be made to be present at all times. There would seem to be a conflict too for the woman in so far as she has got to play a role with someone, which means attempting to see what they mean by their actions, when this someone is so obviously helpless and dependent on her for everything, including meaning.

Isaacs goes on to divide her study of the child into intellectual aspect and emotional aspect. He can apprehend "reality", in terms of the former, but tends to distort "reality" in terms of the latter. On this point Isaacs distinguishes her approach from that of Piaget. While the child is egocentric with regard to feelings and fantasy, according to Isaacs, he is not so in his thinking. The child is egocentric to the extent that he has not learned to think. For our purposes the argument, though interesting, is unacceptable because of its inherent assumptions about an external reality directly knowable etc. and the consequent separation of cognition and emotion. What we can accept and make use of is the description of the young child, whose experience of many events in his life is likely to be emotionally more "highly charged" than that of the adult. In our terms we might say that a child's experience of events in his life is more likely to result in "imminent or active construct changes" than is likely to be the case with an adult. As we saw before when discussing the meaning of egocentrism in construct theory terms, a child's constructs are likely to be few, loosely organised and relatively unstable. In such circumstances we can readily understand the occurrence of some kind of change as a child is faced with a fairly constant stream of new events. In similar terms, we might expect that such a child, in trying to anticipate events, will many times be aware that the events with which he is confronted lie outside the range of convenience of his construction system—i.e., he will experience anxiety.

In describing the behaviour of an infant waiting for his mother to come and feed him, for example, we might say that the potential exists in the situation of his becoming aware of not having constructs to make sense of what's going on. Alternatively he might cry more vigorously, kick his feet, and maybe even shake his cot, in which case we might describe him as "actively elaborating his perceptual field", i.e., aggressive behaviour. We would expect to be able to distinguish other instances of "imminent or active construct changes" in the behaviour of the young child, like fear, guilt etc. Furthermore, we would expect to find these occurring most frequently in the most significant context of his life, i.e., his relationship with his mother.

Some effort has been made in the above discussion to suggest lines of approach to the emotional context of the mother–child relationship in terms of construct theory. While we consider it undesirable to introduce text-book categories into this discussion, we have done so as a temporary measure in order to illuminate certain aspects of the subject which we felt were in danger of being overlooked. We hope that this deviation will be considered worthwhile in having directed our attention to the process of construct change as a useful construction of mother–child interaction. We have tried to work out some of the implications of assuming this to be a reciprocal relationship by construing general dimensions which may exist for both people in being so involved with each other.

Overview

In this paper we have argued for more theoretical consideration to be given within construct theory to those aspects of human development specifically associated with childhood, i.e., to the "creative exchange" of "people making people". The starting point of our discussion was provided by the related writings of such construct theorists as John Shotter, from whose work the last quote has been taken, and Phillida Salmon. The central features of the framework we see emerging from their combined approaches, may be summarised as follows: Men have created for themselves a means of making sense of their world, and in the process, a means of becoming what we recognise as human. A child born into a human community gains access to this system through his relationship with a caring adult. Shotter makes use of Vygotsky's concept of instruction (Vygotsky, 1962) whereby the child's spontaneous actions become deliberate, in describing this relationship. The process is seen as reciprocal, however, in that while the mother is the one who helps the child *select* among his activities, she is not responsible for the activity itself. The knowledge of himself and the world provided by this activity constitute a significant factor in the interactional processes between mother and child which may not therefore be seen simply as one-way guiding of child by adult.

Both Shotter and Salmon individually lay stress on this aspect of the relationship. Shotter talks of the "idiosyncratic patterns of interaction", while Salmon emphasised the importance of "sociality" in understanding the interaction between mother and child. This brings us to the issue of role relationships with which we have been concerned in one guise or another throughout the paper but particularly in the last two sections. We discussed the question of whether or not it was theoretically possible to conceive of a role relationship in the mother–child context, in the light of Piaget's influential formulation, cognitive egocentrism. While such a notion seems to preclude the occurrence of socialised behaviour in the child under seven, the situation is otherwise in construct theory terms because of the difference between the two theories in the way they view the nature of social relationships. Since the "success" or "failure" or a social interaction cannot be said to depend solely, or even primarily, on its resulting in intellectual cooperation, as is the case in Piaget's terms, construct theory provides a construction of what happens between mother and child consistent with its construal of all other relationships. That is, it construes the interactions in terms of the mutual playing of roles, and the "launching of unprecedented behavioural undertakings" (Salmon, 1970, p. 213), by two people.

Although the particular meaning attributed to it by Piaget is denied, the phenomenon of egocentrism as seen in the behaviour of young children still remains a factor to be taken into account. Construct theorists have set themselves the not insignificant task of making sense of a role relationship between two people, one of whom is acknowledged as possessing few, loosely organised and relatively unstable construct systems. What constructions might be made about such a relationship and what might be involved for the two people concerned, bearing in mind our commitment to a notion of persons and of meaningfulness in personal terms, is discussed in our section on "emotion". The construction of "emotion" in terms of "imminent or active construct changes" (Bannister, 1962) is applied to the interaction. At a general level, the events constituting birth and childhood are seen as demanding changes in the construct systems of mother and child of a type and magnitude not found in any other relationship. The awareness of the need for such changes must be considered an important factor in the development of mutual construing systems. That is to say, the role relationship that develops or does not develop between mother and child must reflect how each person makes sense of the demands for change that arise as a result of knowing the other.

In summary, it could be said that the implication of a construct theory approach to what happens between mothers and their children, what is involved when we talk of socialisation, is the construal of a relationship which is somewhat more robust than our psychological theories would sometimes have us believe. In turn, we may construe our child clients, not as passive victims of their environments, nor as demons of primary process thinking, but as persons, a construction which, in truth, may include one or other of these unfortunate construals but which essentially involves us with subjects who can "transcend the accepted parameters of the situation by inventing new ones" (Salmon, 1970). It seems important that those of us who are involved with children should have access to a theory which allows us to perceive ourselves and our clients as capable of transcending the past.

References

Bannister, D. (1962). Personal construct theory: A summary and experimental paradigm, *Acta Psychologica* **20,** 2, 104–120.

Isaacs, S. (1933). "Social Development in Young Children", Routledge and Kegan Paul, London.

Kelly, G. A. (1969). "Clinical Psychology and Personality: Selected papers of George Kelly", (Maher, B., ed.), Wiley, New York.

Kelly, G. A. (1970). *In* "Perspectives in Personal Construct Theory" (Bannister, D., ed.), pp. 1–29. Academic Press, London and New York.

Newson, J. (1969). Comments on the nature of pre-verbal communication between mothers and their children in the age range 10–20 months. Mimeo, Department of Psychology, University of Nottingham.

Piaget, J. (1932). "The Moral Judgment of the Child", Routledge and Kegan Paul, London.

Piaget, J. (1951). "Play, Dreams and Imitation in Children", Routledge and Kegan Paul, London.

Piaget, J. (1962). *In* "Comments on Vygotsky's Critical Remarks" concerning "The Language and Thought of the Child" and "Judgment and Reasoning in the Child", M.I.T. Press and John Wiley & Sons, Inc., New York and London.

Piaget, J. (1968). *In* "Six Psychological Studies", (Elkind, D., ed.), pp. 3–73, Vintage Books, New York.

Salmon, P. (1970). *In* "Perspectives in Personal Construct Theory" (Bannister, D., ed.), pp. 197–221 Academic Press, London and New York.

Shotter, J. (1970). *In* "Perspectives in Personal Construct Theory", (Bannister, D., ed.), pp. 223–253, Academic Press, London and New York.

Shotter, J. (1973). *J. Theory Soc. Behaviour*, **3,** 2, 141–156.

Vygotsky, L. S. (1962). "Thought and Language", M.I.T. Press and John Wiley and Sons, Inc., New York and London.

Living on the Horizon

Alan Radley

> "What we think we know is anchored only in our own assumptions, not in the bedrock of truth itself, and that world we seek to understand remains always on the horizons of our thoughts." George Kelly, *The Psychology of the Unknown.*

There have been several attempts to categorise George Kelly's personal construct theory alongside other schools or approaches to psychology. For example, Holland (1970) makes out a case for the theory being strongly influenced by Kelly's existentialist thinking, although Kelly himself was at pains to remain outside this, and indeed all categories in which other psychologists tried to place him. One reason for Kelly's concern might be that he produced in "The Psychology of Personal Constructs" a theory which approximated to a position which he never lived to express fully. This, of course, we can never know. What we can know, however, is the extent to which Kelly's later ideas, as expressed in the papers published after his death (Kelly, 1969), go beyond the formal structure of the theory published ten years earlier. It would appear that while others were seeking to draw parallels between contemporary schools of psychological thought and personal construct theory (and being criticised by its author for so doing), Kelly was at the same time trying to elaborate his thinking concerning the model of man upon which his theory was based.

The formal theory of personal constructs was based upon the metaphor of *man-the-scientist*, who seeks to predict and control the course of events in which he is involved. The exposition of the theory contains many references to these important aspects of scientist-like behaviour, and Kelly embodied prediction and control in his idea of the personal construct. The idea of the construct can be seen as the crystallisation of Kelly's standpoint regarding man in his scientific aspect. Introducing the theory, he says of this standpoint:

> "Here is an intriguing idea. It stems from an attempt to consolidate the viewpoints of the clinician, the historian, the scientist and the philosopher. But where does it lead? For considerable time now some of us have been attempting to discover the answer to this question. The present manuscript is a report of what has appeared on our horizons thus far". (1955, p. 5).

The key words here are "what has appeared on our horizons thus far". For in the years which followed the publication of the theory Kelly continued to extend actively the horizons of this intriguing idea. In his later writing he elaborated considerably the metaphor of man-the-scientist so that it extended well beyond the processes of prediction and control.

In "The Strategy of Psychological Research" (1969c) he presents an account of the scientific enterprise based upon the experimenter's willingness to be *personally involved* in the field of his enquiry, to be *committed* to that undertaking, and to reappraise what happens as a result. The importance of involvement and commitment is re-asserted in "The Psychology of the Unknown" (published in this volume) together with the role of *faith* in man's own efforts to make closer approximations to the nature of things which he is trying to explain. In one of the last papers which he must have written ("Ontological Acceleration", 1969a), Kelly uses the standpoint of constructive alternativism to argue his position regarding the role of psychology in the broader enterprise of human understanding. He says:

> "In the perspective of this viewpoint the task of the psychologist is not simply to tabulate the categorial ways in which men presently behave, and the circumstances in which each occurs, but to engage himself with his fellow men in exploring the uncharted realms into which the human quest may be expanded. And in this mutual enterprise one may well wonder what outcomes will ensue and to what unexpected vantage points their joint experience will lead. So now, in order to survive, psychology must invent as well as discover. That will make it a discipline disconcertingly alien to the one in which most of us were once trained". (1969a, pp. 16–17)

It would seem that Kelly is using the psychology of personal constructs to point towards aspects of man-the-scientist which are not fully expressed in the original theory. Rather than being a theory about ways of seeing the world, Kelly is indicating that construing is also about a person's involvement *in something*, his faith in *what is not yet actual* and the commitment of *himself* to his actions. It is these issues which form the core of this essay, which is an attempt to make room for the broader and deeper conception of the term "personal construct" implied in Kelly's writing. We can relate these issues to questions about the relationship of the person to his world, each issue having its extension in a particular concept of importance to Kelly's theory. We can relate faith in what is not yet actual to *anticipation*; involvement in something to the person's *behaviour* in his environment; and the person's commitment of himself to the question of *self-control*. In doing this we are not trying to know more about the theory of personal constructs as such, but are attempting to find out what a psychology of personal constructs might be about. That is to say, our purpose is not to support or to deny the formal theory which Kelly presented: it is to enquire further into the possible ways in which we can extend and reformulate our ideas about what we mean when we talk of a personal construct psychology.

Anticipating the Future

"Twice or thrice had I loved thee
Before I knew thy face or name."
John Donne

In a certain sense, there is an element of the future in everything which we do. Inasmuch as our actions are directed towards something which is not yet attained, then we are working with an image of some future situation which we wish to encourage or which we wish to avoid. Where we are clear in our minds as to what it is that we want to achieve, then the problem becomes one of how to set up and execute plans that will bring us to our goal. Hammering nails and posting letters are good examples of activities with easily specifiable ends which lend themselves to this kind of analysis (Miller, Galanter and Pribram, 1960). However, there are many occasions when the goal or end-point which we seek is not so readily specifiable, and sometimes is most difficult to make explicit. Quite often these situations require us to act before we are able to specify exactly where our behaviour will lead us, and part of our efforts will have to be devoted to clarifying our exact purpose on the way. For example, the person who sets out to paint a picture or to write a poem is unlikely to be able to say at the outset exactly what he is trying to achieve, beyond painting the picture or writing the poem. If pressed, the most he may venture is some description of the "tone" or "feel" or "sentiment" which he has, and which he hopes to convey. The scientist who undertakes a new direction in his research may not be able to tell us why he is pursuing this line of questioning, beyond his expectation that it will lead to a discovery in the end. These examples point towards the question of how it is that we, like scientists, undertake one course of action rather than another in the belief that it will be most fruitful in the end. Unlike hammering nails or posting letters, when the goals are easily specified, we are here pressed beyond the problem of plans for answering questions towards the deeper problem of "how are questions posed initially"?

Let us note, in passing, that we are involved with similar issues to those discussed by Kelly in "The Psychology of the Unknown" (this volume). We are interested in how people commit themselves to the future with a certain faith that their efforts will produce worthwhile results. These results, be they for artist, scientist, or layman, will be worthwhile if, through their efforts, they have "transcended the obvious". But how can one know whether the question one is asking is worthwhile? We have already stated that a person cannot know this in any precise way, and yet we have throughout acknowledged that somehow choices are made on the basis of hunches or feelings which lead towards actions or lines of inquiry which later bear fruit. Polanyi (1967) reports Plato's description of this paradoxical situation, in which he stated that either you know what you are looking for, and then there is no problem; or you do not know what you are looking for, and then you cannot expect to find anything. This paradox is brought home to us most forcibly when we are at the outset of our inquiry and other people ask us to explain what we are doing or what we are looking for. Then must we resort to descriptions of our feelings, our hunches and often our despair at being unable to explain further. At such times our feeling of subjective certainty seems inversely proportional to our ability to provide explicit justification for our actions.

The problem then is one of how a person anticipates the future. Anticipation is central to the theory of personal constructs, and in the original presentation of the theory Kelly was explicit about his view of how we go about predicting events.

> "What one predicts is not a fully fleshed out event, but simply the common intersect of a certain set of properties. If an event comes along in which all the properties intersect in the prescribed way, one identifies it as the event he expected. For example, a girl in her teens anticipates eventual marriage. There are few men in her life and the system of constructs which she uses to keep them arranged is fairly simple. The predicted husband does not exist for her in the flesh, but simply as the intersect of a limited number of conceptual dimensions. One day a young man comes along and plumps himself down more or less on this waiting intersect. Her prediction, like the navigator's, is confirmed and, before anyone realises what is happening, she marries him." (1955, p. 121)

As it stands, this description is suggestive but provides no convincing solution to the paradox. It suggests that we know our problems in terms of the intersect of abstract properties which point towards an, as yet, unknown event. What it does not deal with is the question of the specifiability of the problem. If the girl knows (as a navigator knows) the co-ordinates of her anticipated husband, could she not specify fairly accurately the kind of person he will be? If so, then she "has no problem" in the terms of our paradox, because she knows what she is looking for. For in our description of anticipation the person is hard put to say exactly what question he is asking, let alone specify the kind of answer at which he might arrive. Here, however, is a clue to the problem. For the crucial difference between these two kinds of expectation is that the person can *articulate* one while he cannot do so for the other. He knows it in a different way, wherein it can be felt but not seen.

Polanyi (1967) has offered an explanation of this paradox by proposing that we conceive of the scientist as possessing a *tacit knowledge* of the things as yet undiscovered in his inquiry. Tacit knowledge is that knowledge which we cannot specify, for the reason that it is comprised of particulars *from which* we attend towards other things. Polanyi gives the example of learning to use, as a blind man might, a probe to feel our way in the dark. At first we are aware of the stick in our hand as the "thing" in our consciousness, but as we learn to use the stick it becomes transformed in our experience. We gradually become aware of the objects which we touch with its point, and now our attention is directed towards these objects by means of the stick. The sensations in our hand take meaning in terms of the objects which we touch with its point, and the stick becomes an extension of ourselves, so to speak. The knowledge which we have in the use of the stick is tacit insofar as it becomes something which we attend from towards something else. The blind man may not be able to specify this knowledge in detail because he knows the world *through* it and is not aware of these particulars, just as we may not be aware of the knowledge upon which we depend when making judgments about others.

Polanyi has emphasised that in making particulars function in this (tacit) way so as to know from them, we *interiorise* or internalise them. So, for example, does the blind man initially come to incorporate the stick within the use of his own body by making it an extension of the muscular system through which he knows the world. He comes to *dwell in* the stick in using it to know the physical world. Using this analysis, Polanyi concludes that in all scientific enquiry we are guided by such tacit knowledge, from which we attend towards further particulars which are connected in a way which we cannot define. This accounts for the sense of conviction, or faith that there is something there to be discovered, even though we are unable to state exactly what it is.

At this point we are able, with the help of Polanyi's analysis, to make a distinction which will clarify the problem we have posed, and also to shed further light on the issue of anticipation as discussed by Kelly. The distinction is between *anticipation* and *prediction* as we shall use these terms in the remainder of the essay. At first glance, Kelly appears to use these words interchangeably, as if he meant the same thing when using either one. However, we have seen that his original statement on prediction does not seem to encompass the experience of the expectancy of the unknown which was described earlier. In using, at one point, the example of a navigator charting the co-ordinates of the North Pole, Kelly is describing a clearly specifiable goal or problem to be solved. The navigator may *predict* the event of his arrival there. That is, he may *say beforehand* (pre-dict) the time and place which defines the achievement of his goal. The scientist who has a hunch about a line of inquiry is unable to say beforehand what he will discover although he may feel convinced that his research will be fruitful. Let us, for the remainder of this essay, reserve the word *anticipation* for use with regard to such tacit knowledge from which we attend towards other things which we cannot yet specify. Anticipation is therefore an aspect of the way in which we know the world through the interiorisation of particulars. Like the blind man's stick, these particulars are transformed from being things which are distant from us, to being extensions of ourselves through or from which we attend towards the world. To anticipate something then, is to *live through* or to dwell in particulars which point towards those things which we may as yet be unable to specify.

Having arrived at this distinction, we now need to show its relevance to the broader question of how people approach the future. We shall need to ask about the place of prediction and anticipation in behaviour, and about the relationship between the two. Personal construct theory can be of help here, for while Kelly did not make the distinction outlined above, his writing points towards a similar meaning for the important place of anticipation in his theory of man's behaviour.

> "Always the future beckons him and always he reaches out in tremulous anticipation to touch it. He lives in anticipation; we mean this literally; *he lives in anticipation*!" (1969b, p. 86)

This statement of Kelly's pinpoints precisely the issue which we have been attempting to make clear, that anticipation is an important aspect of the way in which an individual *lives through* or *dwells in* those particulars which he has interiorised. Since these may be thought of as *constructs from which* he attends towards the world, then they may be considered to form part of his tacit knowledge which he cannot specify. On the other hand, those things which the person has in his awareness, and towards which he is attending, are those particulars about which he can be explicit in his expectation of what may happen. Inasmuch as the navigator can chart the position of the North Pole, or a young girl can specify the characteristics of the man whom she will marry, then they make *predictions* about what will happen and what they will do. They can use words to say in advance what might happen, and to articulate the constructs which define the events which they expect might occur.

In one sense, then, a construct is a process or vestibule (Kelly, 1955, p. 6) through which the person *exists*; it is not just an expression of similarity and difference which he can necessarily hold out for inspection. Anticipation is the basic posture of man towards the world, and it is a description of him at the personal level. By this we mean that, while predictions may be made about all kinds of things (what I shall do tomorrow, who will be the next Prime Minister—in fact, anything I can formulate), they do not necessarily involve us personally. However, my actions grounded in anticipation often do involve me, and this is shown in my total response to the situation. For example, travelling to a station or airport to meet a loved one whom we haven't seen for some time is an occasion where anticipation is in evidence. We know precisely perhaps, the arrival time of the train or aircraft and yet we yearn to be there. All our thoughts and bodily expression tell us this, for we may be unable to keep still or quite composed. Like the scientist on the verge of a discovery, we are involved in what is happening beyond the point where we can justify our excitement in terms of what can be made explicit. Therefore, while we may often think of predictions as expressions of the intellect (rather than of our bodies) we must realise that anticipation is a description of the whole person. The experience of travelling to the station or airport with our heart in our mouth is as good an example of anticipation as any to tell us this.

The preceding paragraphs have been aimed at establishing that our original paradox of acting with respect to a goal which we feel to be correct, but which we cannot specify, is central to the problem of anticipation. From the position which we have outlined, not only artistic or scientific endeavours are to be considered in this way, but all of man's undertakings. If we take this position, then certain issues need to be clarified. Firstly, when do we anticipate and when do we predict? Here is a badly phrased question which we need to get out of the way before we proceed. Anticipation and prediction, like explicit and tacit knowledge, are not separate, sequential processes. The object which the blind man feels with the end of the stick may be a flowerpot or a garden gnome, and if asked he can make a prediction on the basis of a cursory inspection. However, the sensations in his hand (the sensory similarities and differences) through which he knows the object, are operative throughout the excercise. Prediction is the specification of particulars which we know (attend towards) through actions channelised by the ways in which we anticipate events. These two ways of approaching the future are, indeed, different, but are inter-dependent. We tend to experience them in varying balance under different circumstances.

As we have described, anticipation is that which we sense to the extent that we feel involved in undertakings to which we are personally committed. Then we talk of having feelings of perhaps joy or dread which we cannot put into words. These are the constructs which we dwell in, and indeed are living through. That is why Kelly was at pains to point out that constructs are not essentially intellectual or cognitive in nature. At other times we can make a prediction in which we forecast the alternative outcomes of a situation. Though not necessarily so, such predictions may be less self-involving, being a statement of events which are "outside us", described by constructs which we can objectify and hold away from ourselves. Again, however, we must remind ourselves that this basic difference does not characterise different constructs, but refers to the way in which *the person* lives through and attends towards particulars in his experience.

We must now address ourselves directly to the question which we posed initially. How is it that we may act so as to achieve goals which we cannot fully justify, or to solve problems by taking lines of inquiry which we feel to be fruitful? The paradox of "knowing what you are looking for, or failing to find what was never sought" can be seen to arise when we think of ourselves only as trying to predict, objectively, a readymade world waiting to be discovered. It is clear that as we begin to make progress in any effort at solving a problem we can be increasingly explicit about the situation in which we are engaged. Like the scientist, we move from hunch to hypothesis with clearly formulated outcomes. This increase in our capacity to say beforehand what we are looking for is associated with a gradual reformulation of the problem. Duncker (1945) has described this as follows:

> "The final form of an individual solution is, in general, not reached by a single step from the original setting of the problem; on the contrary, the principle, the functional value of the solution, typically arises first, and the final form of the solution in question develops only as this principle becomes successively more and more concrete. In other words, the general or 'essential' properties of a solution genetically precede the specific properties; the latter are developed out of the former." (pp. 7–8)

The general properties of a solution, its functional values, are what Kelly referred to as the co-ordinates of an event, in terms of the intersect of a set of construct dimensions. Thus we seek for a "thing which will prise open a box" before our eyes light upon a penknife, or we seek a word which means "not saying anything" before we hit upon "silent". To a varying extent we will be able to specify what we are looking for, but *in looking* we attend towards the features of possible solutions *through* these ways of organising the problem. The difference here is that Duncker emphasises the reformulation of the problem, its *reconstruction* in the light of our attempts to provide solutions. Thus the constructs which define the original problem are unlikely to be those (though some may be included) which define the successful solution. There is, however, an exception to this process. Duncker also reports that the person may, during what appears to be planless investigation of the situation, stumble on a particular feature which "suggests its functional value from below". In this situation the concrete instance appears to precede the functional value.

In either case, this will lead to an attempt to solve the problem as conceived, such attempts meeting with a greater or lesser degree of success. When our attempt doesn't work, we try to analyse the new problem which has arisen. In doing this, Duncker suggests that the individual "seeks to penetrate more deeply into the nature, into the grounds of the conflict". Let us suggest that this penetration is brought about by what Polanyi has called an *interiorisation* of these particulars, so that they are now included in the essential properties of the solution which is sought. Using our previous analysis, such interiorised particulars comprise the tacit knowledge which, as we use it, remains unspecifiable. However, through dwelling in these particulars we attend towards further features of the problem field, whose meaning is organised in terms of the knowledge which we hold tacitly. In our experience, the problem gradually becomes more and more specifiable (meaningful) in terms of particulars which we can point to in the outside situation. It is to this that Polanyi (1967) refers when he says that "all meaning tends to be displaced *away from* ourselves" (p. 13). This is similar to the blind man with his stick, where the interiorisation of the sensations in the hand leads to their organisation with respect to the things which he touches with the point of the stick.

We can now see how the solving of a problem or the achievement of a goal is attained by a *two-way process* in which we act so as to interiorise certain particulars (penetrate deeper into the problem) and in so doing distance the problem from ourselves by attending towards further things in the situation. That is why as we progress in our inquiries we are able to specify more exactly (predict) what we seek or what we intend to do, while at the same time feeling that we are getting the problem outside ourselves, as it were. This accounts for the way in which scientific investigation can demand, as Kelly (1969c) describes, both detachment *and* commitment from the person involved. We may conclude that while we do, in a certain sense, anticipate that which is yet to be discovered, the solution is not prefigured in our original question. For, as Duncker states, a solution has two roots, one in that which is sought and one in that which is given. It is only through our attempts to come to grips with events outside ourselves that we may reformulate the problem which we have posed. This reconstruction (to use Kelly's term) is seen to be a process with genetic properties, an evolution of our relationship to the problem situation which involves us at a personal level. However, the next question which we must ask concerns the way in which such problems are posed initially. If our strivings are rooted in constructs through which we "live in anticipation", then while we may *see* the problem in terms of events to which we can point, it must derive also from these ways in which our psychological processes have become channelised. On this Kelly (1969a) says:

> "The posture of anticipation, which is the identifying psychological feature of life itself, silently forms questions, and earnest questions erupt in actions." (p. 31)

It is of interest to note that Kelly describes this process as "silent" (of which we cannot tell), and its effect as being "earnest" (having a claim on us, personally). This question of how anticipations arise invites us to look more closely at the relationship between the person and his environment, with a special emphasis upon the way in which he acts within it.

The Person and His Environment

> "We find the same life-process—the self-yielding of organism and environment—on every plane; here in the concrete circumstance is the 'living' truth. Where

then is reality? In the objective situation, or in 'the people'? In neither, but in that relating which frees and integrates and creates. Creates what? Always fresh possibilities for the human soul." M. P. Follett, *Creative Experience*.

Kelly's position is one from which we may view the person as an individual who is able to predict and control events, rather than as a slave to antecedent circumstances. He may do this through placing interpretations of his own construction upon the events in his world. These events are known to him only through his personal constructs, but Kelly assumes that a world does exist "out there" for the individual to construe. His interpretations are open to revision because it is the *person* who construes and (Kelly reminds us of William James' statement), events do not come bearing labels on their backs. The interpretation which we might make of Kelly's standpoint, as summarised here, is that the individual is engaged in evolving a system of constructs within which he can broaden and deepen his understanding of the world. It is in terms of events that the person's system must be judged as having predictive utility and allowing him a measure of control.

> "Thus we hope it is clear that what we assume is that the person makes his choice in favour of elaborating a system which is functionally integral with respect to the anticipation of events. To us it seems meaningless to mention a system qua system. It must be a system *for something*." (Kelly, 1955, p. 67)

The person's means for testing his constructions of events is behaviour, which Kelly calls "man's principal instrument of inquiry" (1970, p. 260). The place of behaviour in personal construct psychology is so important, particularly with respect to our topic of person and environment, that it will be our main concern in the discussion that follows immediately.

From a cursory reading of the opening pages of "The Psychology of Personal Constructs" one might be forgiven for thinking that Kelly was proposing a theory in which man-the-scientist simply behaved in accordance with the hypotheses embodied in his construing of the world. Behaviour, in this sense, *follows from* a person's construction of events, so that if I construe a man as "intelligent", I will address him in a manner which I perhaps reserve for "intelligent" people. On the basis of his reply to my question, my construction of him as an "intelligent" person will either be verified or perhaps shown to be false. In a certain sense, my behaviour is an experiment based upon my hypothesis that the man is "intelligent". Now sometimes this kind of thing does happen, such as when we have doubts about another person and set out to find out what the situation really is, or when a scientist sets out to test a hypothesis which he has derived from his thinking. The special features of these situations are that the individual in question has *made explicit* the nature of his inquiry, and that his behaviour *is designed to serve the end* which he has defined in terms of these predictions.

We have seen in the previous section that *prediction* is not acceptable as an explanation of behaviour which is anticipatory, but is rather that aspect of expectation which involves explication. We shall now see that the interpretation of the role of behaviour outlined above is similarly limited in its applicability to how people act in the world. The separation of construing and behaving, taken to its conclusion, rests upon two fallacies which must be exposed. One is the artificial separation of mind and body, which allows a person to think (construe), and then act and then think again. This dualism is unacceptable, insofar as it makes it impossible for us to act and to know that we are acting. Secondly, it reserves for behaviour the role of serving a preconceived end, decided upon in advance, and (because thought and activity are held separate) controlled by criteria supplied by the mind.*

Such a view as this is unacceptable, if only because we know that we act as we are acting, not in some discrete fashion after the event. We know ourselves as we behave *in relation* to others and to events. Furthermore, our behaviour is not merely a means of verifying what we can already predict, but is the way in which we engage the world and thereby precipitate a new situation. In effect, we can never verify what we believe to be the case because our participation in the situation changes that situation. All we can know is our relating to circumstances; not a knowledge of objects as such, but a knowledge of persons engaged in their environments. In saying this we are rejecting the notion, stemming from the dualist position, that behaviour is a bodily experiment which follows the setting up of a mental hypothesis. We need to rid ourselves of this idea if we are to appreciate fully the interpretation which Kelly came to place upon the role of behaviour in his psychology. In his paper entitled "Ontological Acceleration" (1969a) Kelly points out that behaviour is not the answer to the psychologist's question, it *is* the question. In other words, it is not the dependent variable in human endeavour but the independent variable. This change in perspective is, I think, fundamental to an appreciation of Kelly's viewpoint, and carries further implications for the discussion to follow. He says:

> ". . . behaviour, however it may once have been intended as the embodiment of a conclusive answer, inevitably transforms itself into a further question—a question so compellingly posed by its enactment that, willy nilly, the actor finds that he has launched another experiment". (1970, p. 260)

> "Behaviour puts itself into perspective by exploring its own possibilities, as well as by allowing itself to be the object of our psychological enquiries." (1969a, p. 21)

* This fallacy is reflected in approaches which, for example, explain behaviour as the end of a hierarchy of TOTE units (Miller, Galanter and Pribram, 1960). It is also in evidence in the practice of behaviourism, though divided between two people, or between one person and a pigeon. The pigeon's (or subject's) behaviour is the result of a series of events controlled by criteria supplied by the mind of the conditioning psychologist.

By describing behaviour* as "transforming itself into a further question" and "exploring its own possibilities", Kelly is definitely marking out a psychology in which behaviour is, *in some important way*, independent of the person's way of construing (awareness of) his world at that time. That is to say, behaviour is neither determined by outside events as behaviourists would have us believe, nor is it just the handmaiden of my construing (my awareness) as would follow from the "verification hypothesis" outlined earlier. "Activity", as Mary Follett put it, "always does more than embody purpose, it evolves purpose" (1924, p. 83). By acting in the situation we transform the situation, and it is there for us to reconstrue. But even this may imply a static world to which we come and to which we begin to do things. This is patently false, as we live in a social world in which other people are always acting. Behaviour, therefore, is the living interplay of persons with each other and with their world, which continually creates new variations on what Kelly called the "replicated themes" of our experience. It is the nature of this interplay between persons and their world which we must examine if we are to provide a better understanding of the important way in which behaviour can be an independent variable in the situation.

The self-transforming nature of behaviour, if we accept this description, clearly implies something different from a position which would locate the creative ingenuity of man in his re-conceptualisation of the consequences of his actions. This is, again, to be misled into thinking that personal change only follows from the individual's ability to re-organise his *perception* of the situation in the light of events which his action provoked. This would still leave us with an individual whose behaviour is dependent upon his construing, and who goes around testing his perception of the world. Such a view is similar to the argument that man predicts his world, an argument which we have seen falls short of encompassing behaviour which, though anticipatory, is nevertheless beyond our explication because we attend towards the world through it. Indeed, we do behave in situations in ways which we neither predict, nor can always readily examine afterwards. Let us take an everyday example.

A relative beginner is playing tennis with a friend and is trying hard to keep the ball in court. His attention is fixed upon the ball and, to a lesser extent, the lines delimiting the area where the ball is in play. His partner makes a shot which presents him with a difficult return, but he manages to do this sucessfully. How he was to make that stroke he could not predict, nor can we (but in general terms) see it as behaviour which verifies his construing of the game. This particular stroke, which he may never have played before, nor

* Let us continue to use "behaviour" to refer to the actions of persons without the implication that we contrast it to "mind". After all, the word has long carried this wider meaning for those who are not S-R psychologists.

which he need have isolated in his previous experience, arises out of his attempt to relate to the situation in the ways demanded by his understanding of the game. Using a similar example, Bartlett (1932) made this point:

> "The stroke is literally manufactured out of the living visual and postural 'schemata' of the moment and their interrelations. I may say, I may think that I reproduce exactly a series of text-book movements, but demonstrably I do not; just as, under other circumstances, I may say and think that I reproduce exactly some isolated event which I want to remember, and again demonstrably I do not." (p. 202)

This behaviour is *created* (manufactured) out of the individual's ways of anticipating the events of the game, which are the "living schemata" in which the person dwells at that time. It is this knowledge (of how to make the stroke) which the person holds tacitly, and of which he cannot tell. It does not arise in the player's experience before it has emerged in his organised response to the situation. Afterwards he may reflect upon the outcome of his response and his own experience of hitting the ball, noting the line of its travel and the success of his partner's return.

Therefore, what we may know explicitly, and perhaps even try to repeat, is a particular kind of action in relation to a particular set of circumstances. The person's behaviour (the tennis stroke) is a new way of relating to the situation which was actualised, and through this action new features of the situation *emerge*. For example, the player may abstract from his experience his own style or racket grip, or abstract a feature of play (the topspin of the ball) which resulted from his stroke. Of equal importance, he may note the failure of a stroke, or a near miss, and consider his behaviour in relation to what *might have occurred*. Either way he "knows the event through his own act of approach to it" (Kelly, 1969a).

The behaviour apparently designed only to fulfil his intention of returning the ball "transforms itself" by engaging the player in a new relationship with the situation, from which emerge new features about which he may frame new purposes. That is to say, behaviour is not some mediator through which we disturb events in the outside world in order to perceive (reconstrue) the consequences: behaviour *is* our relating-to-the-situation in which we live. We cannot, therefore, begin with an interpretation of personal construct theory which sees man as primarily withdrawn from a world which he *looks at* through his constructs. If we take this view we are left in the awkward position of having to ask how the person's construing is "converted" into actions. Kelly's (1955) emphasis on constructs as templets, and his concern to show that men give meaning to events through interpreting them, may lead us to imagine ourselves as being "over against" the world, which we classify and conceptualise at our convenience. It is this view which we wish to deny, in asserting that all we can know is our *relating-to-events*, which demands our active participation in the situation. It is this which Mead (1934) refers to when he says:

> "If we accept those two concepts of emergence and relativity, all I want to point out is that they do answer to what we term 'consciousness', namely, a certain environment that exists in its relationship to the organism, and in which new characters can arise in virtue of the organism . . . On this view the characters do not belong to organisms as such but only in the relationship of the organism to its environment." (p. 330)

This latter sentence is an important one, for it asserts that characteristics of the environment do not belong to organisms. This would appear to be at variance with the fundamental idea that constructs are personal—that is, that my constructs are my responsibility. We shall have reason to explore this issue again later, in a different context, but at this point we are now required to be more specific as to how the individual relates to his environment, and to pursue the implications of this for a theory of personal constructs.

Let us return to the example of the tennis player. We have rejected the idea that his behaviour is to be understood as merely a means of verifying his predictions about the situation. (We do acknowledge, however, that a person might play tennis in order to demonstrate his athleticism, or superiority over his opponent: this is however, another matter, and one which is relevant to a later part of our discussion.) Instead, his behaviour can be seen as anticipatory in the sense in which we used this term earlier on. It is an organised (channellised) response to the situation as events present themselves to him at that time. While it is not construing in the sense of articulating a dimension which one can write down or discuss, it is construing in the more fundamental sense in which the player anticipates events within the situation to which he is committed. In the moment of making the stroke he lives through equivalence—difference patterns in eye, body and racket which are directed towards hitting the ball. This is making use of the terms introduced in our discussion of anticipation as dwelling in particulars (constructs) which we know tacitly. It also re-emphasises Bartlett's description of the stroke as manufactured out of *living* postural and visual schemata. We may see that this construing is not the equivalent of the player's previous predictions about tennis or how he will play the game. At the moment of making the stroke *events of that moment* place him in a relationship within which he must act, and it is in the *anticipation* of events that he lives out his response. What we call the person's ingenuity is the creative organisation of the response to meet events as they arise; an organisation which, in being lived, is at that moment beyond prediction or thorough explication. In the very making of the response (acting in anticipation), the person is, to some degree, in-formed by it. The creative aspect of behaviour is in meeting events, not only in reflecting afterwards upon the unforeseen consequences of a considered action. And yet this arises when we commit ourselves to action, when we lend ourselves to the situation in anticipation of particular outcomes. This is little more than saying that the nature of our relating to the situation is one of *participation*. The tennis player makes the stroke (no matter how well calculated) by participating within the larger system comprised by his opponent, the physical situation and the rules of the game.

What then is the nature of this participation? Certainly it is not the state in which we are contemplating or withdrawn from the problem or situation, so that it is an *object* of our attention. In such a state we reflect upon the situation, appraise the various possibilities, and perhaps consider ourselves in relation to it (can I do it? do I want to?). Instead of this, we experience our commitment to action as of ourselves becoming part of something extending beyond us, as one of "absorption into" a problem or activity. Paradoxically, we sometimes also speak of the need to open ourselves to the situation if we are to experience it fully. What we experience are the demands of the situation, and it is in replying to these demands that we are *instrumental* in the evolution of the situation, whether it be a game or a social relationship. Sartre (1962) describes this demand in relation to writing:

> "The words that I am writing, on the contrary, are *exigent*. It is the precise manner in which I grasp them in the course of my creative activity that makes them what they are: they are potentialities that *have to be realised*. Not that have to be realised *by me*—As for my *hand*, I am conscious of it in the sense that I see it before me as the instrument whereby the words realise themselves." (pp. 60–1)

What may help us to clarify this problem is to remind ourselves that in the evolution of any situation—a conceptual problem or a game—new features arise in relation to our actions, and we act in *anticipation* of events as the situation is presented to us. We penetrate more deeply into the grounds of the problem by *interiorising* these new particulars, which become extensions of the basis from or through which we act in the situation. Before his first game of tennis, the beginning player does not feel the exigence of the situation, but knows the game only from a distance, as an object to be considered. During this game, as he makes his first strokes, he is living through (or dwelling in) the organised responses which are created to meet the moment to moment events of the game. These organised responses—his behaviour—are anticipatory in relation to the events which they intend. (The word "intend" is used in the sense of *intentionality* (May, 1969) to refer to the fact that all construing points toward something outside itself.) And as we have seen, the player's stroke transforms itself by taking its meaning in terms of the new events which it is instrumental in setting off. As with Duncker's problem-solving situation, the evolution of his game is reflected both in his acquired skill and in the features of play which he can bring about in the physical situation. Both can be seen as grounded in the two-way process which we have described as "interiorisation" and "distancing of meaning". He gradually understands the game by living through anticipations related to the further events which arise in the course of his activity; this is what we mean by participation in the situation.

This can also be seen to be true of our understanding of each other. Kelly (1969b) says:

> "Thus I, Person A, employ Construct A′, a component construct within my own construction system, to understand Construct B′, a component construct within Person B's construction system." (p. 75)

However, we should not read this to mean that B's construct is apprehended only as an *object* of A's attention. This is clearly not so. When engaged in conversation with a person I do not only subsume his position as an object within my construction system, but rather understand him through an active participation in the standpoint which he presents. I dwell in the particulars which he presents in words, in order to see through them towards the things to which my partner is pointing. Without this participation my understanding of the other person remains superficial; insofar as I retain him only as the *object* of my attention, I do not enter into the standpoint from which he views the situation. As Dilthey (Hodges, 1944) has put it:

> "The fact that we live in the consciousness of the system of the whole is what enables us to understand a particular statement, a particular gesture or a particular action." (p. 136)

While, therefore, it is of interest to know how I perceive others through, say, using repertory technique, this should not be taken as a description of the process through which I come to *know* a person. To make this interpretation is, yet again, to commit the fallacy of limiting construing to a procedure whereby I stand apart from others and look upon them.

We can now draw some conclusions from our discussion and briefly state their implications for personal construct psychology. Behaviour evolves ("transforms itself") through our relating to the world. In that relating we are not individuals who simply do things to objects or to others, but we participate in a system which is larger than the self in our awareness. Macmurray (1957) has described this as follows:

> "It is indeed an integration of the movements of the Agent with the movements of the Other, so that in action the Self and the Other form a unity. This integration *is* the action and its unity is intentional." (p. 220)

Indeed we only know the world as we relate to it, that is, in our behaviour. And our behaviour is our *living* of the ways in which we anticipate events. Because we live in an active world, we are constantly faced with new situations into whose demands we must enter if we are to meet them with a satisfactory response. What emerges are *new behaviours*, and what follows from such actions are transformations of events which in turn demand their own reply. What is primary is that we live out our relating to the world in this fashion, and that this be the starting point for our discussion of personal constructs.

Behaviour, therefore, is the independent variable in the sense that it is not "mine", possessed by a withdrawn and separate "self". By entering into or involving myself in a situation new relationships may emerge (be lived out by me), which I may then make the object of my attention. I may, after the event, perhaps claim to have been changed by the experience. In this sense, the person is a product of the behaviour in the situation; that is what Kelly meant by behaviour being an independent variable. Because we do not see behaviour as just muscular movement, we are in no difficulty in accepting that the lived equivalence–difference patterns involved in making the tennis stroke are *construing*, and that this extends beyond the player's arm to the racket which he holds and the ball he is about to strike. It is on the basis of the *emergent* relationship of his actions to features of the situation that the person may come to predict or to repeat such circumstances in the future. In doing this he will separate out what "he did" from "what happened" or "what followed". He may construe himself, his opponent, or the game in the sense of thinking about them, and be able to conceptualise in a more differentiated manner the turn of events which answer to the various strokes which he might make. Indeed he may plan a whole new strategy in his head, but whether he can achieve it or not cannot be known until he plays again. Behaviour cannot be deduced from the person's construct system *qua* system. Indeed the whole question of action is misconceived if we begin from the standpoint of construing as primarily a perceptual-cognitive process.

The Question of Self-Control

"What stands in your way is that you have a much too wilfil will. You think that what you do not do yourself does not happen." Eugen Herrigel (1953) *Zen in the Art of Archery*.

In the two previous sections we have argued that a person's behaviour is guided by anticipations as well as predictions, and is "of itself transposed" into something which the person did not intend. Both arguments are an attempt to place in context those capacities of prediction and control which are often held to epitomise the inquiring nature of scientific men. A person's constructs are not necessarily for prediction, although he may predict with them; nor are they to be considered only as reflections upon the world in which he lives, although they are his only means of reflection. However, the ability to reflect upon our behaviour and to act in a determined fashion is a capacity which we clearly hold as important. Implicit in this idea are assumptions about what we may term self-control; that the person himself makes decisions about what he will do, and he carries them out. We have seen, however, that people do not always enter into situations or relationships with clear intentions, nor do they simply behave in ways which confirm their expectations or their purpose. To say that a person behaves like an introvert because he construes himself to be an introvert would be to make a statement which explains little, while in certain circumstances being simply untrue. On occasions people do act in ways which, immediately afterwards, appear to them to be out of character, if not inexplicable. At other times individuals try very hard to behave in accordance with their decisions or beliefs about themselves and end up doing something different. Sometimes, even, we try to force ourselves to do something and suffer ignominious failure. What is clear is that there is no immutable control *by* what I construe myself to be in reflection, *upon* how I act in the situation. We have already concluded that to identify construing with thinking (as *opposed* to behaviour) is to enter into a dualism which leads to conceptual difficulties if not to confusion. Furthermore, the question of change in the person's construing of self has been argued to depend upon the creation of new relationships between the person and his world through behaviour. The problem which remains to be explored concerns the nature of the relationship between the person and the construction of himself to which his behaviour refers.

Any construction by the person of "himself" is an abstraction from his behaviour and experience. The self which arises in a person's consciousness is always a self-as-object, and is not to be confused with the person who is construing. The self-as-object is what a person is aware of in terms of his construing at that time. This is most similar to what Mead (1934) referred to as the "me", although Mead's analysis was grounded in an essentially *social* behaviourism. The person who is construing may be likened to Mead's concept of the "I", which cannot be obtained directly in consciousness. In the terms which we have used previously, we may say that the person cannot objectify the constructs which he is living through at that moment in time. It is the individual's response to the situation, his action, which constitutes what Mead called the "I" and we might refer to as the construer. As Mead pointed out:

> "The getting of [the "I"] into experience constitutes one of the problems of most of our conscious experience; it is not directly given in experience." (1934, p. 175)

A further illustration of this distinction is given by Kelly when he says:

> ". . . on occasion I may say of myself . . . "I am an introvert". "I" the subject, "am an introvert" the predicate. The language form of the statement clearly places the onus of being an introvert on the subject—me. . . . Yet the proper interpretation of my statement is that *I construe* myself to be an introvert . . .". (1969b, p. 70)

Kelly goes on to argue that problems may arise when the individual believes that he really is (or really can't be) what he construes himself to be, or would like to be. In such circumstances the individual's construction of himself exerts considerable control because he believes that this particular experience of himself is how he "really is". Paradoxically, this kind of absolutist construing may lead the person to feel himself lacking in control in his relationships with others or to the physical world. This statement requires some elaboration, and to do this we shall turn to some examples to be discussed below. However, at this point we may reformulate the question of self control as concerning not only the way in which the person's conceptualisation of himself may influence his immediate behaviour, but also the wider process whereby he may come to bring more of himself "as construer" into his experience.

The problems most relevant to our question are those involving an individual's attempt to exert or to claim some form of direct control upon himself. These attempts are usually associated with, or stem from, the person's concern to achieve an end which is integral with his conception of how he is or might like to be. Sartre (1957) gives an example of a pupil who is so concerned to be attentive that he ends up not hearing anything which the teacher is saying. Alexander (1932) uses the term "end-gaining" to describe the particular problems which individuals have in carrying out physical or recreational skills with strong competitive elements. In such circumstances the very strength of purpose, of trying, appears to be the factor which most disturbs the person's actions. So it may seem paradoxical to the beginning tennis player that he may make better strokes when simply returning the ball to the other end of the court, than he does when trying to score points against his partner.

The effects of extreme purposiveness such as this are apparent when a person tries to force others or the physical world to submit to what he expects of them. The attempt to force events to fit in with one's expectations after these have been recognised to be inadequate is what Kelly termed hostility. It is even more starkly apparent when the individual turns his hostility upon himself; when he tries to make himself do what he expects himself to do. One imagines a pianist whose playing is not up to his expectations trying to force his fingers to move across the keys as he wishes. Or the sportsman who makes a conscious effort in order to make his arm do what he wills. In both of these cases the individual concentrates upon his fingers or arm during the exercise and these become the focus of his attention. The pianist's fingers or the sportsman's arm are no longer experienced as being extensions of the person, through which he acts, but become separated from him in his consciousness. To do this when skilled behaviour is progressing smoothly is, quite simply, to bring it to a stop. If, at this moment of writing, I begin to concentrate upon the pen and the words emerging beneath it, the flow of meaning at the base of this sentence is disturbed and I falter. So it is in my social behaviour. My spontaneous or natural actions are interrupted by my attempts to "be something" in particular. To the extent that I try to achieve such an image for myself, then I may act self-consciously with reference to this construction of self which I hold. Often this goes alongside conscious attempts not to act in a certain way, and to suppress some particular way of behaving. Again, the effect of this is to isolate some aspect of our behaviour in our experience and, by objectifying it in this way, to try to operate upon it directly. As with the examples given before, this may or may not bring the results which were intended, but the person's experience will not be of having acted in a natural or spontaneous manner. Furthermore, the ways of acting which were self-consciously suppressed may, in other situations, be evoked in spite of the person's wishes. This is not at all surprising if we remember that the person's construing of that aspect of himself, by which it is objectified for him, is not the same as the behaviour itself. As we have seen, the person's behaviour is anticipatory in relation to *events*; when we construe ourselves, we abstract the behaviour from the situation in a way which belies the nature of our relating to the world.

One example of the failure of self-control is the behaviour of the confirmed alcoholic. Bateson (1973) describes the alcoholic's conviction that he can remain sober if he wills it, and that he ought to be "the captain of his soul". His world is polarised in terms of the alternatives "drunk–sober", and he places himself at the "sober" end of this dimension in his endeavours to fight "drinking". His drinking behaviour is distanced from himself in his experience and may at times be emphatically denied. We may say that he *disavows* his drinking behaviour, a term which Fingarette (1969) has used in his explanation of self-deception. As Bateson notes, there is often an element of self-deception in the alcoholic's life, insofar as he may secretly be laying in drink all the while he is professing his determination to stay sober. In the course of events the alcoholic eventually takes one drink which leads to a full scale bout of drinking. At this stage the alcoholic may "hit bottom", by which is meant that he comes to realise that he is in the grip of a system larger than himself, a system which he (the "self" who wills sobriety) cannot control. This perceived powerlessness of self in relation to the larger system as a whole is taken by Alcoholics Anonymous to be the beginning of the road to recovery for the alcoholic. Only by *giving up* the struggle to assert himself at the "sober" end of the "sober–drunk" dichotomy can the alcoholic achieve some measure of control. To be able to say, "I am an alcoholic" is to acknowledge not only the denied aspects of his drinking behaviour, but to place the alcoholic in a position of recognising his relationship to a world containing alcohol. To say that his uncontrolled drinking is the behaviour which finally invalidates the terms in which he construes himself is true, but insufficient.

The important element of this reconstruction is the alcoholic's *giving up* his self-control, and his acknowledgement of his newly perceived relationship as part of a larger system.

This example is useful in pointing up issues which, when we look further, are applicable to many less extreme situations. Illusory self-control depends upon the identification of oneself in terms of a construction which bisects one's own behaviour. It is the affirmation of oneself as this and not that. Most importantly, it is to continue to believe one can act in terms of this abstraction in the face of repeated disconfirmation from one's own failure to behave in a particular way.

Simpler examples can be found in our daily experience. The man who flies into a rage and desperately pulls at the drawer which jams is trying to "control" his environment. "He" is pulling at "the drawer". Only when he *gives up* this attempt can he begin to analyse the problem. He notes the swelling of the wood, or the way in which the drawer checks relate to the upturned handle of the hammer which he casually threw in there on some previous occasion. This analysis is achieved by *changing his relationship* to the drawer. It is no longer merely an *object* to be pulled at *by him*, but the slight movements and resistances transferred through his fingers take on a new meaning in pointing towards other possible reasons for the drawer to jam. In the analysis of the problem the person lives through these tactile transformations of sensory difference (constructs) so that he becomes part of the larger system ("person plus drawer") which he is instrumental in changing. We are calling attention to a similar process in the case of the alcoholic or any person who gives up the attempt to force his behaviour between the alternative "selves" which he defines in his awareness.

The illusory self-control to which we have referred depends upon the person attempting to apply or to verify that abstraction of himself, in its necessarily narrowed form, in his relating to others. In our attempts to "try to do" or to "make ourselves do", we are divided in our experience. We may feel that we have a better and a worse self, or that we know what's right in our "minds" but our "bodies" won't submit to our will. We are led again to a dualist position in which the person may experience himself as needing to be goaded or pushed to do something. As individuals we do indeed employ a kind of "reward" or "need" psychology when in this condition of being divided upon our own constructions. And yet, in spontaneous action, in which we enter into or dwell in the features of the larger system of which we are a part, we feel no such division, nor do words like "reinforcement" or "motive" have much relevance. In such situations the person as a "self" does not come into it: we are involved in a situation in which construing happens without categories being placed by "me" on the situation. In their study of the role of the ego in work, Lewis and Franklin (1944) wrote:

> "Experiments . . . make it apparent that, on certain occasions, man's selfish needs are so little a part of the motivational system which guides him that participation of his 'self' in a task is not even necessary for the achievement of his goal. The goal is reached when the task is done; the agency of doing need not be the self."

What then is the agency? At this point in time we can only say that it is the *person in relation to the situation*, or in Macmurray's words, the unity constituted by the interaction of the movements of self and other. Put another way, it is the larger system of construing which is lived out by the person in his relating to others or to events. For constructs, as Kelly wrote of them, are not dead axes of reference, but are living processes constitutive of an organic system:

> ". . . anticipation, which is the identifying psychological feature of life itself, silently forms questions, and earnest questions erupt in actions."

To involve oneself in a situation, to give up *looking at it* in terms of a particular construct can be more aptly described as an *endorsement* by the person of his ways of anticipating events. To become involved in a problem, a relationship or task is to be instrumental within it; to let it happen rather than to "try to do, or not to do" it. There is nothing unusual about this; it is the mode of action which makes up much of everyday living. As William James described it:

> "'*Will you or won't you have it so?*' is the most probing question we are ever asked; we are asked it every hour of the day, and about the largest as well as the smallest, the most theoretical as well as the most practical, things. We answer by *consents or non-consents* and not by words. What wonder that these dumb responses should seem our deepest organs of communication with the nature of things!" (1950, p. 579)

To consent to an "earnest" or "probing" question about oneself is, therefore, to involve oneself in the situation through acting within it. As Kelly says:

> "Knowing things is a way of letting them happen to us." (1955, p. 171)

This does not imply that individuals are passive in the face of events, but that they discover things through being instrumental in realising new situations. "Letting" is a word which implies this instrumental role which we play; the "dwelling in" the larger situation which leads, as we have described, to the establishment of new relationships in our behaviours with others. Characteristic of this mode of activity is spontaneity, in which the individual experiences the larger system of which he is a part working, so it seems, of itself. From the previous section we have seen that construing in action cannot be understood if we limit our analysis to the individual within his skin, and fall into a mind–body, person–environment dualism. Spontaneous action is the interplay between persons, or between the individual and his environment, in which the person lives through the construing which is his relating to the environment. The situation becomes more one of the play between person and environment than one of the contest between self and other.

From this discussion it would not be an error to state that what "I" am is beyond "me". The "I", the construing organism, is what we refer to when we speak of ourselves acting in anticipation, or of living through our constructs. It is the tacit basis of action, which is disclosed through our relating to events. We place interpretations upon, and reflect upon our relating to others or to the world. These constructs allow us to articulate the alternative ways in which we might act in future situations. But it is not, as we have seen, only the degree to which we can *predict* and *control* our actions which assures us self-control in a world of ever changing activity. Instead, "the key to man's destiny is his ability to reconstrue that which he cannot deny" (Kelly, 1959).

There are two senses in which this statement sheds light upon our discussion. It acknowledges that the person may be faced with placing new interpretations upon himself in the light of his spontaneous behaviour in a situation to which he has committed himself. And it also poses succinctly the problem of the person identifying himself as being "this" and "not that". Like the alcoholic, it is what he denies about himself, what he does *not predict* (e.g., "drinking"), which remains to be reconstrued in the light of its possible disclosure in his subsequent behaviour.

The control to which Kelly is pointing is, again, seemingly paradoxical in nature. It is achieved by the person transcending, through his behaviour, the alternatives expressed by his bipolar construction. Self-control is not to be achieved through trying to see whether one is really an introvert or an extrovert, nor through the attempt to verify in one's behaviour the self which the individual wills himself to be. The "better self" which we may desire is not already there simply to be willed into existence: it is still to be gained in fresh activity. The question turns upon the person's ability to give up trying to embody himself in his constructs (his self-image) and through his behaviour to reconstrue his relating to the situation.

On Significant Questions

In a sense, all of what has gone before in this paper has been about "transcending the obvious". The nature of anticipation, behaviour, and identity can be brought to the point at which all of these topics bear upon the issue of how the person deals with what he can but partly know. Part of the reason for this must also be that these topics are not, in themselves, separate but exist together in the life of the individual.

The position we have described repeats, and in some cases extends, the arguments put forward by Kelly (1969a) most clearly in his paper entitled "Ontological Acceleration". However, it is to be contrasted with any interpretation of personal construct theory which would make it a theory of the intellect or of a "rational self". The main task in this essay has been to make room for a broader notion of construing than that suggested by our usual understanding of prediction and control. While we do not predict events in order to control them, we can only do this on the basis of organised construing which is beyond the "reason" which appears in our consciousness at that time. Perhaps we should re-phrase this to say that such construing is beyond the reason which we can *articulate* or spell out at that time. For much of our effort is devoted to making sense of, or trying to formulate, our experience of living in the world. It is in this way that, through our activity, we are always coming to know ourselves in relation to others and to our environment. What this means, in fact, is that out explicit knowledge, where we can say which constructs we are using, arises from our lived experience. We may consider a scientist working towards a discovery which he cannot predict, a novice beginning to get the feel of playing tennis or a person asking questions about himself. In each case the interpretations which he places are not, strictly speaking, upon events "out there", but upon the construing through which he is relating to his world. This latter construing, however, is *not* at the level of interpretation. It is the construing which we have described as "indwelling" or "living through", expressed in activity. Therefore, the questions which the scientist asks, the aspirations of the tennis player and the good intentions of an individual are not conceptualisations which spring into being at a distance from these individuals. What appears on their horizons is, indeed, what they can formulate, describe and to a greater or lesser extent know explicitly. And yet all of these interpretations (for such is what they are) are placed upon their own lived experience *within which they already dwell at that time*.

What we ask questions about, particularly questions which are important to us, are those features of the world to which we are already relating in some significant way. We sometimes call this "experience", in the sense of things which happen to us or otherwise grip us in ways which command our attention. For a psychology which takes its stand in cognition or perception, this presents the problem of how the perceiver knows significance *in* what he sees, before he sees it. Consider, however, the position which we have been assembling throughout this essay. The person is actively engaged in a construing system which extends beyond the limits of what we might call his self; his actions are his indwelling in constructs which relate him as a person to features of an active world. We have used the examples of the tennis player and the problem-solver to illustrate how the person is engaged in creating, through his behaviour, new relationships between himself and features of the environment. The path and speed of the tennis ball, or the handle of the hammer obstructing the opening of the drawer are new features of the situation which emerge in relation to the person's behaviour which he dwells in at that moment. The creative or self-transforming nature of this activity arises through the new relationships which emerge. When we say that they emerge we mean that such relationships are not created by the interpretations which the individual may later come to place upon them. Our interpretations, the constructs which we place upon the world and which we can communicate in words, are our attempts to know explicitly that which we have isolated as significant, as a question. This significance—the important questions which men ask—cannot be conceived from nothing. We, as self-conscious individuals, do not create significant questions: they emerge within us through our active participation in the world of which we are a part. That is why the posture of anticipation *silently asks questions*, to which we may or may not respond as self-conscious individuals. How the scientist deals with his question, whether perhaps he shirks the responsibility of pursuing his hunch, is an issue which arises *after* the question has arisen within him. These are questions concerning how he (or any man) *finds* what he is seeing after he has long seen significance in it.

We conclude that new questions are posed by behaviour. But they do not come from our perception of "things" which "we" disturb. They arise within activity which is imminent in all of the parts of the system. That is why Mead (1934) spoke of new features arising *in relation* to the organism: they do not strictly belong to the person involved. The significance of an important question, then, initially lies beyond the intellect, or the constructs subsequently placed upon it, because the question arises in activity. And yet we are not to imagine such activity as the tennis stroke or drawer-manoevring to be either trial-and-error or habit. Activity is lived construing, which is anticipatory, directional and organised. In that sense the significance of a new question is to be found in a rationality to be sure, but not in a reason which we can articulate. As Pascal said: "Le coeur a ses raisons que le raison ne connaît point".

What then do we construe or know explicitly? We can answer that our inquiries, being directed by such questions, are about the world of activity of which we are a part. We can only know others and the world in relation to ourselves, and these relationships are always changing because activity *evolves*. Comparing this to Mead's (1934) analysis of the evolution of mind (p. 134), we can say this. The ontological appearance of a significant question takes place when the whole process of activity in which the person is involved is brought within that individual's experience, and when his adjustment to the activity is modified by the awareness which he thus has of it. We are always coming to know a world in which we live and which, in a very real sense, is living through us.

The End of Personal Construct Theory

In this essay we have not attempted to make personal construct theory the focus of our inquiry, but have instead referred to Kelly's work in the context of the issues which we chose to discuss. In so doing, however, it has become clear as to the kind of perspective which we have taken on the theory, and the type of interpretation of it which we reject. We have criticised the position that constructs are *essentially* interpretations placed upon events, or "transparent patterns or templets" (Kelly, 1955, pp. 8–9) which order our perception of the world. In effect, we have rejected the standpoint of construing as primarily reflecting upon the world, and embraced the idea of construing as primarily acting within the world. Our suspicion is that, while Kelly fully understood both positions, the explication of the theory of personal constructs was carried out mainly from the position of construing as interpretation. This is not at all surprising, as much of the work was directed at psychologists who were trying to help clients to articulate their problems in the clinic. And yet Kelly's writings are punctuated with warnings that we are not to interpret constructs in this way only:

> "Not only is a construct *personal*, but it is a *process* that goes on within a person. It thus invariably expresses anticipation. It is easy to forget this point and to start thinking again . . . of a concept as a geographic concentration of ideas." (1955, p. 1089)

> "A construct owes no special allegiance to the intellect, as against the will or the emotions." (1969b, p. 87)

> "[The psychology of personal constructs] . . . is also taken to apply to that which is commonly called emotional or affective and to that which has to do with action or conation." (1955, p. 130)

In this discussion we have argued that the interpretation of construing as a perceptual-cognitive process cannot explain certain things about human behaviour and experience. To take this perspective is to make Kelly's contribution into little more than a theory of individualised conceptual systems, grounded perhaps in repertory technique measurement but inapplicable to the wider and deeper issues of human behaviour and experience. Within such interpretations of the theory, a personal construct has been held to be something which the individual makes up (rather than it being determined by the environment), and which is "mine" rather than "yours". I believe that our analysis leads to a further interpretation of the word "personal", which is adumbrated by Kelly:

> "Construct theory, or better, *personal construct theory*—a term which implies that a construct is as much a personal undertaking as it is a disembodied scheme for putting nature in its place—suggests that human behaviour is to be understood in a context of relevance." (1969a, pp. 11–12)

A "personal undertaking" might be taken to refer to the "dumb consent" by which James described our commitment to action, in which we dwell in our construing. Construing-in-action is a personal involvement which, paradoxically, implies our commitment to a question or task which may not, at the time, be felt to be ours. To be absorbed in a personal relationship or in some activity is to participate within construing which is neither placed upon the other person nor self-consciously possessed by me.

The idea that constructs are templets to see through perhaps devolves from the narrow interpretation of man-the-scientist attempting to predict and control. And yet it would seem that in Kelly's later writings he is trying to make explicit those other aspects of scientific endeavour which make it a personal undertaking. In such writings (particularly "Ontological Acceleration"), Kelly appears to have almost emerged through the metaphor of man-the-scientist to realise scientist-as-man. In his discussions of faith and commitment in science Kelly is outlining an earnest question which was never, for him, to be crystallised into a formal theory. His later writing is, I believe, clear enough for us to see that what was appearing on his horizon then was a more fully articulated statement of what underlay the writing of the "Psychology of Personal Constructs".

Our position on this is as follows. While it is futile to seek for qualities in the things to which we attribute them, so is it mistaken to seek for constructs "in the heads" of the persons whom we are studying. To do this is to confuse the intellectual, communicable constructs of George Kelly's theory with the processes which psychologists try to study in human behaviour and experience. Personal construct *theory*, to be communicable, has to be expressed in an intellectual way if it is to be understood by others. Yet we must not then believe that, if we wish to study persons, we must only seek to find the constructs in terms of which those individuals make their own lives rational. While Kelly may at times have argued for the latter point of view, he was aware (1955, p. 130), of the crucial distinction between the two.

To make constructs *per se* the object of psychological study is to reify personal construct theory; it is for me to make the mistake of seeking in persons the constructions which *I* attribute to them. As we have tried to show in this essay, such an approach limits itself unnecessarily to a study of conceptual systems, while missing the larger context within which construing is going on. We have described the individual as coming to know, not things, but the system of activity of which he is a part. In a similar fashion, the personal construct psychologist is coming to know something of the larger system of human behaviour and experience within which he participates as a person. But here are no "personal construct things" to be discovered, measured and set aside. Rather there is the process of construing, in which persons are continually engaged in steering their lives within a world of ever changing activity. It is the world of activity, of which persons are necessarily a part, which forms the basis of the approach which we are suggesting here. It is what we live today that we might tomorrow come to know—when, of course, the situation (and we within it) will have changed. Constructs as man-made interpretations are always, then, approximations to the living "truth" which it *appears* is always just beyond the horizon. To study Constructs is, like the study of the Unconscious in Freudian theory, to commit a form of psychological idolatry. Bakan (1966) has pointed out:

> ". . . as soon as the notion of the unconscious is taken not as that which is unconscious, but as something manifest by itself . . . then even psychoanalysis is idolatrous. The intellectual task which psychoanalysis sets itself is that of making what is thus unconscious conscious. To the degree that it tends to fix upon that which it has made conscious, it is idolatrous. But to the extent that it stresses the existence of that which is still (un-conscious), it avoids being 'idolatrous'." (pp. 11–12)

In similar terms, personal construct psychology may be seen as having as its end the making of that which is un-known, known. However, it is not a psychology of the unknown in the sense of tracing our progress from "the known to the unknown" (Kelly, this volume). Instead, it might be a psychology of the un-known, in the sense of affirming that what is there to be found and made explicit are the lived relationships which comprise a world of ever changing activity.

References

Alexander, F. M. (1932). "The Use of the Self", Methuen, London.

Bakan, D. (1966). "The Duality of Human Existence: An Essay in Psychology and Religion", Rand McNally, New York.

Bartlett, F. C. (1932). "Remembering", Cambridge University Press.

Bateson, G. (1973). The cybernetics of "self": a theory of alcoholism. *In* "Steps to an Ecology of Mind", Paladin Books.

Duncker, K. (1945). On problem solving, *Psychological Monographs*, **58,** whole no. **270**.

Fingarette, H. (1969). "Self-deception", Routledge and Kegan Paul, London.

Follett, M. P. (1924). "Creative Experience", Longmans, Green and Co., New York.

Herrigel, E. (1953). "Zen in the Art of Archery", Routledge and Kegan Paul, London.

Hodges, H. A. (1944). "Wilhelm Dilthey: An Introduction", Routledge and Kegan Paul, London.

Holland, R. (1970). George Kelly: constructive innocent and reluctant existentialist, *In* "Perspectives in Personal Construct Theory", (Bannister, D., ed.), Academic Press, London.

James, W. (1950). "The Principles of Psychology", Vols. 1 and 2, Constable, London.

Kelly, G. A. (1955). "The Psychology of Personal Constructs", Vols. 1 and 2, Norton, New York.

Kelly, G. A. (1959). Personal construct theory and psychotherapy. Lecture for the second Los Angeles Society of Clinical Psychologists' Post-Doctoral Institute.

Kelly, G. A. (1969a). Ontological acceleration, *In* "Clinical Psychology and Personality: Selected papers of George Kelly", (Maher, B., ed.). Wiley, New York.

Kelly, G. A. (1969b). Man's construction of his alternatives, *In* "Clinical Psychology and Personality: Selected Papers of George Kelly", (Maher, B., ed.), Wiley, New York.

Kelly, G. A. (1969c). The strategy of psychological research, *In* "Clinical Psychology and Personality: Selected Papers of George Kelly", (Maher, B., ed), Wiley, New York.

Kelly, G. A. (1970). Behaviour is an experiment, *In* "Perspectives in Personal Construct Theory" (Bannister, D., ed.), Academic Press, London.

Lewis, H. B. and Franklin, M. (1944). An experimental study of the role of the ego in work. II. The significance of task-orientation in work, *J. exp. psychol* **34,** 195–215.

Macmurray, J. (1957). "The Self as Agent", Faber and Faber, London.

May, R. (1969). William James' humanism and the problem of will, *In* "William James: Unfinished Business", A.P.A., Washington.

Mead, G. H. (1934). "Mind, Self and Society", University of Chicago Press.

Miller, G. A., Galanter, E. and Pribram, K. H. (1960). "Plans and the Structure of Behaviour", Holt, Rinehart and Winston, London.

Polanyi, M. (1967). "The Tacit Dimension", Routledge and Kegan Paul, London.

Sartre, J. P. (1957). "Being and Nothingness", Methuen, London.

Sartre, J. P. (1962). "Sketch for a Theory of the Emotions", Methuen, London.

Personal Construct Theory: An Approach to the Psychological Investigation of Children and Young People

A. T. Ravenette

This essay presents an opportunity of putting into writing some of the thoughts and experiences which have arisen out of the application of personal construct theory to the problems associated with children and with the people who work with children. Inevitably the essay will be personal in the sense that it will reflect something of my own development as a professional psychologist, and implicitly it will reflect something of my own personal prejudices and biases. I shall not, however, make this an occasion for the confessional nor for excessive self-examination, although personal construct theory might sanction both of these activities as basically growth promoting.

As a professional psychologist my concern is practical rather than theoretical and pragmatic rather than academic. This needs to be said in order to give a general direction to the reader. Having said that, however, I must also say that theory and practice should go hand in hand and that the detachment and involvement which each implies may be seen as a rhythmic redirection of energies as part of a growth process. A practitioner must be prepared to act, and therefore to err. He cannot wait until the niceties of theoretical ambiguities are resolved, nor can he wait for the findings of academic research. He is involved jointly with his clients in the ambiguities and dilemmas of life itself, and this is the context within which personal resolutions and personal research findings are created. Nonetheless, out of the dynamic of successive engagements between psychologist and client, ideas emerge which may lead to revised ways of working and these in turn may provide material which can enrich the basic formulations of theory, both as theory and as practice. Although my own involvement is with adults as complainants about children, and with the children who are complained about, there may be some original notions in this essay which are relevant in other therapeutic contexts. In the light of the very nature of personal construct theory it would be surprising if this were not so.

This essay will first explore something of the historical perspective leading up to my own present thinking and practice. This perspective will provide the basis for presenting some ideas which have not previously been made clear, ideas which I think may have important implications for practice. I shall then work out these implications by describing a number of techniques, each of which illustrates one facet of a single theme. Illustrative data will be drawn from children either having problems or presented as problems. I shall not offer an exhaustive account of interviewing techniques, that would be well beyond the scope of this particular essay, nor is the presentation concerned directly with developments in grid methodology which itself has now become an independent field of study (Bannister and Mair, 1968; Bannister, 1970; Landfield, 1971; Bannister and Fransella, 1971; Fransella, 1972). I shall, however, refer to them as necessary stages in the development of interviewing techniques. The sequence of the essay therefore runs from a practice to theory and back again, a rhythmic pattern which is embedded in personal construct theory itself, and which seems to reflect some of those universal rhythms which underly development and growth.

Historical Perspectives

In this part of the essay I propose to elaborate two themes which, although quite separate, are yet linked in an interesting way. The first theme is that of my own developmental steps in the use of personal construct theory, the second is the queries and difficulties of research students anxious to use grid techniques in the planning of their research, together with the difficulties which trainee psychologists experienced in adopting a personal construct theory approach to their work.

The decision to embark on a personal construct theory approach to children stemmed from a dissatisfaction with the irrelevance of traditional psychometric approaches and with a distrust of the assumptive framework underlying the use of projective techniques. Neither provided me with a basis either for understanding the troubles which children presented, or for helping teachers make a more useful sense of those very children who were causing them anxiety. Personal construct theory, on the other hand, immediately offered the promise both of purpose and relevance. There were of course no guidelines to show how the theory could be made to work with children. Kelly (1955) himself says little or nothing about his own work with children although he describes in graphic terms some aspects of his involvement with people who did deal with them. Under the circumstances it was natural to make a start with the techniques which Kelly devised to elicit constructs and to elucidate their organisation (i.e., grids). After all, these techniques were easily understandable and were easily assimilated within a conceptual framework of numerical description and statistical analysis.

Needless to say, however, this direct transposition of grid techniques from adults to children did not work (Ravenette, 1964). The reason is clear. The population for which Kelly invented his construct elicitation procedures and his grid techniques was made up of university students, not children, and what was within the grasp of students certainly far exceeded the understanding of children. The world in which Kelly moved was manifestly not the world of working class children within which I moved.

Bannister's work (Bannister and Mair, 1968) provided a key for the first development of grids with which children could cope, i.e., by using photographs in association with ranking techniques. It was a long time, however, before the question of children's constructs and their elicitation was resolved. Practically the issue was dealt with by providing constructs which were concerned with what significant people might expect and what these people might feel about children. (Ravenette, 1975 gives a résumé and an illustration of this.)

When I started developing a grid methodology which worked with children I was convinced that this was personal construct theory. Fortunately the falsity of this belief did not matter and it was possible to elaborate a variety of different grids with which it was possible to explore some of the ways in which children were able to make sense out of things. All the time, it was clear, however, that the child himself was given little scope for providing his own observations within a grid framework. This weakness was resolved when situational pictures were used as the elements for a grid. Under this condition, the ways in which constructs clustered together made no real sense unless the child described what was happening in different situations. In this way the grid procedure provided a means whereby the psychologist's prescribed constructs acted as keys to unlock the doors of the child's own world. The formal use of a grid with prescribed constructs therefore now becomes a technique whereby the child's own construction of his world can be investigated. Clearly we can use the same procedure when the elements are photographs of children, and the further description of those elements can therefore provide a knowledge of some of the child's store of words whereby he discriminates his peer group, and perhaps himself. Where, for Kelly, the personal construct was essential to the grid itself, the use of the prescribed constructs within a grid structure provide a powerful means for eliciting the child's constructs and constructions. The procedure is illustrated by the following case, using a two way analysis of an 8 × 8 grid.

THE CASE OF J.B.

J.B. was aged 15 years 10 months. He was of bright average intelligence and was seen in a Remand Home. He had had a history of school refusal and had recently left a succession of jobs. As a result he had been labelled "work shy". His father had killed J.B.'s older brother, probably whilst drunk, and it was known that there had been tension between husband and wife. Father had been in gaol for manslaughter. There was trouble between mother and son, and recently mother had turned J.B. out of the house. Previously J.B. had been in a children's home in the hope that he would attend school from there.

The prescribed constructs were derived from an awareness of the family relationships and also from a knowledge of the boy's behaviour in the Remand Home. They were:

A. Least likely to have friends.
B. Most likely to get on well with mother.
C. Least likely to get on well with father.
D. Most likely to understand other boys.
E. Frightened of what he sees in the family.
F. Feels he must keep away from other boys.
G. Most likely to be the same sort of boy as himself.
H. Most likely to be the sort of boy his mother would not like him to be.

(Constructs A, C and H were presented in the opposite direction.) The spontaneous constructs, i.e., his descriptions of the boys in the photographs, were:

1. Fairly quiet and reserved. Probably more interested in staying home than going out.
2. Probably very shy. Probably very good in school. Very interested in his work.
3. A little bit shy, not very much. Not particularly interested in school work.
4. Probably likes to go out. A little bit quiet. Very tactful.
5. Likes to go out with his friends all the time. Very outgoing. Probably untactful.
6. Likes to go out some of the time but helps at home. Probably quite interested in school work.
7. Very outgoing. Fairly tactful. Probably likes to stay in quite often. Prefers friends to come to him.
8. Not very bright. Probably stays in quite a lot. Probably sees a few friends regularly.

The analysis of the correlational data indicated two construct clusters and these are reproduced with rearranged rank order in Table I.

TABLE I. *Re-arrangement of basic rank order data in relation to construct clusters for the case of J.B.*

		Elements								
		2	3	5	4	8	1	6	7	
Least likely to have friends	A	1	2	3	5	6	4	7	8	Most likely to have friends
Mother would not choose him to be like	H	2	3	1	6	5	4	7	8	Mother would choose him to be like
Frightened of what he sees in the family	E	2	3	1	5	4	8	6	7	Not frightened of what he sees in the family

		Elements								
		2	3	1	4	8	6	7	5	
Most like himself	G	1	2	3	6	5	4	7	8	Least like himself
Understands other boys	F	2	1	3	5	4	6	7	8	Least likely to understand other boys
Feels he must keep away from other boys	D	1	3	2	5	4	6	7	8	Least likely to feel he must keep away from other boys

CONSTRUCT CLUSTERS FOR J.B.

Construct cluster (A, H, E)

This cluster brings together the following ideas: Boys who are "least likely to have friends" are boys who "mother would not choose him to be like" and who "would be most likely to be frightened by what they see in the family". The opposite pole of the cluster would be defined as the opposite of this statement.

The corresponding element clusters for the two poles are (2, 3, 5) and (6, 7). An examination of the boy's spontaneous constructs for these two sets of elements reveals a lot of differences which cut across the two sets but one clearly defined pair of opposites, namely "untactful—tactful". If we broaden cluster (6, 7) to include element 4 (which is more like it than it is like (2, 3, 5)), "tactful" again appears and this lends some support to the idea that "untactful–tactful" is a central theme for the definition of this particular construct cluster.

It is interesting that, during the interview, the boy himself introduced the importance to him of being tactful in the presence of at least one member of his family. Psychologically, therefore, the construct seems particularly meaningful for this boy and within the context of his family.

Construct clusters (*G*, *F*, *D*)

This cluster brings together the following ideas: Boys who "are most like the sort of boy he is" are also "most likely to understand other boys" and "would be most likely to feel that he must keep away from other boys". The opposite pole of the cluster would be defined as the opposite of this statement.

The corresponding element clusters are (2, 3, 1) and (7, 5). Examination of the spontaneous constructs related to these elements shows a fairly clear contrast between "shy, quiet and reserved" and "very outgoing". This might be labelled as a form of social introversion–extraversion. His behaviour in the Remand Home suggests that the boy was, in fact, beginning to experiment along such lines, and his previous history of school refusal and not staying long in the outer world of work is consistent with self-characterisation along this dimension.

Useful as grid procedures may be, it is also necessary to develop other questioning methods in order to explore the child's construction of his world. One specific technique involved elaborating with the child such causes of complaint that he might have with other people. Kelly (1955) provides a set of questions (originally formulated by Maher) whereby the child can state what he sees to be the trouble with boys, girls, teachers etc., how he understands why they are like that, what he would wish to happen, and how such changes might make a difference. Later I shall present this technique in more detail, but at the moment it is sufficient to say that even six and seven year old children are able to respond to such enquiries in a meaningful manner. Other sets of questions were invented to sound out a child's identifications within the family and his expectations of school.

These developments lead to the formulation of certain principles which seem to me to derive from, and be extensions of, personal construct theory as a means of working psychologically with children and young people. The basic tool of the psychologist is the question, and a part of his professional skill lies in his ability to invent better and better questions. Better in this context means facilitative for the child and penetrating for the interviewer. The investigation itself can be seen as a process in which *elicitation* of constructs comes first, and the *operationalisation*, or use, of constructs comes second. The grid is one way in which this second phase is carried out, but it is possible to invent different ways and later in this essay I shall present one such alternative.

The second of the two themes in this historical account is concerned with my contacts with students. These students were of two kinds, the first kind were research psychologists aiming to acquire higher degrees, the second kind were trainee professional psychologists who were trying to broaden their interviewing techniques by using grids in particular, and perhaps a personal construct theory approach in general. The research psychologists had two questions: How could they use grid techniques in their research? and how could they elicit children's constructs?

As my own experience increased I found it more and more difficult to help research psychologists with their questions since in order to do so I had to challenge some of their fundamental assumptions about children and personal construct theory. Quite correctly they were seeking test instruments which fitted in with the methodological cannons of academic research, but they also seemed to have an implicit notion that measures should or could be simple. At the same time they seldom saw the possibility of involving children themselves in resolving some of the researcher's own difficulties almost as though it must be the psychologist who knows best. I cannot go into all the issues which these questions pose but I would like to raise just a few. The first of these is in relation to grids and their interpretation. Clearly a grid which is completed without any spontaneous involvement of the child other than in the rank ordering of elements will be restricted by the sensitivities of the psychologist. It will not say much about the child and his psychological processes. Any statement about reliability and validity can only refer to the extent to which the child fitted himself to the task and intentions of the experimenter. If we invite the child himself to contribute to the grid we are faced on the one hand with the difficulty of standardising the procedures and on the other arriving at general observations. These remarks are about grids in relation to meaning. Other remarks are about grids in relation to statistics.

It is, of course, easy to refer basic grid data to a computer and then abstract out what we want. I must confess, however, to a feeling of unease when confronted with print-out from the computer of one child's eight element grid. The amount of material thus presented hardly seems justified by the basic data. More importantly, however, the use of a computer removes the psychologist from his data and this may cut him off from the possibility of making important discoveries for which the computer was not programmed. As an example of this I would mention a phenomenon which I call the "joker in the pack". If the reader will return to Table I he will find that element 5 is responded to somewhat anomalously. If this element is removed from the grid the two construct clusters will be positively correlated. Its presence, however, at opposite poles in the two clusters leads to correlations in the original matrix which are nearly zero. This element is the "joker in the pack" because its presence throws awry an interpretation based on assumptions of linearity. In psychological terms this element presumably has a special significance for the client but in statistical terms its presence might well suggest not no relationship, but some form of non-linear relationship. A perfect curvilinear relationship provides a zero linear correlation, but this will only be revealed by an examination of the basic data. Many more reservations might be made about the use of simple grids as research instruments, but perhaps they are best summarised in a more general form. Academic research seems to demand abstractions based on populations of individuals. The grid, however, when carried out as a joint investigation between psychologist and child, may reveal truths only at an individual level. To find that truth requires looking at the fine detail of what happened whereas traditional research demands that the fine detail is ignored, in favour of high order generalisations. In many ways the best use that can be made of the generalisations which statistics provide is in re-ordering the basic data in the light of those generalisations. This may well be true both for formal research as well as for the study of the individual.

When we turn to the research student's enquiries about children's actual constructs the problem becomes much more difficult. At the heart of this question is the issue of what a construct is. I shall present my clarification of this issue in the next section, and make only a few remarks here. The most obvious difficulty stems from the fact that the students seemed to be committed to the notion that constructs were words, and moreover that there might exist, or might be invented, lists of words which were children's constructs. When, however, we took their questions seriously and suggested they asked children for themselves it became clear that this presented even more difficulties. On some occasions when they did this a transcript of the conversation read rather like a lesson in school carried out in the style that the student himself had suffered as a pupil. In other words the students needed to learn how to interview children before they could elicit the material they wanted.

This particular difficulty enables me to bring in my second kind of student, the trainee psychologist learning a way of interviewing. He invariably showed exactly the same difficulty in questioning children as the research students almost as though the whole of his training and experience so far had acted against the idea that if he asked good questions he could trust the child to come up with some worthwhile answers. Even when children had produced a wide range of attributes which might be usable in some form of grid, these students seemed to feel that there were rules whereby they, the psychologists' choose the most important ones, rather than the children themselves. It was as though, somehow, the child was not to be entrusted with the task of making his own choices. This was the psychologist's prerogative, although they may need help and guidance in order to do it correctly. Such a stance is of course in flat contradiction to the stance of the personal construct psychologist. Hopefully we do have some expertise which we can make available to children, but this may be more in the direction of helping him to communicate about himself and his difficulties, and how to envisage alternative ways of thinking and acting.

The point of this historical perspective is twofold. In the first place I think it is clear that my own progress in personal construct theory was in some ways measured by the increasing difficulty in giving answers to those research students, who year after year came up with the same questions. In the second place, by trying to implement a personal construct theory approach it became apparent that I was progressively learning what it was. It is possible that learning the theory from other people's existing practice may have been quicker, but it may not have had the same personal validity. Thirdly, this résumé provides a frame of reference in which I can develop some thoughts about theoretical issues which stem directly from this learning process.

From Practice to Theory

Kelly did not write a developmental theory. Although he was clearly concerned with the problems surrounding children he wrote in greater detail about the work of people who had to deal with children than about the development of children themselves. When he was asked his views about the development of constructs in children he suggested that the earliest constructs were states of the organism. At a later stage people who were important in a child's life acted as constructs, and later still they became represented by verbal symbols. My own experience suggests that he omitted one very powerful basis for children's expectations namely children's own actions and the actions of others in relation to children.*

It was perhaps inevitable that this area of personal construct theory elaboration should be missing in Kelly's work. He was after all primarily involved with a university population and drew some of his inspiration from the similarities he saw between clients with personal problems and students with research problems. When therefore he illustrates constructs and construct systems his samples are drawn from relatively articulate students and therefore appear as rather abstract verbalisations. It is true that he frequently points to the fact that many of our actions are based on anticipations which were developed before we had verbal means of labelling, and he also warns against the dangers of equating constructs with verbalism or literalisms. Nonetheless,

* It would also be my view that this sequence, if it is genuinely developmental, is not restricted to the chronological development of children into adults, but may also be continually recreated at all ages when an individual is confronted with new contexts within which to operate.

the very fact of writing about these things puts a premium on verbalisms and the impact of this on students coming to grips with the theory for the first time seems to be in the direction of their equating *real* constructs with their verbal representation.

There are, of course, a number of important differences between adults and children, and these need to be pointed out. To a great extent the adult is already relatively mature and any psychological change which takes place is on a base of reasonably stable expectations. The years of childhood represent a time during which this relatively stable base is being developed. They are years of transition. Moreover, with children, even if their bases for anticipating acquire some stability the verbal representation of these anticipations is likely to be itself unstable and ephemeral. It is part of the growth process that the child is able to make progressively finer levels of discrimination and to develop a greater hierarchy of abstractions, both of which, over time, lead to richer and more complex construct systems. It follows from this that the elicitation of children's constructs is a far from simple affair, especially when children are young. The words they use should not be taken for the constructs, nor should it be assumed that a child's constructs can be couched at a high level of abstraction. In many ways the constructs which are important behaviourally may be just those which defy easy verbalisation and which in fact exist at a rather low level of awareness.

I think that some of the difficulties and ambiguities which these last observations imply stem from a single failure in semantic discrimination. We need to recognise a distinction between "scientific" and "spontaneous" or ordinary, language. When Kelly talks of constructs and construct systems he is offering a "scientific" language in which the terms have very precise meanings within an articulated theory. Sadly, however, we so easily use scientific expressions as a part of ordinary discourse and frequently it is not at all clear, when these expressions appear, whether they are to be taken in their "scientific" status or as ordinary words. The expressions "construe", "construct" and "construct system" all share this double usage. I think that Kelly himself sometimes fails to discriminate between these different usages, or at least he fails to point out the nature of the trap. Before presenting my own resolution of the problem I would like to say a little more about what I think constructs are meant to be or to do.

We can go along comfortably with Kelly in seeing the construct as a two-ended affair whereby we have a basis for discriminating the phenomenal world of people, of objects, of internal states and of all their manifold inter-relationships. Discriminating in this sense also carries something of the meaning of anticipations or expectations. It is also easy to go along with the idea of hierarchical arrangements of constructs and a progressive development of system complexity over time. These statements are all of course couched in scientific language within personal construct theory. When, however, I have worked with children and young people trying specifically to elicit constructs both they and I have found the work extremely difficult. We have struggled hard, frequently ending up with "one-ended" constructs and on occasions I have had the suspicion that the client has merely gone in for "word-spinning" in order to finish the ordeal. It has not been unusual for the client to challenge the whole enterprise as boring and irrelevant.

Gradually it dawned on me that when I was asking my clients to produce constructs I was inviting them to *do something they had never done before.* That being the case they were committed to finding expressions for abstractions which they had never before verbalised, or they had to generate abstractions where previously their experiences, their thoughts and feelings had been relatively fluid. Under these circumstances it is not surprising that they found the task very difficult and exhausting. If for the moment we forget the scientific language of personal construct theory and revert to everyday language the interview begins to flow. I say to them that I am interested in how they make sense of things, and ask questions to that end. They are then able to respond fluently, meaningfully and often with considerable perspicacity. I cannot say that, at that stage, they had produced constructs. They have talked of what they saw, what they did, what they felt, what they said, what other people said and so forth. They did, in fact, produce material out of which constructs might eventually be fashioned, and, if they were invited to reflect back over all they had said and done, they were sometimes able to produce abstractions which were novel and meaningful and which might indeed be called constructs in their own right. The transition from a scientific terminology to an ordinary terminology therefore enabled interviewing to become a live possibility instead of an arid exercise. Clearly I am advocating a further level of description from that which Kelly formulated in the theory of personal constructs. The justification for this is that it loosens up the whole process of interviewing by allowing the invention of questions which enable clients to talk about themselves. And this of course is but the first step in the resolution of their dilemmas. There is, however, an implication from this argument which is both theoretically important and clinically useful.

I would like to put forward the view that the knowledge of a person's constructs is, of necessity, inferential, both for the client and for the psychologist. This follows from the argument that the client does not "know" his constructs until we provide a situation in which he is asked to produce them, and when he does produce them, he produces them for the first time. Although we may have a personal hope for fact rather than inference there are certain advantages in this formulation. The very speculative nature of inference forces us to go back to the client for his observations on the validity of our thinking. Further the acceptance of inference allows us to think and think again if our first inferences seem unproductive. The retreat from a factual basis for asserting what a person's constructs are to an inferential judgment about what a person's constructs might be offers a far greater range of therapeutic options whilst at the same time reducing the risks of wasted time and energy on discussions of what the truth of things might be.

None of this is, of course, out of line with Kelly's own practice. His extended chapter on self-characterisation (Kelly, 1955, Chap. 7) seems to me to be an essay in inferential analysis. In this task the client is not invited to produce his constructs but rather to write a self-characterisation sketch as by someone who knows him intimately and sympathetically. Under these circumstances a knowledge of his constructs can only be inferential, based on a study of the ways in which the material is composed, the inter-relationships between its various parts and the internal consistencies and inconsistencies of the whole work.

I now want to extend the argument to construct systems as such. I have already said that people do not know what their constructs are, and indeed are frequently puzzled when asked formally to produce them. The notion of a construct system must therefore be of the same order. People do not appeal to their construct systems in order to act. They are their construct systems and always have been. Life goes on reasonably smoothly for most people without over much conscious deliberation, and problems are usually taken in their stride. That being the case it follows that construct systems as such are built up and maintained at a low level of awareness. If conscious choices are made which lead to a proliferation or simplification of construct systems, the choices once made cease to be matters for conscious concern. Our construct systems therefore are essential parts of ourselves which we, perforce, take for granted, do not need to formulate and seldom, if ever, need to refer to.

This formulation has certain implications, the most obvious of which is that in order to discover something of a person's construct system we have to invite him to explore and communicate that which he is already taking for granted. It also means that we, as interviewers, must be careful not to take for granted those things which are self-evident to the client, but not perhaps to us. If we fail to do this we are likely to fail in understanding what he has to say. I can illustrate this best with an example taken from an interview with a mother. She described an incident where her son was "showing off", an expression which to me (and to most people whom I have asked) meant boasting or unnecessarily drawing attention to oneself. This meaning, however, did not make sense in the context of the mother's report so she was asked to say what her son actually did. She reported the following sequence of events: he pulled a face, ran over to the door, threw it open, ran out slamming it behind him, and ran down the stairs into the street. This description, seemed to me more likely an expression of anger (which did fit the context) and the mother agreed that this was indeed the case.*

In this particular case we chose not to take the mother's abstraction for granted and invited her instead to elaborate the concrete referents underlying that abstraction. As one outcome we were able to agree a different abstraction to cover the concrete data, an abstraction which had important implications in relation to the mother's relationship with her son. The same principle applies also to the reporting by a client of events. The events which clients report have meaning only in relation to their own construct systems. Just as we take our constructs for granted so many events in a sequence will go unreported because they also are taken for granted. What is reported in relation to what is left unsaid is comparable to the relationship between figure and ground. When a client gives us the salient facts as he sees them we tend to provide our own "ground" in order to understand what he is telling us. This may be good enough in ordinary conversation, but in psychological work with clients we frequently need to investigate the ground against which the client's own report has meaning. In other words when we are offered the spoken events in a sequence we must ask for the unspoken events. I can illustrate this with a small part of an interview with John who is aged fifteen. John was seeing me because he had been threatening to commit suicide. He was having violent angry outbursts at home. He had not been able to go to school. His father had died in an accident some five months previously. John himself had complained of his bouts of violent anger and as a therapeutic task I had asked him to observe and report one of them to me in detail. In this particular interview I asked him to report back on the last outburst. He said that he had asked his friend to join him so that they could do something together. Everything John suggested his friend turned down. This made John very angry and he stormed out. This was his report. Whilst the sequence does indeed carry a consistent logic it says very little about the detail of the events. So he was asked to close his eyes and recreate the whole sequence from the time he went to knock up his friend to the final angry outburst. He was to report what he said, did, felt and thought, what other people did, what noises he heard and so forth.

The very ordinariness of the task made it difficult and John needed considerable prompting in order to do it. What did come out, however, was extremely important for an understanding of many aspects of John and his place in the family. The revised account now included many more events: his friend was not ready to go out and would not be for thirty minutes. John

* Subsequently I have always asked my clients (all of whom might loosely be described as "working class") what they have meant when they said "showing off" and more often than not they do in fact mean being angry. This observation has of course considerable implications for a proper sensitivity to language in cross-cultural researches.

spent that time with his mother and with Paul, (an adult) who, as John said, would marry his mother when he, John, was "straightened out". Between them they discussed the possible things that John might do with his friend, but it was mother who made most of the suggestions. In fact the suggestions that John's friend turned down were very much those that John's mother had put forward in his absence. It seems to me very clear that this elaboration now includes a number of issues which are important in understanding John's relationship to his family and the bases for his anger. One might specify his mother's intention to remarry so quickly, the onus placed on John to "straighten himself out" and the conflict in loyalties his friend inevitably imposed in unwittingly turning down the various suggestions for joint activity which the mother had offered.

I have been stressing in this section the view that people are, by and large, ignorant of their constructs and construct systems, and in this sense their construct systems are part of that large aspect of living and experience which is taken for granted. At the most general level of practice therefore I would suggest that when things in a person's life, or in a family's life, go so wrong as to call for outside help, the problems are likely to be located in just those areas which the person takes for granted rather than on the outside events with which the client claims to be failing. People usually cope remarkably well with new tensions and new demands. When they fail it is my guess that the existing constructions which a person is using must be looked into rather than the events which the person presents. To do this of course is rather daunting since it requires the psychologist to question what is taken for granted and to look again at the obvious. This is daunting too for the client, who may well prefer to explore the dramatic rather than the commonplace.

From Theory Back to Practice

As I said at the outset of this essay my main concern is a continued involvement with children and young people and with those adults who are worried about them. Thus my excursion into theory (which stems directly from work with children) becomes important if it leads to improvements and changes in interviewing techniques and therapeutic strategies. Before illustrating this, however, it is necessary to present a framework within which interviewing techniques can be described.

We are unlikely to make much progress in understanding our clients unless we can induce them to talk about themselves. Thus a prime function of interviewing techniques is that they make it easy for a child to respond. It needs to be remembered that *children do not present themselves as having problems for which they require help*. It is rather the case that they are presented as the focus of complaints by adults. Under these circumstances there is no reason why a child should see a psychological interview as other than meaningless and irrelevant.

One of the greatest inhibitors of communication about oneself is that we do not know what the enquirer is wanting to get, nor what he will do with it if he gets it. This is even more likely to be true for children. It becomes important therefore that children know why they are being interviewed, and what the purpose of the psychologist's questions are. If, therefore, we specify the topic of the enquiry and the nature of the questions which will be asked, the child is given freedom, within limits, for him to communicate his understanding of himself and others. An enquiry which is contained within a systematic structure allows a wide range of thoughts and feelings to be explored with relative safety for the child and with a considerable economy of time for the investigator.

A second principle indicates that if we want to know someone well we should explore the areas in which he is expert. In relation to children, granted a state of trust, we should be prepared to investigate the child in relation to school and in relation to his family, and with delinquents their expertise in delinquency.

A third principle suggests that we must be wary of assuming that we know what a child means by his descriptive labels. We do not necessarily share common ground with children, nor is the recollection of our own childhood a guarantee that what he says matches our own experiences. Thus we must be prepared to ask and ask again. If, for instance, a child says that he is upset when he is involved in a fight we need to ask if this is because of the possibility of physical pain, or because of psychological embarrassment in case he loses, or because the edicts of his partents are being flouted, thereby testing his loyalties and so forth. He takes for granted that we know which aspect of the situation is upsetting, but, of course, we do not know, unless we ask.

In summary, therefore, I am suggesting that if we are to maximise our chances of helping a child to talk to us we must take him into our confidence about the issues we might both be interested in, we ask questions in a systematic manner and we enquire in areas in which the child feels relatively safe in his own expertise. At any point we can choose to question the basis for the child's response and we may need to help him in this by offering a number of possible answers from which he alone can pick the one which is personally relevant. Structure frequently enables the verbally inhibited to talk and can also be used to constrain the garrulous. Within this framework the point of my theoretical exposition becomes clear. We do not, in the first place, seek for constructs, instead we try to find out how a child makes senses of himself, and people and their inter-relationships. This is the overall aim of the interview. We shall arrive at a child's constructs as a result of inference, and hopefully this will be a shared activity. In the second place we shall be concerned with exploring those aspects of a child's perception and awareness which he takes for granted, that part which operates at a low level of awareness. In a way we can call this an exploration of ordinariness since it is in the everyday experiences that we seek a child's construct system, not in fantasy nor in the dramatic incidents of life. It must be conceded that to talk about the ordinary, and the part of oneself which we take for granted is not easy.

There is, however, a way around this difficulty. Those aspects of life which, in one way or another, we label trouble represent ways in which our expectations are invalidated. In fact the variety of troubles we experience may be valuable pointers to expectations which we take for granted. If therefore we invite a person to talk about troubles we are automatically exploring his unverbalised expectations of life. Troubles are in fact a rich source of enquiries. Individuals themselves may represent troubles as in delinquency. Individuals may complain of the ways other people are a trouble to them. Troubles may be those inner feelings as when we are ourselves troubled or upset. We can use these different aspects of "troubles" as themes through which we can explore with a child or young person his ways of making sense of himself, of others and of situations. It is a series of techniques based on "troubles" which will form the last part of this essay.

Delinquency Implications Matrix and Polygon

This technique was developed in work with delinquent boys and its use assumes that the boys are to some extent already fairly conversant with delinquency. "Trouble" in this context is the fact that boys brought before the court are already "in trouble". A boy is invited to give his knowledge of eight common delinquent activities, and only if he does this satisfactorily is the implications grid given. The eight activities are recorded on separate pieces of numbered paper (an activity which the boy shares). The main task is for the boy to consider each delinquent activity with every other, and say if, in his view, boys who commit the one are likely, by and large, to commit the other. His responses are recorded in a matrix and when that part of the task is completed they are analysed through the implications polygon. Every step of the task and the analysis is explained so that he sees the pattern of his own thoughts reproduced in diagrammatic form. The pattern of clusters which is implicit in the polygon is usually difficult to see until the linkages are teased or "shaken out". When this is done clusters of delinquent activities emerge which can then provide the basis for further elaborative questioning.

The following example was worked out with a 15 year old boy (Joseph) who had been in trouble with the police for "going equipped to steal" and "housebreaking". He had also been a pupil at a residential school for maladjusted children from which he had frequently played truant. The eight delinquent activities appear at the left of Table II and the symbol "0" in the body of the grid means that the boys who commit delinquent activity numbered on that row are likely also to commit the activity numbered by the column.

The implicative links in the body of the grid are represented in Fig. 1. One way linkages are shown by a dotted arrow and a reciprocal linkage is shown by a solid line. The pattern emerges much more clearly in the "shaken out" diagram in Fig. 2.

TABLE II. *Delinquency Matrix Implications for Joseph*

		1	2	3	4	5	6	7	8	
1. T.D.A. (Taking and driving away)	1				0	0		0	0	1
2. G.B.H. (Grievous bodily harm)	2			0				0	0	2
3. Mugging	3		0		0	0	0		0	3
4. Housebreaking	4					0	0	0	0	4
5. Going equipped to steal	5	0			0		0	0	0	5
6. Receiving stolen goods	6	0			0	0			0	6
7. Vandalism	7		0	0					0	7
8. Truancy	8	0	0					0		8
		1	2	3	4	5	6	7	8	

Three possible forms of interpretative analysis are possible in terms of hierarchies (c.f., Hinkle, 1965), in terms of mutual exclusiveness, or (and this is the simplest form of analysis) through reciprocal implications.

The reciprocal implications for Joseph's responses appear in Fig. 2 and it can be seen at a glance that "housebreaking" (4), "going equipped to steal" (5) and "receiving stolen goods" (6) form one cluster and "truancy" (8), "G.B.H." (2) and "vandalism" (7) form a second. These two clusters are linked by "T.D.A." (1) and "G.B.H." (2) has also reciprocal implications with "mugging" (3). We can now ask more questions to elaborate the bases for these clusters. He says that the (2.8.7) cluster is made up of boys who would be *hard nuts* whereas the (4.5.6.) cluster is made up of boys who would be *crafty*. In response to my question about who would be criminals for life he says the (4.5.6.) boys but of those at the opposite end (2.8.7.) he says *as they grow older they will grow out of it, they realise what they are doing*. His attention was then drawn to the other axis represented by "vandalism" at one end and "mugging" at the other. Of boys who do "mugging" he says *they know what they will do, they wait for a person.* Of boys who commit vandalism he says *they go in big groups, they don't know what will happen until they get somewhere.* He was finally asked where, in this pattern of delinquency activities he would place himself. His answer was unequivocally as the "T.D.A."/"truant".

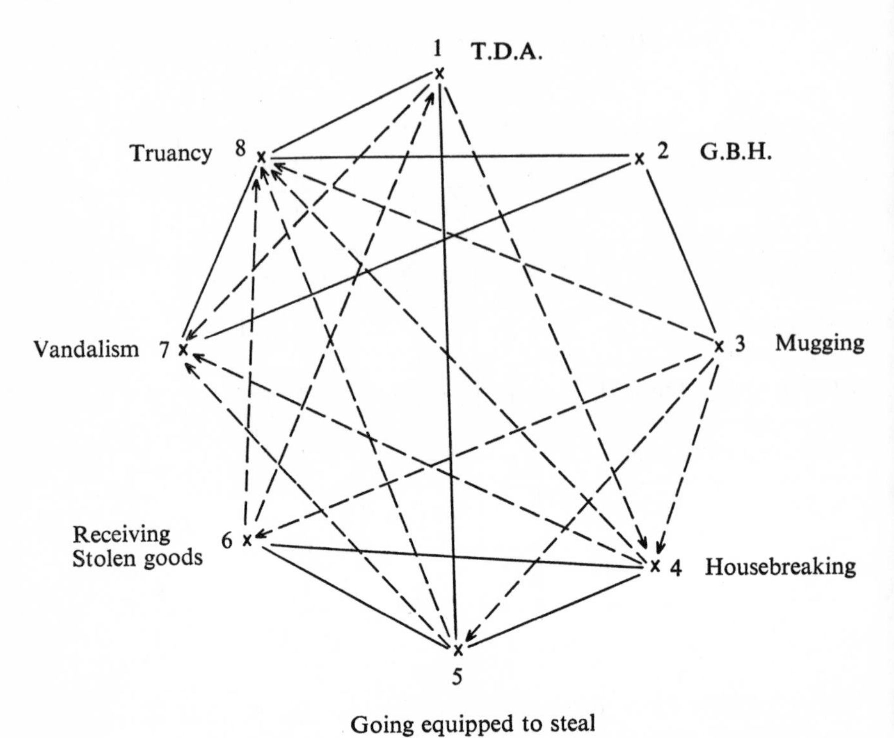

Fig. 1. *Implications Polygon for Data in Table II.*

If we wish to infer constructs underlying Joseph's conceptualisation of delinquency and delinquent boys it would seem that two should be sufficient to encompass the data. The first would have reference to aggressive, unthinking adolescent delinquency as opposed to crafty, near professional delinquency. Included in this construct would be correlated ideas of delinquency as a stage of development as opposed to delinquency as a final stage of development. The second construct would be concerned with violence and sets criminal intent together with a degree of social isolation against spontaneous delinquency as an aspect of group membership. "T.D.A."/"truancy" forms a linking

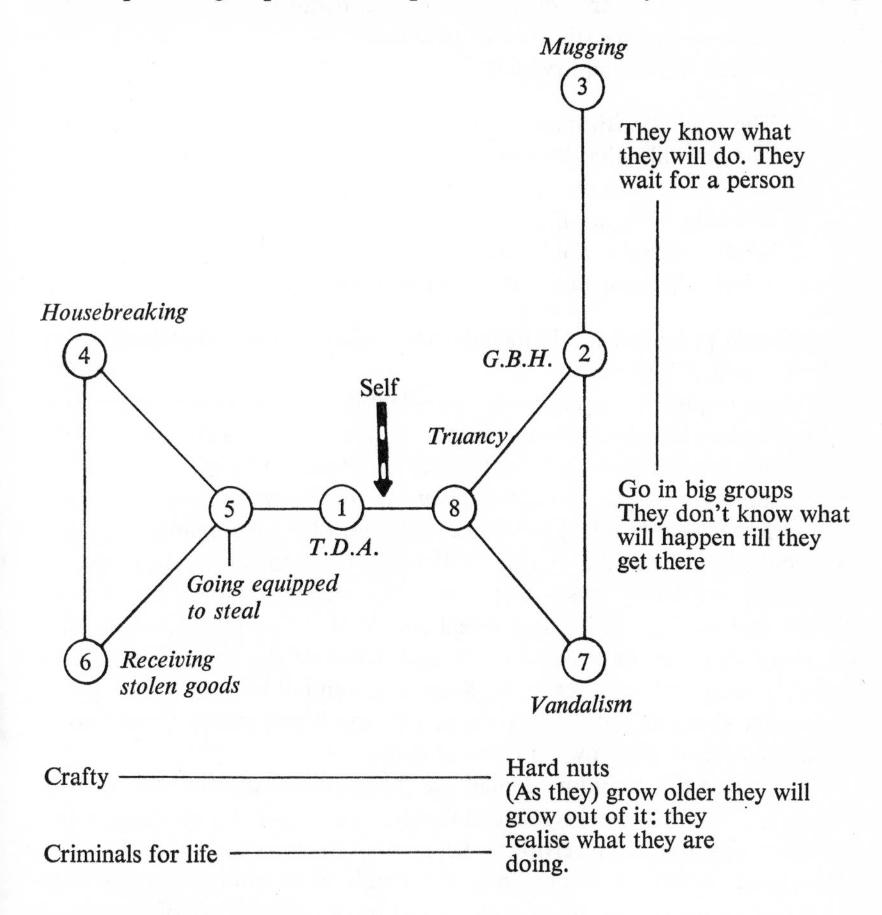

Fig. 2. *Delinquency Implications for Joseph* ("*Shaken out*" *version*)

concept and since this is where Joseph places himself seems to imply for him a choice point between a growing out of delinquency on the one hand, and a life of professional delinquency on the other. The more important question of whether or not Joseph can abandon the delinquent choice goes beyond this presentation.

The Elaboration of Complaints

The focus of this technique is, quite simply, the complaints which an individual voices against people who might be important in his life. I quoted the technique earlier in the essay but present it here in greater detail as part of the theoretical stance based on the investigation of troubles.

The pattern of the enquiry runs:

1. The trouble with most . . . is . . .
2. They are like that because . . .
3. Another reason they are like that is . . .
4. It would be better if . . .
5. What difference would that make? . . .
6. What difference would that make to you? . . .

(The people presented in (1) include boys, girls, teachers, brothers, sisters, fathers, mothers and self).

In order to present the child with an orientation to the task he is reminded of the way in which for instance teachers will say "The trouble with that boy is . . ." or a mother will say "The trouble with that family is . . .". He is then invited to say what *he* considers, *from his own point of view*, the trouble with different people might be. Some children, of course, immediately offer adult complaints, but this can be challenged. If for instance a boy says that the trouble with most boys is that they fight we need to know if he is echoing parents and teachers or if this is a real complaint of his own. Other children will deny that they have any trouble with some of the persons, and this is perfectly acceptable as a response. Some children fail to answer each part of an enquiry about an individual person. This too is acceptable. Omissions are as informative, in their own way, as answers.

To illustrate this technique I shall use the responses of a 16-year-old West Indian boy who had been remanded for the two apparently unrelated crimes of indecent assault and receiving stolen goods. (In connection with the charge of indecent assault we do not know the details so we cannot establish either the extent to which there was provocation or the seriousness of the assault. As frequently happens, the technical language of the law obscures rather than reveals the behaviour of the delinquent.)

In reporting James' responses I shall reduce them to consecutive prose by omitting the numbering of each question in the sequence and italicising his actual words. I shall also insert an interpretative comment where I feel it appropriate.

1. The trouble with most BOYS is *they most likely want to go round with each other in a gang*. They are like that *because they have nothing to do,*

they follow each other, and *because they like to have friends.* James could give no way in which things might be better.

2. The trouble with most GIRLS is *they also like to go round in gangs.* They are like that because *they like to follow the way of the boys*, and *boys and girls have to get together sometimes, that's when it begins, when they start liking each other.* James could give no way in which things might be better.

James is here pointing a contrast between solitariness, which he seems to find painful, and being in a gang, which by definition is a trouble. A sense of the emptiness of life permeates these answers, yet the resolution of this emptiness implies trouble. It is noteworthy that James can give reasons but he can give no implication for how things might be better.

3. The trouble with most BROTHERS is *the old brothers and sisters like to put the young ones down.* They are like that *because they have to stick together* and *because there might be some disagreement.* It would be better if *they stuck together and tried to work together.* If this happened *as they worked together they would really understand each other's feelings*, and *if they understood my feelings they might be able to help me.*

James here points to disharmony with his older siblings and the barriers which he feels are put up in the way of understanding. He points to the need for solidarity, presumably against a hostile world, amongst siblings as opposed to group membership which merely fills an emptiness of life. In the end, however, he points to the primacy of his feelings and a need for these to be understood, a need which could be met by working together, rather than talking together, with his siblings.

4. The trouble with most FATHERS is *they most likely favour the girls more than the boys.* James could give no reasons. It would be better if *they not only liked the girls but liked the boys as well.* If this happened *they would be able to help the boy in his feelings* and if this happened to him *if my father showed me that he really liked me he would help me not to get in such trouble.*
5. The trouble with most MOTHERS is *they most like the boys.* James could give no reasons. It would be better if *they liked girls as well as boys, if girls got on with them as well as the boys.* James could not say what difference this would make.

James describes a family system in which the alliances are across the sexes rather than within the sexes. This can obviously create problems for an adolescent boy, especially if he sees the outside world as hostile. It is perhaps not suprising that he documents the case against his father rather more fully than the case against the mother. This is in line with his complaint that his brothers also have failed him.

6. The trouble with JAMES himself is *I like my way too much and don't get it*. He is like that because *I suppose my mother likes me more than my dad* (*does*), and because *I was brought up that way*. It would be better if *I was liked by both mum and dad*, then *my dad would understand my feelings*.

Although James puts forward the issue as one of having his own way, almost inevitably he points back to his feelings and the failure of his father to understand him as the real problem. With the family constellation as James describes it, perhaps homosexuality might be one logical next step for this boy. We do not know what happened.

The remaining two "troubles" techniques I shall illustrate with material from one case. The boy in question is Mark, aged 10 years. He presents a serious problem in school both to teachers and to other children because of his behaviour. He is said to have been difficult from the time of starting school. He alternates between being "good, hard working and conforming" and being a "troublemaker". Other boys are suspicious of him because of his unpredictability and the instability of his interpersonal relationships. He is said to have acquaintances rather than friends. It is known that there are considerable tensions within the family, and between the parents and school. Mother is dominating and unpredictable, father tends to be self-effacing. In the first part of the interview (not reported here) Mark showed an awareness of many tensions at home. He indicated that he was closely identified with his mother but seemed to regret that lack of involvement with father. A younger brother in fact was closely identified with father, and, as a boy, was aggressively masculine, assertive and daring.

Perception of Troubles in School

In this technique the child is offered eight pictures of ordinary situations in school, drawn with some ambiguity as to detail, but otherwise quite straightforward. He is then invited, within a sequence of questions, to isolate and describe the child who might be troubled or upset. If the child says that no-one is troubled he is invited to consider "*if* someone is troubled, who would it be?" If the child gives an adult as the one who is troubled this is accepted but the child is then invited to give an alternative version in which it is a child who is upset. The sequence of questions runs thus:

1. What do you think is happening?
2. Who might be troubled and why?

3. How did this come about?
4. If you were there what would you do and why?
5. What difference would that make?
6. What kind of boy is the one picked out in Q.2?

Out of the responses to this sequence it is possible to gain some idea of how a child actually perceives various situations in school and how he understands some of the interactions which take place there, how willingly he identifies himself with these situations and the extent of his understanding of different ways of coping. The final question in each sequence presses the child to a level of abstraction which summarises the troubled character whom he has presented.

This sequence represents a complete technique in itself, and as will be seen from the data obtained from Mark, the information is extensive. We can, if we so wish, proceed to some method of systemising the data. The therapeutic value of this rests on the fact that the child is thereby committed to a serious review of his own understanding of things together. At the same time the possibility arises that he might make some important personal discoveries which could influence his future behaviour. With Mark we used the formulations arising from his answer to question 6 as the material for an implications matrix (as previously described).

Before giving the details of Mark's responses I should make some comments on the differences between the aims of this technique and the more commonly known projective techniques, e.g., T.A.T. and C.A.T. (c.f., Ravenette, 1973). In the first place the child's imagination is turned to the reconstruction of ordinary things rather than to the creation of "fantasy" stories. The telling of stories would, in fact, be a disadvantage since it avoids the questions which are posed, and uses up a great deal of time in doing so. In the second place the investigation takes the form of an active dialogue, not only in setting the structure, but also in collaboratively clarifying the meaning of a child's responses. In the third place we invite the child to be an observer and reporter on incidents which he shares, and also to take some responsibility in imagination for his own involvement in school. Fourthly, at the interpretive level, a premium is put on the mapping of conscious awareness, rather than the deliberate exploration of lower levels of awareness. That aspect of a child's functioning is taken up either inferentially as in analysis of the data, or as a basis for feedback and further exploration with the child himself during the interview. Mark's responses to the eight pictures are given in Table III.

The amount of information provided by the questions in Table III is very great, so great as to make interpretation itself very difficult even if such an analysis were necessarily called for. (Should we wish to undertake such a task, Kelly's suggestions for analysing self-characterisation sketches might well be relevant. In this he assumes that the client seldom moves far from his starting

TABLE III. *Mark's responses to perception in schools*

	What do you think is happening?	Who do you think is upset and why?	How did it come about?	If you were there what would you be doing and why?	What difference would that make?	What sort of a person is the one who is troubled?
1.	Playing a game	The 2nd from left would be upset because he is always getting picked on.	This is a game when they are fighting. The weak one is upset.	If Paul were there he would not mix, as they would pick on him.	This would probably lead to trouble.	The central figure is the boy who *always gets picked on.*
2.	They are coming out of school, mother is waiting for them.	The mothers are upset, they did not realise the time. Housework not done.	Probably they did not start their housework until late.	Paul would do nothing, he cannot do the work.		
2a.	The boy had a bad day at school, his sums were wrong, and he was told off.		He probably got up late, he was still tired.	Paul would say to the boy "it is not your fault."	He would do this to please him.	The central character here would be *not very bright*.
3.	A lesson in school—arithmetic.	Boy at blackboard is upset trying to do a sum. He may be unable to do it.	He does not like doing sums, he does not listen.	If Paul were there he would say "don't worry", nothing he can do.	This would make no difference.	The central character here would be a *trouble maker.**
4.	They are playing on some apparatus.	The boy does not like the apparatus, he is scared of heights.	He had been up a hill before and fallen down.	If Paul were there he would say, "have another go."	It might make some difference to the boy.	This boy would be a *quiet and shy boy*.

5. A lesson.	One of the boys does not like the lesson, he is not doing it.	He does not like the lesson.	Paul would tell him have a go at it. He may get it right.	The boy might like the lesson if he gets a few right.	The boy does not like the lesson, *cannot stand making errors*.
6. A game of football.	The boy who cannot play, they won't let him play.	They do not like him.	He would ask the boys whether I could play and maybe be could.	It is worth asking.	This boy would be *no good at football* (or outdoor games generally).
7. This is a presentation.	The boy who gets nothing is upset. He was not in to pass the things.		Paul would say to him "you can have another go next year maybe."	The difference is the boy would think I will try and not get colds.	The boy is *untidy* he hasn't a dad. He would be a good sportsman.†
8. Two boys getting the cane.	They were not in the wrong, but were getting punished.	The boy who picked on him gets in trouble and retaliated.	Paul would say it was not his fault.	He may not get the cane.	This boy *does not like getting into trouble*.

* *Did not like doing work*.
† This sequence is extremely puzzling because it seems to bring together very contradictory ideas, probably indicating extreme anxiety.

point, and when he does it is usually by means either of contrasts or through the elaboration of what he has already presented.) At a very simple level of analysis we can take all the descriptions given in response to Question 6 and see in them the characterisation of at least one boy who would find school a difficult place in which to exist. We should not assume of course that this

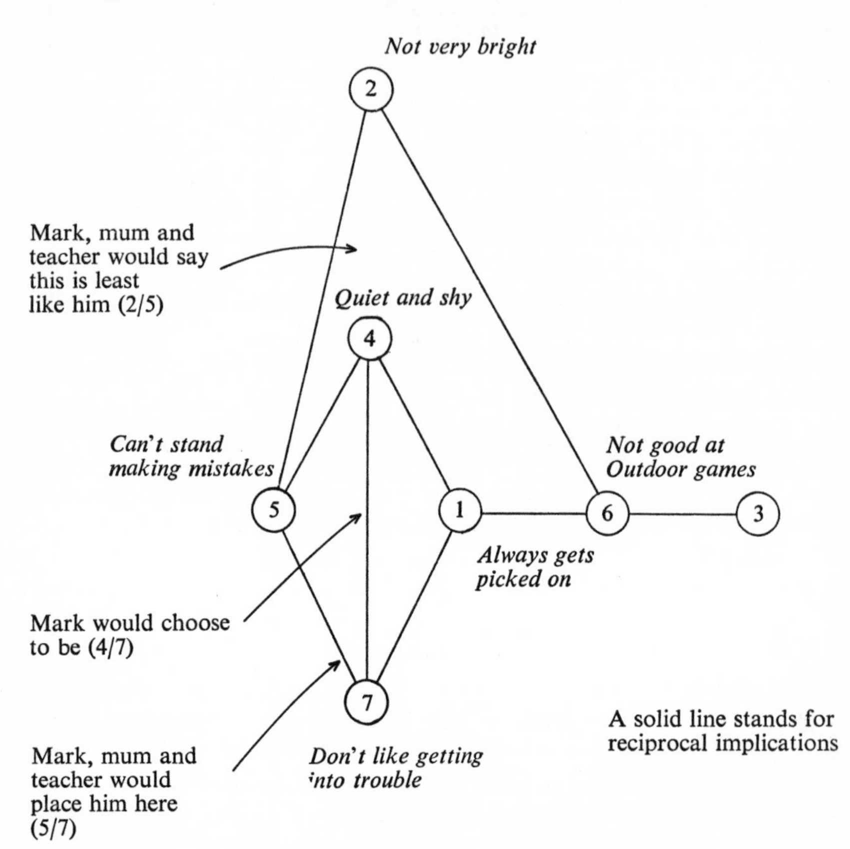

Fig. 3. *Personal Attributes Implications for Mark* ("*Shaken out*" *version*).

characterisation is Mark himself, although it might be one version of him. The version he, in fact, offers must be inferred from his answers to the Question 4, and his view of his effectiveness from the answers to Question 5.

In practice we used seven of the characterisations given in response to Question 6 as the material for an Implication Grid Matrix (as previously described). The graphical analysis was carried out with Mark and the "shaken out" version appears in Fig. 3. The elaborative choice after this graphical analysis was simply to ask Mark to indicate where he fitted in the diagram where his parents and his teacher would put him, and, if he could be different, where he would choose to be.

A study of the content of the clusters suggests that three dimensions (or, inferentially, constructs) may be necessary to understand the pattern of implications. The first dimension could be intellectual brightness with only the "not very bright" pole named. Mark confirmed this himself. A second dimension would be concerned with trouble with the two extremes labelled "*not liking to get into trouble*" and "*troublemaker*". The third dimension is defined by only one end, "*quiet and shy*". Later in the interview Mark does in fact define a boy who is rough, able to beat people up and not scared. Presumably this could be the missing pole, in which case this dimension would have important implications for the style of interpersonal relationships open to him. The use of an implications grid procedure, provides a way in which the ideas generated by the "Perception of Troubles in School" technique can be reduced to manageable proportions.

The Exploration of Personal Troubles

The fourth technique explores a child's personal troubles as expressed through the medium of his drawings. He is reminded that (all people) boys, girls, men and women, have times when they are troubled inside themselves. They feel hurt, angry, ashamed, embarrassed, worried and so forth. We would like him to draw pictures to show five occasions in which he would be troubled or upset. While the invitation is being made a sheet of paper is folded into six rectangles and a mark is put into five of them. It is pointed out to him that these marks are merely to help get him started on his drawings and he does not have to use them. He is given the pencil and invited to carry on. If he says he does not understand what he has to do an example is given of what another boy has done. This is usually sufficient for him to understand. Some children produce less than five drawings and this is acceptable. In the sixth space the child is invited to draw a situation in which everything would be fine, he would feel good, and people would seem good. When he has finished drawing he is asked to say what is happening in each picture. This completes the basic task. He is then invited to respond to a rather challenging elaboration. He is asked to think of a child who, in all of these troubling situations would not in fact be troubled. Specifically he is asked to give three descriptions of such a boy. Two more questions are asked, when would this new character be troubled or upset, and when was the child himself accurately described by this new character. The answers to these two questions frequently show a depth of understanding and an originality which is surprising both to the interviewer and to the people who are familiar with the child. In many ways this particular enquiry forms a dramatic high point in an interview.

The material which illustrates this technique is drawn from the last part of the interview with Mark. He was able to give only four occasions in which he would be troubled or upset.

In the first picture boys are kicking footballs into his face. The physical hurt, and possible nosebleed would be the upsetting aspect. The second picture shows a boy spitting at him. The spit would go down his clothes and his mother would tell him off. In the third picture a boy is throwing stones at him and it would cut his head open. The fourth picture shows a boy ducking him in the swimming baths, pushing his head under. The fifth picture now shows the occasion in which everything would be fine. Mark would be winning a game of table tennis against another club.

The pictures illustrate the theme of personal inferiority as opposed to personal superiority in relation to the peer group. In essence, when Mark can demonstrate superior skills and when interpersonal relations are governed by clear rules, he will feel good. When he lacks skills and situations are unstructured he will expect to be in trouble.

Mark's description of the boy who would *not* be troubled in these situations is of someone who is. *Rough*, 2. *Could beat them up*, and 3. *Not scared of anything*. Implicitly therefore Mark sees himself as troubled or upset when he is not rough, when he feels physically inferior and when he is scared. This newly invented character would himself be troubled when someone *older* was doing these things to him (i.e., the events in his pictures) and this character would best describe him when *younger* children started on him. Mark's age status therefore becomes an issue against which his own behaviour needs to be understood.

It is tempting to see these alternatives as having some relationship to acceptable masculine and feminine roles, a formulation which also fits the cluster of attributes already described in the previous section. Such a formulation also receives support from the father's relative inaccessibility at home, Mark's identification with his mother and feminine activities at home, and his brother's aggressive masculinity. It is probably of some significance that both Mark and his parents set great store by his imminent transfer to a secondary school, and have specifically requested a place in an all boys' school.

And what Difference does All of this Make?

I have presented the illustrative material in a rather detached, if not academic, manner, but it can be of little value unless its practical utility and therapeutic potential is indicated. I shall conclude by showing two ways in which this kind of investigation has the possibility for generating change. The first way relates to the client himself. The second way relates to the people with whom the child or young person is interacting and to whom the child presents a problem.

When a child is confronted with this style of interview he is, perhaps for the first time, invited to think seriously about himself and his ways of making sense of things. We could call this a stocktaking, or a mapping exercise. Either analogy is useful. Out of this exercise he may develop alternative ways of making sense, or, at the very least, become aware of the sense he has traditionally been making of people and things. Whilst neither of these experiences is necessarily therapeutic, each contains the possibility of generating some change of view when the old events reappear.

A new awareness of habitual choices at least opens up the possibility of a new response. The interview itself presents the child with an opportunity of being treated with seriousness, and not being given pat responses to what he himself offers. In many ways Kelly's Sociality Corollary is relevant here.

> "To the extent that one person construes the construction processes of another, he may play a role in a social process involving the other person."

The techniques and strategies, of which those in this essay are examples, are implicit communications to the child of the possibility that at least one person might be capable of understanding him. If he can be understood, perhaps he can pose his behavioural questions in ways which generate less distress. This is one aspect of therapeutic change.

The second practical value rests on the fact that the complaints which lead to our involvement with children are those of the adults rather than the complaints of children themselves. In this sense the problems reflect a failure in mutual understanding between adult and child. But it is the adult who bears the greater responsibility, and if the communication between them bears no relationship to the child's construction of himself and others, then misunderstandings and problems arise. It is sadly the case that when the adult, in his communication with a child, does not understand the child's construction of himself the relationship between adult and child will probably generate friction. A small example will illustrate this. A very disturbed boy raised a fist to teacher. The Head Teacher took the boy away to his office and playfully smacked his bottom. He then asked the boy if the treatment was fair. The boy immediately said that it wasn't. Rather puzzled by this the Head Teacher asked why, and the boy replied that he wasn't a baby, and that to smack him on the bottom with his hand was to treat him as a baby. It would have been fair if the teacher had used a cane. In this way the boy himself was given an opportunity of communicating his own construction of himself, thereby reducing the risks of further interpersonal failure.

A personal construct approach to interviewing children should, by definition, lead to the finding of those constructions whereby the child makes sense of himself and others. The communication of this to teachers, and others who are involved with children improves the chances that the child's outlooks are taken into account by those significant others who so often complain of difficulties. The Sociality Corollary is again very relevant here together with the concept of *hostility* as defined by Kelly:

> "Hostility is the continued effort to extort validational evidence in favour of a type of social prediction which has already proved itself a failure."

When children and teachers continue to maintain their own constructions of each other, in the face of continued interpersonal difficulties, we see this as hostility in the classroom. When a teacher can abandon, if only for a limited time, his existing constructions of his problem children he may provide room for growth for each of them, and for himself. When a child, if only for a limited time, can become aware of his own constructions of himself and others, he too may enjoy a breathing space in which more harmonious relationships can develop. Kelly, through personal construct theory, invites the psychologist to reconstruct *his* role in such a way as to help him to promote just those kinds of change.

References

Bannister, D. (ed.) (1970). "Perspectives in Personal Construct Theory", Academic Press, London.

Bannister, D. and Mair, J. M. M. (1968). "The Evaluation of Personal Constructs", Academic Press, London.

Bannister, D. and Fransella, F. (1971). "Inquiring Man", Penguin, Harmondsworth.

Fransella, F. (1972). "Personal Change and Reconstruction", Academic Press, London.

Hinkle, D. N. (1965). The change of personal constructs from the viewpoint of a theory of implications, unpub. Ph.D. thesis, Ohio State University.

Kelly, G. A. (1955). "The Psychology of Personal Constructs", Norton, New York.

Landfield, A. W. (1971). "Personal Construct Systems in Psychotherapy", Rand, McNally, Chicago.

Ravenette, A. T. (1964). Some attempt at developing the use of repertory grid techniques in a child guidance clinic, (Warren, N., ed.), Proc. Brunel Symposium

Ravenette, A. T. (1968). "Dimensions of Reading Difficulty", Pergamon, Oxford.

Ravenette, A. T. (1973). Projective psychology and personal construct theory, *Proj. psychol. J.* **18,** 1, pp. 3–9.

Ravenette, A. T. (1975). Grid techniques for children, *J. Child. Psychol. Psychiat* **16,** 79–83.

Core Structure Theory and Implications

Charles Stefan

Carl Rogers remarked that if one did therapy according to the dictates of personal construct theory, one would be busy in intellectual and unemotional enterprises which lack depth. These impressions led him to wonder whether therapy modelled after Kelly's dictates could be effective. My first reading of Kelly (1955) left me with a similar impression. It seemed the theory focused on man's cognitive efforts and dismissed his passions. Where, for example, was the chapter on the unconscious? Terms like hostility, aggression, and guilt were re-defined and reinforcement was transformed into a broader, more inclusive concept and re-named validation. Terminology familiar to me, which served as cognitive pegs to make sense of events and ideas was summarily dismissed. No data was offered to support the fundamental postulate and eleven corollaries upon which Kelly based his theory. It seemed an uninspired philistine theory, lacking continuity with established personality theories, as if Kelly did not appreciate what had gone on before him. Had I questioned no further, I would not be writing this paper; for me to adopt a personal construct orientation, the theory must provide a way to understand a broad range of phenomena usually associated with "depth approaches". My understanding of these phenomena has come to rest on what Kelly terms "core constructs", and I plan to provide an elaboration of this aspect of personal construct theory and to consider the implications of core construct theory to both personal growth and suicide, areas traditionally reserved for "depth" interpretations.

There is a subtly compelling message within the theory: the object of personal construct theory—individual man—is seen as earnestly observing events and construing or replicating these events in some private fashion in order to predict future events. As an observer, the individual must know as much as he can about the instrument he uses to make observations, and that instrument is himself. Therefore, he must construe himself; that is, he is in the business of construing an image not only of outside events, but also of himself. He is, in fact, engaged in a construing process which results in a self-creation. At any given point in time, the construing person may stop, reflexively view himself as a finished product and elaborate an identity as if this identity had certain immutable features and certain degrees of freedom.

Individuals—who themselves are elements of the real world—continuously monitor and construe the events of the real world. The construing process involves interpreting events in terms of the similarities and differences among them and developing a finite repertoire of dichotomous constructs which are organised into a hierarchical structure. Once generated, constructs provide the basis for the prediction and anticipation of future events, and they also provide, according to Kelly (1955), the channels for human behaviour. The individual as a real event becomes the object of his interpretive process and becomes construed as a self by observing similarities and differences between himself and others. Those constructs which include the construer as an element are referred to as self-constructs. At a theoretical level, what is said to be characteristic of constructs in general applies to self-constructs as well. At the level of direct experience or first order, self-constructs are verbalised in the form of attributional statements considered by the construer to be "like me" statements. Examples are: handsome, honest, intelligent, hard working, rich, etc. It is noted that such constructs are based on comparisons of at least triads of individuals; one of whom is the construing person. Although they have been referred to as first order constructs, they are not the original constructs of the individual. The original are, according to Kelly, pre-verbal, and if communicated at all, they are either symbolised or behaviourally enacted. Many, but not all, pre-verbal constructs Kelly sees as falling into the realm of convience of physiology and not psychology:

> "If a person is asked how he proposes to digest his dinner, he will be hard put to answer the question. It is likely that he will say that such matters are beyond his control. They seem to him to be beyond his control because he cannot anticipate them within the same system which he must use for communication. Yet digestion is an individually structured process, and what one anticipates has a great deal to do with the course it takes." (Kelly, 1955, p. 51)

An individual's repertoire of first order self-constructs is organised into patterns of interdependence, which are then structured hierarchically into subordinate and superordinate relationships. Those constructs in the structure which identify a pattern or identity for a loose grouping of constructs are referred to as core constructs. Core constructs are characteristically superordinate (subsuming first order constructs as elements), less permeable than first order constructs and more comprehensive, possessing a wider range of convenience. The core constructs are those constructs by which the construing individual comes to posit an identity or meaning to his behaviour. Because of them, he is free to act in some ways and is blocked from taking action in other ways. It is a system which the individual has defined for himself and which he controls. For example, first order constructs a person may attribute to himself are that he is rich and handsome. These may be then subsumed under the superordinate or core construct of playboy by the individual, whereby he achieves a distinguishable character. Core constructs are more difficult to validate or invalidate than are first order constructs, because they operate at a less direct order of experience. As constructs, core constructs are more abstract. They are inferential statements based upon first order constructs and are designed to facilitate the organisation of diverse data into simple generalised interpretations. A first order construct may be invalidated by experience. In order for core constructs to change, the elements subsumed must be affected.

It is important to stress that the degrees of freedom which exist from any particular reflexive view of oneself are not to be viewed as haphazard alternatives, but are the consequences of the individual's self-construction as well as of his construction of events outside himself. Therefore, not only is man creating himself, but in some sense and to some degree he is determining his future. If this is the result of the construing process, one needs to be wary lest by his own constructions he becomes trapped, circumscribed, or confined by his machinations. Freedom is gained when an individual defines himself in terms of attributes and achievements such as being able to walk, being able to run, being able to drive a car; however, freedom is diminished when an individual defines himself simply as a walker, runner or driver. To define oneself implies a setting of boundaries—definitions of what one is and of what he is capable and is not capable—the former implies freedom; the latter, constraint. Freedom means having a choice of all alternatives; to explore or experience more of one's environment, and to obtain validating information regarding the accuracy of one's system. A personal construct system allows one to predict and control events. An inadequate system, however, can leave one feeling at the mercy of fate, circumstance, or the will of God. Without an adequate validational process, the theory becomes stagnant and inaccurate. Thus, what is meant by the self in personal construct theory is a person in motion, constantly developing and revising strategy to deal with the world as accurately and effectively as he can.

The core construction of one's self can be exemplified by using the game of chess as an analogy. At the start of the game, one is black or white, on one side of the board or the other. You have at your disposal sixteen pieces that move in particular ways and, like being born, there is no influence one can bring to bear at the beginning point. It is an event outside the realm of personal determination. During the game, one is free to move in specific ways at specific times. There is never total freedom; movement is determined or limited by past moves and future objectives. Early moves are designed to facilitate subsequent movement, while later moves are determined by the strategies and structure developed. At any point in time, the set of moves made forms patterns or strategies. To succeed in chess one must be able to develop and elaborate effective strategies, and to revise or discard strategies to meet the exigencies of the situation. It is a major objective of the game to gain as much freedom of movement as one can. One needs freedom to arrive at more effective strategies. The more one can anticipate and control the moves of his opponent, the more freedom he has to develop an alternate strategy and thereby expand his freedom base. It is a circular scenario. The player is never without a strategy until the game is over; at this point, freedom is no longer applicable. The game has ended.

The "self", "personality", or "life style" can be construed as analogous to chess strategies. In initiating action or in responding to perceived situations, the person develops strategies which take into account previous actions and future objectives. A child is concerned with developing skills and abilities, while an adult applies the skills and abilities learned to accomplishing objectives. During this activity, individuals typically describe themselves with adjectives that acquired their meaning from patterns of action; the descriptive adjectives imply permanence, while patterns of action may change abruptly.

A major proposition of this chapter is that man is creating a self or a set of core constructs which determine his range of freedom by acting "as if" certain personal attitudes, judgments, and evaluations were reflections of some factual and permanent core within him rather than the subjective psychological events which are either empirically untestable or have no empirical validity at all. The very attribution of factual status to these subjective psychological events subsequently determines the person's behaviour in a way which affects his relationship with his social and empirical environments, in ways which confirm or validate the original attitudes, judgments, and evaluations. The created self is then a mental construct. Yet, if taken seriously, it becomes reified, stultifying future growth and ensnaring the individual in a psychological role which may be ineffective.

The institutions which make up the social environment are based historically upon systems of thought which do not foster self-reflexive criticism, and they may operate as well to prevent the individual from questioning the substance of his self-generated identity.

Admittedly, there exists a reasonable need to preserve the integrity of the social system; that is, there must exist a system or structure that provides some context for the individual to incorporate novel events. Anecdotal accounts in the literature suggest that recalcitrant attitudes may operate within the social environment which support the continuance of systems even when they are no longer effective. The system of language and the principles of logic upon which it and further western culture are based, can also lock the individual into thinking about himself as if he were one and only one particular kind of person for once and always.

Man inherits a language which fosters the development of personal convictions of uniqueness and permanence. The structure of language can lead one to think about people, objects, and oneself in particular ways. In fact, because the logic of language is so deeply ingrained, it has been argued that "the structure of our language does our thinking for us" (Sapir–Whorf–Korzybski, 1960, p. 339). Skinner (1971) provides many excellent examples; an act of aggression comes to be viewed as an individual's trait of aggressiveness.

A person's views and descriptions of himself may restrict him from engaging in actions which otherwise might disconfirm self-descriptions. Examples are statements such as: "I cannot get through a Math course; I was never good at talking to girls; I will always love you; I am impotent or frigid." A student was so positive of his inability to succeed academically that even after winning a scholarship, he threw away the letter which announced the award, since it could not be his—some mistake had been made by the granting university.

Hardin (1961) offers an analysis of the self-labelling process; he asserts that there are three classes of truth: "Class I—those truths that are unaltered by the saying of them" (p. 5). These are the truths of the real world which are not dependent on my thinking (for their existence). For example, lightning strikes a tree whether I know it or not; two plus two equal four; and apples fall to the ground. "Class II—those truths that are made true by being said" (p. 5). This is the truth of rumour, suggestion, or anticipation. For example, everyone knows that fat people are happy-go-lucky. It is also the truth of self-images: "I cannot do this or that because I am too old." In medicine, the Class II truth is the force behind placebos. Sugar pills cure, once someone says they do. "Class III—those truths that are destroyed in the saying of them" (p. 5). Hardin tells us that history's impact is gained through Class III statements: If it is mentioned that an event is bound to happen, people will ensure that it will not; or one can interpret Orwell's novel, *1984*, as an attempt to prevent the conditions described from happening by describing them.

The point is that there are individuals who believe that there are only Class I truths. Further, when they encounter human events which are characteristic of Class II and III truths, they erroneously treat them as if they were immutable fact, i.e., Class I. Thus, when assigning subjective attributes to themselves which may have no status empirically, these attributes are treated as Class I truths and are consequently considered to be out of the realm of personal control—hence, self-fulfilling prophecies.

It is not just the self-labelling that is the culprit. The labelling process is merely part of an overall pattern of thinking which is designed to manage unrelated events by simplifying them into an organised system. This system can lead one to the conviction that once a person behaves in a particular manner, he will always and forever perform in that way, no matter what the context. He cannot go beyond the limits of his core structure as a person. His opinion of himself is apt to be couched in exclusive terms: I am always honest; I never lose my temper; I am a professional. That is, if he views himself and is viewed by others as capable in one area, he runs a strong probability of being viewed as capable in all other areas. Needless to say, if the person labels himself negatively, it is a strong probability that he will see himself negatively in all other interrelated areas until he views himself as a particular kind of person. This pattern of thinking serves as the basis for the well-known phenomenon of homosexual panic. The person, after one bit of evidence, no matter how weak, decides to construe himself exclusively as a homosexual. It is the assumption of exclusivity that is the causal agent of panic. There is no way out of the quandary; there are no alternatives. One concludes that he is or is not a particular type of person. One does not go about changing one's style of life or one's self-concept with ease or without consequences. Once developed, a past style of life (strategy) becomes a commitment.

One criterion in evaluating a theory is its ability to accommodate information other than that used to develop the theory; this principle applies to personal construct theories of individuals and to public theories of human behaviour. Can core construct theory accommodate current theory and research in the areas of depression and suicide?

Current experimental work in depression and suicide does admit cognition as a major determinant. Characteristic, self-determined patterns of behaving are viewed by Neuringer (1961, 1964) and Schneidman (1957, 1961) as factors in accounting for suicide. Levenson (1974) summarises a literature review, stating:

> "The findings suggest the emergence of a group of serious suicide attempters with a cognitive style which furnishes them with a view of the world, which is highly undifferentiated, inarticulate, and global." (p. 155).

On the basis of his knowledge of suicide, Farberow (1950) notes:

> "A logical question intrudes, despite the fact that it has implications which are contrary to our mores and ethics. What would happen were the threatening patient provided with the opportunity, taking all due precautions to attempt suicide? For example, would it be therapeutic to allow a suicidal to swallow some harmless placebos or other substance provoking a brief disagreeable illness? One important fact provides at least a partial answer to such a question. It is known that most suicides are intent on commission of their act by one particular method at a certain time or place and that they will not accept or attempt any other method." (p. 67)

Farberow appears to be arguing that methods of dying as well as methods of living must be consistent with one's psychological core structure. The actions of men move along the channels which are open to the individual as a result of his construct system and they are, more pointedly, expressions of what one considers to be the essence of his identity. The suicidal person is not carrying out actions as experiments to determine who he is; but instead, his identity is announced and foretold by his suicidal actions.

The non-experimenting person operates from a fixed, tightly defined core structure which subsequently leads him to a reflexive view of himself as an accomplished, completed product. In contrast, the experimenter operates from a core structure less rigidly defined and more permeable, which results in a reflexive view of himself as incomplete and engaged in an ongoing process. It would seem the former must look for security in the product, viz. himself. Further, the non-experimenter is consigned to a course of action designed to validate his self-concept, while the experimenting person must find security in trusting in the process and the future outcome. These styles represent two quite different orientations. Kelly (1965) distinguishes between these two orientations in the following manner:

> "Now as for self-concepts: . . . I suppose I could start out with the sage comment that it should be called a self-percept rather than a self-concept. But I'm not sure I concur. The self may have the character of a construct as well as that of a construed event or object."
>
> "Assuming the self I am talking about when I refer to myself as an object, I am led to look for the dimensions in terms of which I suspend myself in psychological hyper-space. The more I follow this line of thought the more identity I have, the more static I feel and the more isolated I feel hanging over there all by myself. I'm not sure I like the idea. Identifying oneself in terms of his construct system can have this effect of making him feel immobilised, particularly if he uses constructs designed to take care of individual differences."
>
> "Of course I do want to be different from others—I think. But the implication is that I dare not change lest I slip into someone else's shoes. But suppose I used constructs that opened up for me channels of movement. Now what? Does this mean that I have relinquished my identity—my fixed identity—in order to live and be different from myself? I think so. But now, what is my 'self'? Is it an object fixed in space, or is it not the system of pathways I have opened up to movement? If it is the latter it is nearer to being a concept, or system of concepts, than it is to being an object to be perceived. Perhaps the self-concept is not a concept about the self but rather the set of concepts perpetrated by the self. How's that for confusing the issue?"

Seligman (1975), working in controlled quarters with animals, coined the term "learned helplessness" to describe a psychological state related to this discussion. The term "learned helplessness" refers to a condition which is observed most frequently in animals which have been exposed to a situation where they could not avoid noxious stimulation. When placed in a situation where escape is an alternative, they will generally not learn to escape. Seligman's experiment utilised a shuttle box of two compartments; one side of the box had an electrified grid and was separated from the other side (which was free of shock) by a barrier. Dogs placed on the grid side were free to jump the barrier to the shock-free side. Dogs which were first subjected to unavoidable shock and then placed into the shock side of the shuttle box whined, urinated, yelped, and finally accepted the shock condition passively. If the dogs so treated did manage to inadvertently pass the barrier to the non-shock side of the box, they did not appear to benefit or learn from the experience, so that on successive trials they accepted shock without attempting escape. Two-thirds of the dogs treated to unavoidable shock failed to learn a response to avoid shock. In comparison, naive dogs (dogs not subjected to the unavoidable shock conditions) learned very quickly to escape the shock. Ninety-five per cent of the naive dogs avoided shock by escaping. Further, once they successfully avoided shock, they could not be reduced to a state of helplessness by being trained to unavoidable shock.

Seligman concluded that the dogs of his experiments had become helpless by learning at some time that they had no control over the trauma they had experienced; thus, they had learned "helplessness". This condition, according to Seligman, is analogous to depression in humans:

> "The depressed patient believes or has learned that he cannot control those elements of his life that relieve suffering, bring gratification, or provide nurture—in short, he believes that he is helpless." (p. 93)

Miller and Goleman (1970) view the potential suicide from a pointed view involving the interaction of three psycho-social concepts called commitment, communication, and crisis:

> "Commitment is a process of role-embracement in which the self is identified as a special kind of person, a person epitomised by a hero image and concomitant personal ideology and value system. Over the course of time, an individual may become committed to such a self-concept to the exclusion of other self roles." (p. 72)

It would seem reasonable to assume that, if a person relies upon an exclusive core construct system, he is vulnerable to recurring crises. Miller and Goleman regard this state of crises as stemming from an "over-commitment" to one's role. "Communication" within the Miller and Goleman model is a process by which the person receives evidence which he utilises to change, validate, or make innovations in his self-image. Finally, "crisis" is an impeding need for change in one's self-definition which cannot be easily incorporated without damage to one's self-esteem.

The Miller and Goleman model (1970) is strikingly consistent with Kelly's interpretations of psychological role enactments, validation, and threat respectively. They conclude that in order to avert further suicidal actions "... it would seem necessary to somehow change this 'dead end' state—to change the self-definition to one offering more potential and different alternatives for action" (p. 76). This conclusion is essentially the same as the conclusion reached regarding change in personal construct theory, i.e., to change behaviour one must begin by changing personal constructions, to accommodate the current situation.

Beck (1967), writing on depression and suicide, arrives at a similar position regarding the importance of the individual's conceptualisations, especially those which refer to the self. Beck believes that underlying all pathology are disorders of thought or "impaired" reason. The individual misconstrues events in such a way as to produce a negative view of the world, himself, and his future. Beck's writing on suicide concludes with a proposition that the self-evaluation of oneself as "hopeless" is central to suicidal behaviour. For Beck, "hopelessness" stems from impaired reasoning which plays a crucial role in most cases of depression and, consequently, in suicidal behaviour. The main thrust of Beck's argument is that the suicidal behaviour of the depressed patient stems from specific cognitive distortions which are characterised as pessimistic. The person systematically misconstrues his experience in a negative way and anticipates a negative outcome to any attempts to attain his major objectives or goals. Those essential conceptualisations which serve as a base for the individual's personal view are referred to by Beck as schemas. Beck's (1967) definition and stated function of "schema" parallels Kelly's (1955) notion of a construct. A schema is:

> "A structure for screening, coding, and evaluating the stimuli that impinge on the organism. It is the mode by which the environment is broken down and organised into its many psychologically relevant facets. On the basis of the matrix of schemas, the individual is able to orient himself in relation to time and space and to categorise and interpret his experiences in a meaningful way." (p. 283)

Kelly and Beck both emphasise the individual's construction of his environment and himself within it. Both note an implicit ability of the individual to change his view of the world, and both see a person as responsible for changing it. However, it should be noted that they do differ on how schemata or constructs originate.

Bakan's (1967) view of the suicidal individual's construct system is quite different:

> "The act of suicide is an act which must be conceived and planned and carried out on the basis of one's own initiative, without consultation with anyone else, and without anyone else's cooperation or moral support, and sometimes over the obstacles that others might place in one's path. It is an act which involves an independence of enterprise and an alienation from others extremely rare in our otherwise cooperative society. It is highly private, self-determined." (p. 120)

As Bakan's description implies, suicide is the act of a *dreamer*—someone who is so egocentric as to construe himself as the hub of the universe. Such a person is operating from a construct system which gives a pattern and meaning to his world that he arbitrarily accepts as ineffably true without questioning or testing out its assumptions for a goodness of fit with the real world. In this instance, the self is based upon a core construct system which marks the individual as independent of sources of validation other than his own. In short, he sees himself as distinctly different, as a self-contained closed system. Thus he is elaborating a self-myth without awareness that other constructions of himself are possible.

What has been described so far is not to be mistaken for the style of thinking that characterises only suicidal or neurotic individuals. Rather, it is the style of thinking of so many people that it can be considered characteristic of western culture. Defeatist, fatalistic statements can be found in all quarters and age groups. Defeatist views are exemplified in such folk statements as: "You cannot change human nature"; "As the twig is bent, so it will grow"; while fatalistic attitudes are implied in the following: "When my ship comes in"; "My luck ran out"; or, "It's God's will". Conventional beliefs such as these carry with them attendant negative attitudes regarding the potential of the individual to become other than he thinks he is, or to change. Further, inasmuch as attitudes like these are familiar and customary, they are treated as "right" sounding phrases; consequently, such attitudes are charged with the neutral valence of acceptance, and are to an extent the trademark of the conformist. Individuals who fit this situation refuse to challenge their abilities, identities, or to explore uncharted areas of personal development. As children they are usually considered to be polite; to know their place; and to be good students. As adults they are good citizens—indistinguishable from thousands of others, they are considered non-curious and not inquisitive intellectually. Parents perpetuate these cultural attitudes and beliefs by serving as models for their children and by demanding that their children behave and believe as they do. Thus, the notion that change is not possible and that new and different ideas are threatening or dangerous is communicated.

In the realm of education, it is a common belief among educators that students are successful or unsuccessful academically because of their particular genetic endowment. Educators who operate out of this bias will not seek alternative methods of instruction which might have the effect of uncovering hidden talents, because the reasons for success or failure are seen to lie in the students and not themselves. Thus, their students' original assumptions are never disconfirmed or challenged.

In other areas involving people and their growth, as in mental health, professionals spend much of their time attempting to label behaviour correctly, rather than attempting to look for ways to foster change in behaviour. Agnew and Bannister (1973) implied the unfruitfulness of such endeavours by showing that diagnostic language is not really a technical specialised language, since it offers no more structure on reliability than lay language when used to describe human behaviour. Further, Bannister *et al.* (1964), pointed out that even after a differential diagnosis was determined, the treatment for different manifestations of psychopathology was most often the same, viz, chemotherapy.

Mental Health Agencies are for the most part represented in the United States by bureaucratic mental hygiene departments that believe in the *status quo*. The attitude that has to be changed here is not necessarily that of the client, but the attitudes of those who work with clients; namely, the professional mental health worker. Instead of fostering alternative ways of living, State systems usually direct their efforts to maintaining the existing structure. Any new behaviour in which a client engages is interpreted pathologically as reported by Rosenhan (1973). Once interpreted as such, tranquillisation is usually viewed as a panacea to knock out behaviours judged negatively by someone. Even more incredible, tranquillisers are often the treatment thought able to produce new behaviours which are viewed as appropriate or desirable even when the target behaviour was never or has not been in the repertoire of the person for several years.

Bannister summed up the state of affairs clearly:

> "Psychiatry has had more than enough time to produce a cure as well as has psychology, but both have failed. The ability to effect change lies with the patients themselves. All we can do is to create the proper environment where that change can come about." (Personal communication, 1975)

Because the condition of being resigned to one's fate appears to be socially communicated, I am inclined to believe that Bannister's appeal for the creation of environments which foster exploration is basic to positive therapeutic effectiveness. Fairweather's (1964, 1974) work with chronic mental patients exemplifies the approach of attempting to create an environment wherein an attitude toward experimentation on the part of patients is encouraged. Although his work has been considered by some as a method of behavioural control, this characterisation misses the point of what Fairweather is attempting. His programme provides patients with opportunities to develop positive first order constructs; for example, to see themselves as actively responsible, capable of negotiating with other patients. Fairweather's "hospital–community lodge" programme provides for the development of a more appropriate personal construct system within an institution and for continuity as the patient enters the community.

Similar negative attitudes and self-evaluations are held by students. Further, these negative attitudes and self-evaluations lead to the development of core constructs which, when stated, describe the individual as failing, struggling, and as unsuccessful academically. For example, they take the following form, "I don't have a head for mathematics", "I have no talent for art", "I'm dumb", or the negative evaluations are the result of comparisons such as: "He does well in school, because he has a high I.Q. My I.Q. is low, so I must do worse than he". Rosenthal's (1966) Oak School Experiment was an attempt to directly influence student behaviour by changing the expectancies teachers had toward their students. The effect of such a change in expectation was demonstrated by measurable differences in standardised test scores. How a change in teacher expectation caused students' behaviour to change is unclear. One interpretation might be students' concepts of themselves changed because of the change in the teachers' expectations for them. However, Rosenthal did not attempt to assess any change that might have been produced in the students' core constructions, so this hypothesis was not tested. In a similar vein, Hudson (1970) demonstrated that a temporary change in "roles enacted" would result in changes on standardised tests. He conducted his experiments by asking students to take standardised tests as if they were "scientists" and to re-take them as if they were "Bohemians". He reported significant differences between the test results. However, he, too, was not interested in looking for core structure change.

In contrast to these studies, Stefan (1973), designed a study to tap core structure change. He used a rank-ordered grid technique employing culturally held attributes of an idealised good student as constructs. The grid constructs were statements like: Does well in Math; likes Art, Science, and Politics; gets good grades, etc. Statements which are usually thought of as characteristic of poorer students were also used as constructs. In addition, the following two self-statements "like me" and "like I would like to be" were employed.

High school students were asked to rank ten student peers on the above constructs. Two indices were derived from the grids: centrality, reflecting the extent to which the individual's "ideal student" agreed with cultural definition; and discrepancy, an indicator of where the individual positioned himself subjectively in relation to his student ideal.

Two groups of students were defined. The first was characterised as having a definition of an ideal student that agreed with that of their culture, and perceived themselves as wanting to become like that ideal. Experimental Group-I came from this population. The second group were those whose concept of an ideal student was discrepant from the definition of their culture and who none-the-less wanted to become more like their discrepant ideal. Subjects for the Experimental Group-II were drawn from this population.

The experimental procedure involved having both experimental groups enact a role of a good student for five weeks. The role was defined as an aggregate of those behaviours which the subjects believed were characteristics of good students. At the end of the five-week period, the grid was repeated and pre- and post-grid scores were compared. It was hypothesised that as a result of role enactment, the centrality score would increase while the discrepancy index would decrease. In other words, the results should indicate that those students possessing a cultural equivalent definition of an ideal student should, as a result of the role enactment, develop a construction of an ideal student more like the culture's and see themselves as behaving more in keeping with that ideal; while on the other hand, those students who possessed a culturally discrepant construction of an ideal student should see themselves as further away from their ideal as a result of role-playing. Further, their definition of the ideal student in comparison to the experimental Group-I should be less like the cultural definition . The power of a role enactment to influence such changes rested on the assumption that good student behaviour is reinforced by the environment.

The statistical analysis indicated no significant differences between groups. However, an interesting, if surprising, trend in the opposite direction developed. Subjects who had a discrepant image; that is, they did not have a cultural equivalent definition of the ideal student, saw themselves functioning closer to their conception of an ideal student while those with a culturally equivalent definition became less like their ideal. These were unanticipated results. The students who took part in the role enactment were interviewed at the end of the experimental procedure. The interview was conducted specifically to assess each subject's personal reaction to his experiences while playing the role. These interviews provided additional information with which to interpret the findings of the experiment. The theme most often heard was negative. Especially negative views were held by those who had the culturally equivalent definition of the ideal student already in their repertoire, and who then played the role. In essence, they felt that role enactment created a situation of dissonance for them. What they reported was as follows: "I played the part, but my friend asked what's going on with me"; or "I did what you told me to do, but stopped around the third week because my buddies were not hanging around with me". One girl reported she felt funny doing things that were not like her, and another boy claimed his buddy thought he was "nuts" by asking questions in class. Others were less articulate and claimed they did not want to change, or "Who wants to be a good student in the first place?" Although the experimenter placed pressure on these individuals to change in the direction of their stated ideal, there was also pressure in the students' environment for them to remain unchanged. It became clear that other environmental pressures were acting to prevent such a change.

It does seem that the greater magnitudes in discrepancy indices (becoming less like one's ideal) observed on post-testing may have resulted from some perceived change in social allegiance and the accompanying subjective threat which was reported in the interviews. It suggests a hypothesis that to express an ideal or future goal does not necessarily imply that one is attempting to reach that ideal, or that one even really wants to reach it. An ideal may be regarded simply as a possible alternative, or as an expression of someone else's standards, or as a reference point from which one operates and defines himself at a given point in time. Any attempt to move a person toward his expressed ideal should take into account the fact that one's stated ideal may not be a powerful influence or, for that matter, even a desired objective. An attempt to make changes in the direction of a stated ideal may, ironically enough, lead to rejection of the stated ideal. For as the stated ideal becomes closer it becomes more defined, and thus it may lose its positive valence by virtue of its more precise definition. From the experimenter–therapist perspective, this may look like failure, while from the subject–person perspective, valuable information is gained.

What may have been gained is knowledge that the goal was not really wanted. How this happens is not clear, but one possibility suggests itself: in order to enact the role of the stated ideal, it is necessary to define the ideal in discrete behavioural terms, which by themselves carry no special merit. Thus, while "good student" is an ideal with a certain initial positive valence, "asking questions" is a behaviour that may not carry any positive valence, and since adding up a lot of zero valences results in zero, one may conclude that the original goal, in this case an ideal student, was not worth pursuing after all.

In this instance, the individual merely reconstrues the goal and his relation to it. On the other hand, the goal may not have been there to be reached; instead, it may have served as a symbol to express certain aspects about one's core structure to others, as well as oneself. In this case, the knowledge gained that the goal was not there for attainment requires one to restructure his core constructs.

What I have attempted to develop in this chapter is the proposition that men do create themselves by construing themselves and investing themselves with a degree of permanence and uniqueness. It is also true that the individual is influenced by his environment and culture to think of himself as being basically constant. It is also convenient, if not desirable, to view one's self essentially as a basic core identity, which can be elaborated, rather than to hold a view that one is continually in the process of structuring a new and updated set of core constructs. An example is the habit of kissing one's mate upon returning home. Suppose, one day one arrives home troubled, distraught, and headachy—not really in the mood for interpersonal niceties. To miss the arrival embrace is to bring down suspicions, questions, concerns—all much more time-consuming than the kiss. It would seem much more expedient for one's frame of mind to keep the habit, kiss one's mate, and get on with the business of taking care of one's concerns. The point is that one's self-system is an established tool which is used with some degree of efficiency. To give it up—even parts of it—is somewhat costly. On the other hand, the cost of thinking of one's self as if one were not responsible for one's life style and situation is to give one's freedom up to others, to events, to the environment, to the mercy of fate, as it were. Colin Wilson (1956), in his postscript, to *The Outsider* sums up this type of thinking. He states:

> "I was born in 1931 into a working-class family in Leicester; my father was a boot-and-shoe operative who earned £3 a week. This meant that education was hard to come by. I realise this sounds absurd at this point in the twentieth century. But what has to be understood is that English working-class families—particularly factory workers—live in a curious state of apathy that would make Oblomov seem a demon of industry. My own family, for example, simply never bother to call a doctor when they feel ill; they just never get around to it. One family doctor—an old Irishman, now dead and probably in Hell—killed about six of my family with sheer bumbling incompetence, and yet it never struck anyone to go to another doctor.
>
> "This explains why, although I was fairly clever at school and passed exams easily enough, I never went to a university. No one thought of suggesting it." (p. 289)

He continued to describe the situation as vague, brainless, cowlike and drifting.

This attitude of formless drifting which occurs when constructing the self as hopeless is what Beck sees as the core of depression. It is also what Seligman would term "learned helplessness". As illustrated by the case of a "poor student", to change, to adapt to crises, implies giving up dearly cherished myths of permanence and constancy.

The individuals characterised by Wilson seem to typify a core structure system which incorporates a basic mythical assumption that they are not influential in controlling their fate. Associated with this type of assumption is a fixed and static self-reflexive view. One is born, grows up, and dies. The channels of activity open are not seen within the realm of personal influence. If such an individual's core structure is threatened, disorganised, suicide may be considered as an alternative to prevent further anxiety from being experienced. It is this view of suicide upon which Landfield (1971) has focused. He sees suicide "as an assertion of meaningfulness" and also as an act which . . . "may pre-empt man's journey into total chaos by preserving ways of interpreting, valuing, and living". Beck (1967), has a similar view of suicide in that suicide seems to be a stress related act. That is, the person attempting suicide is anxious and thought-disordered, attempting to express via his actions the message that all is not well. Consequently, he continues to operate in ways to extort from his environment confirmational evidence for a personal construct system that has been disconfirmed. In this sense, suicide becomes what personal construct theorists term a crisis of validation.

On the other hand, suppose we elaborated man as the creator of himself to its fullest extension. What would the consequences be? To some degree, others have also been in this area before us. To see man as a self-creator is to posit "a being in control" nature to man; which is equivalent to seeing him as having an independence with no bounds. In a sense, it is to see him as Plato did—the determiner of truth and ideals, or like the romantic poets as the possessor of an untainted God-like spirit which was corrupted by the social environment. The key features are: just as man can come to believe in the myth of himself as constant with sustained consequences; so, too, can he come to believe in himself as having such independence and creative capacity as to be in constant ultimate control of his environment, himself, and his destiny. Seen in this way, the act of suicide may serve to prove to the individual that he has control over his core structure.

If a crisis occurs for the person who operates from the mythical viewpoint that he controls his environment, then suicide is an attempt to gain control over his environment in a manipulative way; that is, to extort evidence for a disconfirmed system. Hence, the types of systems described thus far which may lead to suicidal acts are viewed as pathologic. The former involves thought disorder, while the latter seems to be sociopathic in the sense that it is designed to bring about environmental change—and not a change in one's personal construct system. Further, suicide seems to represent, in this type, a validational crisis also.

Yet, there is a need to describe a third type of core structure which may lead one to consider suicide as an alternative. One's personal construct system may be quite functional and effective. It may be permeable enough to accommodate a personal crisis. Yet, as with all systems, a notion of the system's limited range of convenience in regard to time may not have been built in. In terms of personal construct theory, Kelly's (1955) "constructive alternativism" has not been incorporated into the personal construct theory of the individual. Here, situational factors become important. The individual may be working toward a particular goal. As he gets closer, the goal becomes re-defined and its value is changed as occurred with the high school students in the Stefan (1971) experiment. Working towards retirement is another example of a system with "faulty constructive alternativism". In this case, suicide is viewed as a crisis of transition, flowing from the fact that the individual does not foresee that his system needs to be viewed as effective for only a limited span of time, after which it needs to be replaced.

Finally, when there are no alternatives preferable to suicide for the active, controlling person, he may—after carefully reviewing alternatives, including suicide—be *more* likely to commit suicide. This type of suicide is similar to Bakan's (1967) conception of suicide as an act of independence, enterprise, and alienation. Suicide is not necessarily the pathological effect of cognitive disorganisation; it may be required by the application of one's construct system to a particular situation.

In a very real sense, the core structure myth which each individual operates and maintains has a limited life span. Over the course of time it has been elaborated in order to deal with the events of the world. We develop objectives and ideals. By experimenting on the world, we pursue some ideals and reject others as they acquire definition and meaning. We develop a sense of self from applying our strategies to predict and control events. Crises occur; loved ones die, we fail repeatedly, we decline in ability. Individuals respond by suicide or by changing their core construct systems.

Personal construct theory teaches a spirit of leaving behind the old and creating the new. A client once said:

> "I would like to go through my past as if I were travelling from town to town, and place to place; I would stop at all the places that made up my life. I would keep the memories of those that were good and useful, and forget those that were not".

References and Related Reading

Agnew, J. and Bannister, D. (1973). Psychiatric diagnosis as a pseudo-specialist language, *British Journal of Medical Psychology*, **46,** 69.

Bakan, David (1967). "Suicide and the Method of Introspection", Jossey-Bass, Inc., San Francisco.

Bannister, D. and Fransella, F. (1971). "Inquiring Man", Penguin Books, Harmondsworth.

Bannister, D. and Mair, J. M. M. (1968). "The Evaluation of Personal Constructs", Academic Press, London and New York.

Bannister, D., Salmon, P. and Leiberman, D. M. (1964). Diagnosis—treatment relationships in psychiatry, *British Journal of Psychiatry*, **110**, 726.

Beck, Aaron T. (1967). "Depression, Causes and Treatment", University of Pennsylvania Press, Philadelphia.

Becker, Joseph (1974). "Depression: Theory and Research", V. H. Winston and Sons, Washington, D.C.

Fairweather, George W. (1964). "Social Psychology in Treating Mental Illness", John Wiley and Sons, Inc., New York.

Fairweather, George W., Sanders, D. H. and Toratsky, Louis G. (1974). "Creating Change in Mental Health Organisations", Pergamon Press, Inc., New York.

Farberow, Norman (1950). Personality patterns of suicidal mental hospital patients, *Genetic Psychology Monographs*, Vol. 42.

Farberow and Schneidman, eds. (1961). "The Cry For Help", McGraw Hill Company, New York.

Hardin, Garett (1961). Three classes of truth, **ETC.**, XVIII.

Hudson, Liam (1970). "Frames of Mind", Penguin Books, Ltd., Middlesex, England.

Kelly, G. A. (1955). "The Psychology of Personal Constructs", Vols. I and II, W. W. Norton and Company, New York.

Kelly, G. A. (1961). Suicide: The personal construct point of view, *In* The Cry For Help, (Farberow, I. N. L. and Schneidman, E. S., eds.), McGraw Hill Company, New York.

Landfield, A. W. (1971). "Personal Construct Systems in Psychotherapy", Rand McNally and Company, Chicago.

Landfield, A. W. (1976). A personal construct approach to suicidal behaviour, *In* "Explorations of Interpersonal Space", (Slater Patrick, ed.), Wiley, London and New York.

Levenson, Md. (1974). Cognitive correlates of suicidal risk, *In* "Psychological Assessment of Suicidal Risk, pp. 151–163, Charles C. Thomas, Pub., Springfield.

Miller, D. and Goleman, D. (1970). Predicting post-release risk among hospitalized suicide attempters, *Omega* **1,** 71–84.

Neuringer, C. (1961). Dichotomous evaluations in suicidal individuals, *Journal of Consulting Psychology* **25,** 445–449.

Neuringer, C. (1964). Rigid thinking in suicidal individuals, *Journal of Consulting Psychology* **28,** 54–58.

Neuringer, C. (1967). The cognitive organisation of meaning in suicidal individuals, *Journal of General Psychology* **76,** 91–100.

Neuringer, C. (1974). "Psychological Assessment of Suicidal Risk", Charles C. Thomas, Pub., Springfield.

Rodgers, C. R. (1956). Intellectualized psychotherapy: reviews of G. Kelly's psychology of personal constructs, *Contemporary Psychology*, **1,** 335–358.

Rosenhan, D. L. (1973). On being sane in insane places, *Science* **179,** 250–257.

Rosenthal, Robert (1966). "Experimental Effects in Behavioral Research", New York: Appleton-Century-Croft.

Sapir, E., Whorf, B. L. and Korzybski, T. (1960). In Donna Woral Brown: "Does Language Structure Ifluuence Thought?" **ETC.,** XVII, p. 339.

Schneidman, E. S. (1957). The logic of suicide, *In* "Clues to Suicide", (Schneidman, E. S. and Farberow, N. L., eds.), McGraw-Hill, New York.

Seligman, M. E. (1975). "Helplessness, W. H. Freeman and Company, San Francisco.

Skinner, B. F. (1971). "Beyond Freedom and Dignity", Alfred Knopf, New York, 1971.

Stefan, Charles (1973). The effect of a role enactment on high school students' performance and self-image, *Unpublished Doctoral Dissertation*, Ohio University.

Wilson, Colin (1956). "The Outsider", Dell Publishing Company, Inc., New York.

Participating in Personal Construct Theory

Peter Stringer

One of the most attractive, but also most confounding, features of Kelly's work was the variety of levels at which he presented his ideas. People have been variously attracted to Kelly for the advantages that the repertory grid technique appeared to offer over scaling instruments like the Semantic Differential, for the heady excitement of constructive alternativism as a fledgling philosophy of science for psychology, and, somewhat more rarely, for the explicit elaboration and coherence of the theory of personal constructs itself. Confusion often arises from our uncertainty of the level at which we are exploring Kelly's ideas. Rep grid studies commonly have little congruence with the theoretical or metatheoretical framework. We may wish to adhere to constructive alternativism, but find ourselves willy-nilly accumulating fragmentary invalidation of psychology's latest theoretical whipping boy. Within psychology Kelly's ideas present a rare intellectual challenge for those who appreciate the value of operating simultaneously and congruently at all three levels.

In this essay I shall offer no prescriptions as to how the challenge might be met. Underlying what follows is my own sense of having failed to respond to it. But I hope readers will be able to project out of my confusions what some of the implications might be of trying to meet the challenge. There are, I believe, important political and moral implications as well as technical and theoretical ones. I shall describe some of the uses I have made of what I took Kelly to be saying, in the fields of planning and architectural design. The context is accidental. I happened to be working in a school of environmental studies when I became interested in Kelly's ideas. But it turned out for me to be a stimulating context and the notions presented in this essay have crystallised with me over the past six or seven years.

Environmental Perception and Evaluation

Westow Hill "Triangle" in 1970 was a decaying Victorian shopping centre in South London adjacent to the site of the old Crystal Palace. When the

Palace burned down in 1936 much of the Triangle's reason for being there evaporated. Its post-war dilapidation and economic decline can be attributed largely to that event. The building stock and shopping provision was deteriorating; and the centre's main street suffered considerable congestion by through traffic. Against this background there was a general feeling among residents in the areas served by the centre that some form of redevelopment should be attempted as soon as possible. There were two very energetic local amenity groups, one of which persuaded two of the responsible local planning authorities to prepare some redevelopment proposals for inclusion in a public exhibition.

It was the proposals which Croydon and Lambeth planners came up with that formed the elements in a fairly large rep grid study which I carried out the following year. The study was intended to serve as my initiation into the newly developing field of architectural, or environmental, psychology! Although I was at that time wilfully and greedily hooked by the mighty trident of constructive alternativism, personal construct theory, and the repertory grid technique, there were several more particular reasons why I chose that orientation.

Even if the field has become accepted as environmental *psychology* much of the most interesting work done during its first decade or so, as in so many other fields of psychology, has been done by non-psychologists. In 1970 the issues of environmental perception and evaluation, for example, had been established primarily by geographers and planners. My own study was designed to examine people's perception and evaluation of alternative environments through the medium of planning proposals. In deciding on the methodology to use, I was interested to see on what basis these other disciplines had proceeded.

Broadly there seemed to be at least three types of reason why environmentalists were inviting lay people to evaluate parts of their physical environment. Firstly, they might be looking for a justification of action already taken by environmental decision-makers, or testing whether their goals had actually been achieved. Sometimes they had a proleptic purpose: discovering what people value now, on the assumption that repeating it and emphasising it would lead to satisfaction in the future. And thirdly they might be attempting more generally to find out what kind of value systems operate in relation to the environment and in what ways they correspond to other and higher-order value systems.

The "justificatory" approach I found highly unsatisfactory. Sometimes it involved observing people's behaviour and estimating the extent to which it matched the planners' expectations. But this is not equivalent to having people evaluate a planned environment. It is misleading to infer values from the observed behaviour. No consonance between the two realms can be guaranteed. For people to be held to be evaluating the environment, there must be a direct and explicit ascription of positive or negative value on their part.

Congruence or lack of congruence between planners' and lay persons' value systems may equally cause problems for the justificatory approach. On the one hand, if there is congruence the planners have no need to look for lay evaluations, except perhaps to establish the congruence—and that comes under the third of the approaches outlined above. If it is lacking, and if the planners' decisions have been made in terms of their own rather specialised values, lay people will be simply unable to make an appropriate evaluation. Their evaluation could not justify or refute decisions taken on other grounds.

The second or "proleptic" approach which I detected in the literature, depended on one of two assumptions: either that people's values once ascertained should be maximally actualised; or that one has rules to enable one to decide when to ignore lay values, or when to aggregate values with other considerations if values cannot be or are not allowed to be superordinate. One has either to by-pass or make assumptions about higher-order values. However, an individual's or group's statement of values, their attachment of value to elements, or their ordering of elements by relative preference, cannot be treated as a "vote". Their statement does not implicitly confer authority on another party to use or interpret it as a legitimation of any future action which may affect them in any way. Similarly, any rules that limit the extent to which lay values are heeded are necessarily independent of the simple expression of those values.

The third approach, which aimed at a general understanding of people's values, at first sight looked too academic and abstract to be useful in solving the practical problems of environmental policy- and decision-makers, which was the context within which I was to do my study. But it did offer an acceptable method of obtaining some of the information which might clarify their problems, while avoiding the assumption of the two previous approaches. It assumed nothing about the identity or otherwise of the values of different groups, nor about what one might do when one has attained some understanding of values. These points will, of course, have to be decided at some time if problems are to be solved. Ideally one should decide on them in advance of seeking the evaluations. But the important point is that they can be treated as external matters which carry no assumptions with them.

I would criticise the justificatory and proleptic approaches, not only for the assumptions they make, but also for the methods they have commonly adopted—which, of course, reflect those assumptions. The justificatory approach commonly uses a technique in which subjects are asked to rate a number of elements, that is parts of the physical environment, on a number of descriptive and/or evaluative scales. The elements will include one or more in which the investigator is particularly interested, which he probably hopes will receive more favourable ratings than the others. I call the scales "descriptive and/or evaluative" because there is no way that one can guarantee in advance that they will be used evaluatively as intended. It is a difficulty of this approach that one cannot know how people will interpret the scales. It is even a matter of conjecture that one is offering them scales that are at all personally meaningful. One might use an instrument like the Semantic Differential which has some claim to universality. But it would still be necessary to show that the structure of environmental meaning was similar for lay people and experts. The experts would have to accept the Semantic Differential as an adequate expression of their environmental semantic structure. These procedures are unlikely.

The proleptic approach of asking people to compare a number of environments, or aspects of the environment, generally proceeds by way of preference rankings. But a preference ranking of an appreciable number of elements is an unlikely operation. Typically one may express preference for one or two elements over the remainder of a set. Over a long period, one might come to express simple preference orders within many of the subsets of a larger set. Although these would be transformable to a rank order, they do not justify the straightforward production of a rank order within the total set as a meaningful psychological operation.

The third approach which I distinguished above avoids the problems of the other two. In looking for a general understanding of environmental value systems, there is no need to confine oneself to responses which are assumed to be purely evaluatory. It is, in fact, more reasonable to integrate evaluatory with perceptual studies in the light of the consistent finding from investigations of person or object perception that one of the major components, and often the largest component, in subjects' judgments is evaluatory in tone. This seems to occur even when care has been taken to avoid suggesting to subjects that their judgments might (or might not) be evaluatory. Such highly specific response modes, therefore, as preference rankings are misleadingly specific.

One's success in arriving at a *general* understanding of people's environmental value systems will depend on not pre-empting the focus of inquiry too soon. It is necessary to give play for individual differences to emerge. A representative and comparable set of environmental elements which are salient for individuals will enable them to give a clearer picture than one element only or a disparate set. If a range of responses is elicited, rather than the simple preference orderings which have often been used, their meaning can be more richly defined by their observed relation to one another. Fewer assumptions about their meaning need be made by the investigator.

This, then, was one set of considerations which guided me to approach environmental perception and evaluation through the rep grid technique and in the context of the proposed redevelopment of the Triangle. Though I would not wish to deny that my pre-existing attraction to personal construct theory may have led me to formulate the considerations in the way I did. The rep grid preserves individual construct systems. It integrates perception and evaluation. Meaning is, in part at least, defined explicitly within the grid structure. Alternatives are construed, and these were supplied by the alternative redevelopment proposals for the Triangle whose salience for local people was already established. The planning context gave a future orientation to people's construing which is highly conformable to the spirit of personal construct theory. For these mainly technical reasons the rep grid looked like the right instrument to use.

It is not the purpose of this essay to give the substantive results of the Triangle study. They have been discussed fully elsewhere. I will go on to discuss a second set of reasons for taking a personal construct theory orientation.

Public Participation in Planning

The initiative of the amenity groups I referred to earlier, who persuaded the local planning authorities to prepare a range of alternative redevelopment proposals and to exhibit them to the public, is an indication of the wider socio-political context for the study than the somewhat technical considerations above. It is an instance of the public participating in planning matters.

Public participation has become something of a political rallying cry in recent years. It is in the field of planning, rather than education or the social services, for example, that most examples of it can be found. It has been enshrined in the 1971 Town and Country Planning Act as a statutory obligation that planning authorities should ensure that "adequate publicity" be given to their proposals, and that an opportunity to make representations on them should be given to those people who might be expected to wish to do so. The Skeffington Committee reported to the government on ways in which the public might be enabled to play a part at the various stages of plan development. One of their specific suggestions was that alternative proposals should be prepared and that the public might make comparative evaluations of them or state their relative preferences for them.

Participation, then, formed the socio-political context for the Triangle study. In addition to examining the public's reaction to the planning proposals, I decided to try to answer two simple and specific questions about what might constitute "adequate publicity". This was done by showing to sub-groups of my sample the planning proposals drawn up in different map formats. The maps differed both in chromaticity and in the extent of Ordnance Survey base included. It is important to determine whether coloured maps are significantly better communication devices than black-and-white ones because they are so much more expensive to produce. Participation budgets are generally very meagre. When redevelopment proposals are published overlaid on an Ordnance Survey map, it is usually possible for people to identify which individual properties may be affected. To avoid the "blight" which may fall on the affected properties, it has been suggested that the Ordnance Survey base should not be incorporated in that part of the map where redevelopment is proposed or should be omitted altogether. In my study the criterion of the adequacy of the maps for publicity purposes, in whatever format, was taken as their power to enable people maximally to differentiate between alternative plans when construing them.

Examination of the two hundred grids I collected produced a rich and useful set of results.* Coloured maps turned out to be better than black and white ones, and those with a full base or none at all, better than maps with a partial base. The map-format appeared to affect people's evaluation of the proposals and the constructs through which they considered them. But despite all this I felt ultimately dissatisfied with the study.

What intrigues me about environmental psychology is that it simultaneously gives one an opportunity to answer some practical questions, and to develop theoretical and methodological aspects of psychology, often in an inter-disciplinary framework. I find that both these activities can be more firmly placed within a moral and political context in environmental psychology than in, say, educational, clinical or occupational psychology. It may be because it is a newer field, rising at a time when moral and political questions also are coming to refresh psychology; or, because more people are more widely concerned today with environmental politics.

But although the political context was clear to me, I failed in the Triangle study to operate within it in at all a thoroughgoing way. The study certainly did produce some sort of operational answers to the practical questions of what might constitute the "adequate publicity" of planning proposals, and of what kind of representations one might get on them from a cross-section of the public. It had something useful to say about personal construct theory and the rep grid technique. In these respects it served the interests of planners, politicians and psychologists, in their professional and institutionalised roles. But it did precious little for Man as such, or for the people who completed the grids. And yet, given the participatory context of the study, that is what the psychological orientation used should have been about. The psychological framework of any study which acknowledges its socio-political context should reflect that context, be isomorphic with it. Participatory democracy has in

* For a more detailed presentation of this information see my "Repertory Grids in the Study of Environmental Perception" *in* "Explorations of Intrapersonal Space" ed. P. Slater, John Wiley, London, 1976.

this way profound implications for psychology, to which I shall return at the end of this essay.

The irony is that my study was perfectly congruent with a socio-political context. But it was the wrong one. It was one which treated participation in planning as just another manipulative exercise for experts, rather than as a means by which citizens could be enabled to move forward to a more understanding and richer relation with one another and with those who work for them. I adopted a methodology which, by following the habits into which I had been professionally socialised, produced useful results, without realising that the socio-political assumptions which attracted me were unusual to psychology and demanded less usual techniques.

In addition to these misgivings I had two entirely unfulfilled objectives in the study. One was to explore ways of enabling the planners involved to make use of the constructions of their proposals which I elicited from the public. Unfortunately, there were over-riding considerations in the whole framework of planning for the Triangle which would have made it an academic exercise. The other was to revisit my respondents after a period of six to twelve months and invite them to reconstrue the proposals in the light of the planners' reactions to their original responses. If investigations of this kind were to be done in future, it would be important to pay more attention to getting planners and public to construe one another's construct systems. By doing this over time there might be an opportunity for learning to occur and for more constructive alternative proposals to emerge. The public should be invited to construe planning proposals in a state of relevant knowledge rather than in relative ignorance, as in my study and in similar surveys which are now increasingly being carried out by planning authorities. They might be asked to alter the elements which they are invited to construe and to formulate their own proposals. They might be encouraged to go out themselves to conduct similar studies, on a small scale, with fellow citizens. In this way, psychology could be of positive help in moving toward a participatory society.

My discussion so far has revolved around a particular empirical study, even if I have not given much indication of its results in terms of "hard data". For the remainder of this essay I shall become even less empirical, in order to speculate on a number of issues arising from looking at public participation in planning and design from the viewpoint of personal construct theory. I shall touch in passing on the problems of creativity in architecture and finish by considering some aspects of the relation between professionals and lay people and between professionals themselves.

Motives for Participating

One of the ways in which I have found a confrontation between the concepts of participation and personal construct theory illuminating is in looking at some possible reasons why people should have come increasingly to wish to participate in the government of their affairs. There are a number of quite distinct reasons why government itself has chosen to encourage participation, though this is not the place to discuss them. But for participation to occur it is not sufficient simply that statutory obligations should be put on local planning authorities and other bodies to engage the public at some point in the decision-making process. Obviously the public must have some inclination to share actively in the process.

There are possibly two major motives. There is a growing recognition that "doing" is more important than "having". And there has been an ever-increasing rate of change in our surroundings and way of life. The two are related. The economic goal of obsolescence and the social goal of mobility lay emphasis upon using an object or situation for a restricted period of time, the end of which one can easily see or anticipate. Objects become impermanent and function adequately for a predictably short period of one's life. It becomes actually difficult now to continue doing the same things day-by-day or in the same way for more than a few years, even if one tries very hard. Change becomes of paramount interest—and change is process not product, doing or being-done-to rather than having.

Both situations and objects are now pregnant with the possibility of their own succession. Objects in our physical environment have lost one of their main characteristics as objects, their stability. Critical distinctions between objects and living organisms are becoming blurred. Objects are taking on apparent capacities of growth, reproduction and death. The processes of development, imitation and decay become more interesting than the products themselves. On the obverse, complaints are raised that living organisms, and especially people, are treated as objects. Ironically, spare-part surgery is introduced at a time when the repair of objects is becoming outmoded.

The most significant thing about the increased rate of change in the objects, activities and ideas which people experience in their lifetime is not the increase in change itself, so much as the agent of change. Whatever relatively small changes occurred in smoother pre-technological days seem either to have been initiated by or within the individual or to have been suffered in direct confrontation with another person. The rare major changes which occurred were usually initiated by a supreme authority or force, or by acts of God. Today a large number of the small and large changes in one's mode of life and one's surroundings are seen to be effected by people like oneself. But they are people with whom one has no direct contact. The alienation which results is heightened by the realisation that nominally and indirectly one does have responsibility for the authority or operation of these change agents through the franchise or through the market-place, and that even small changes, in ways too complex to follow, may have far-reaching repercussions for oneself.

The economic power that one has at the level of final consumption, and the moral authority that one can exercise in the absence of over-riding social or religious dogma or of ultimate legal sanction—for example the much freer decisions one can make about the possibility of birth, marriage, death—also make it irksome to see an equal power to change being exercised over oneself by others. Both small and large changes in one's life, manipulated from without and with no direct personal confrontation, have become a source of irritation.

Three aspects of this account of why people might want participation are particularly conformable to the spirit of personal construct theory. People have come increasingly to realise their capacity to manipulate their own lives and environment, and to resent the irrelevant manipulations of those whose only authority is one conferred by themselves as electors. In being constantly affected by change they are turning their attention from trying to stabilise the past in the present to predicting and anticipating the future. Their manipulations, resentments and predictions are experienced as individuals. They have a personal view of the world, and it is this which is affected by the designs of planners, architects and other decision-makers. Their views should be taken as personal, if for no other reason than because less and less often can one predict an individual's viewpoint on the basis of sex, age, education, social class, race and so on.

A person's view of the world is organised in terms of a system of constructs. A personal construct system is the making sense of events around and within one, ordering them in relation to each other. The system evolves towards an ever more convenient state for making more useful and interesting predictions of future events. A construct system does, of course, also order past events and it is validated by comparing predicted with actual events as they pass. But because of his clinical and therapeutic work, Kelly seems to have been particularly interested in the evolution of construct systems in response to changing situations and to produce a different perception of some part of one's world. A rigid adherence to the validation of a stable construct system and a determination to view the world in a way which led to unvarying and apparently veridical predictions was uninteresting, unhelpful and maladaptive. Personal construct theory is a fitting conceptual framework for an era of change.

An evolving construct system, responding to an internal or external requirement of change, often proceeds by propositions in the form "what if" or "let me look at it as if". They are a device for asking about the implications of construing an event in a particular way. Such propositions may be shots in the dark or be derived from higher-order propositions in the way in which a scientific hypothesis may be derived from a theory. Viewed in this way a plan or a design can be treated as an indication of an evolving construct system. The hovercraft might be an example of a shot-in-the-dark, "what if" proposition. The Boeing 747 or Concorde more clearly represent hypotheses about future travel patterns derived from a theory, however imperfect, of transportation economics. All imply a change in the way one construes transportation. Any design or plan which is not simply a straight repetition of an existing one is a new way of viewing a part of the world.

Architecture as Fiction

If designs can be treated as hypotheses, designing is a matter of formulating hypotheses. Looked at in this way there are some rather intriguing observations one can make about architectural design and particularly about the way in which young architects come to learn to design. These observations are not entirely irrelevant to the issues of participation.

The core of an architectural student's lengthy education is learning to design. He has to handle indeterminate, value-laden and complex design briefs to arrive at an arresting and feasible solution from among the many possible. His solution can be taken to include his hypotheses about how the requirements in the brief might be satisfied by certain building operations. The solution is presented as a drawn plan or model rather than in built form. It is extremely rare for an architectural student to see any of his designs actualised. But it is in the design studio that his creativity is expected to develop.

Even when he graduates, similar limitations on his work are likely to obtain for several years. He may not be given much creative work to do. If he is, the project will often be on such an extended time scale that it will be years before it is built. Many professional design projects never get built. Or he may move to another office before it is. The fee-structure of the architectural profession discourages him from following up on the building-in-use to see whether it satisfies the criteria originally laid down. The distinctive peculiarity of the practising designer, as a creative person, is his customary inattention to the performance of the finished product of his creativity, his hypotheses in concrete form.

The significance of this characteristic is illuminated by Kelly's notion of the "creativity cycle"—the process by which an individual may modify and extend his view of the world by successively "loosening" and "tightening" his construct system. When loosening his system he allows himself makeshift, fluctuating and invalidated constructs, without being concerned about their possible contradictions. They may be vague and unexpected. Subsequently, as the construct system is tightened, they are more closely formulated. Eventually they are put to the test of confirmation or disconfirmation of the anticipations they suggest. A creative individual is one who has actively chosen from the alternative ways of operating his construct system "to encompass and modulate the extremes of loose and tight thinking" in a cyclical process. This process crucially involves forming and testing hypotheses.

Kelly put forward the "language of hypothesis" as one of Man's psychological instruments. He introduced the "invitational mood" as a possible way of using language to orientate one to the future and open up alternative constructions of events. In his paper on the language of hypothesis, Kelly refers briefly to the German philosopher Vaihinger and his philosophy of "as if". He seems to suggest that Vaihinger's position was "that all matters confronting man might best be regarded in hypothetical ways". In fact, Vaihinger was more interested in a class of mental constructs which he referred to as "fictions". He was at pains to distinguish them from hypotheses.

A fiction is an auxiliary construct or device for aiding discursive thought. It has a practical rather than theoretical goal. Whereas an hypothesis tries to discover, a fiction invents. An hypothesis should be confirmed by verification, while correspondingly a fiction is justified. The principle of the rules of hypothetical method is the probability of the conceptual constructs, that of fictional method is their expediency. The distinction between fiction and hypothesis is clear.

Both fictions and hypotheses are important to creative processes. It is the fiction which appears to operate in the loosening phase of the creativity cycle. Other terms for fiction might be phantasy, imagination, invention or ingenuity. Hypotheses are entertained in the tightening phase when the newly formulated constructions are put to the test.

In so far as the young architect cannot concretely test his hypotheses he has to treat them as fictions. Design solutions in architectural school are inventions rather than discoveries. They have to be justified to a teacher or to peers rather than verified. Expediency rules over probability.

Vaihinger warned explicitly against the error of treating hypotheses as though they were fictions, out of laziness, to avoid the chore of verification. The architect rarely seems able or interested to test his hypotheses. He lets them rest as fictions. He embarks upon a creativity cycle in which there may be no rationale for the successive phases of loosening and tightening constructs. If his hypotheses are neither validated nor invalidated within his construct system, his construing is liable to become random, unduly susceptible to outside influences, or to take the form of dogma.

Where dogma takes over, the architect is treating his fictions as though they are of the essence, rather than as a tool for thought. They take on the form of religious fictions whose criterion is their existential validity. If the architect is creative, it is a religious creativity. He is a myth-maker.

Now I believe it is precisely because we appreciate the architect's frequent lazy or frustrated stance as myth-maker that we resent his power to influence our environment without consulting us. The same applies for many other areas of decision-making operated on our behalf by government bodies. The lay person is much less often a myth-maker when he lays out his garden or decorates his living-room. He looks to see whether they work, in part because they are the fruit of his own actions. If professional decision-makers can not or will not take such an obvious step, we may ask to join in their functions and institutionalise processes that will ensure that there is more feedback and that hypotheses are tested. We ask to participate.

Participation as Having, Doing and Being

To return now to the main thread of the discussion I shall look at design participation in more detail and more directly in relation to personal construct theory. It is useful to distinguish three different senses of the word "participation". It can mean having a part of something in common with others—sharing a cake; or doing something in common with others—playing a game of football; or being a part of something, partaking of the essential nature of something.

Design participation, in the sense of sharing something with others which has been designed, involves the individual in accepting the imposition on his way of looking at the world of part of another person's construct system. The imposition is not necessarily undesirable, provided that it is not unwelcome and does not put undue strain on the individual when he tries to incorporate it into his own system or to adjust his system to accommodate it. The disadvantage is that it is a one-way traffic. At its best it is difficult for the designer to anticipate the implications of his design—the manifestation of part of his construct system—for the possibly quite different and numerous systems of other people.

In the sense of actively taking part in the process of designing, participation involves the individual either in trying to fit his construct system to that of the specialist designer or in imposing his system on the designer and denying the professional's need or right to have a specialised set of constructs. The latter position is possible, but very often will be unhelpful. The former is back-to-front. If the designer has a specialised and sophisticated construct system, the layman cannot possibly incorporate it into his own without first construing the world like a designer. But he is not a designer, in the specialised sense at least. The designer should rather be fitting his system to that of the layman, but the difficulty about that is that it might prove inhibiting. It might prevent the designer from looking at innovations in construing which at present are incompatible with lay systems.

The possibly more fundamental sense of "design participation" would entail being a part of a design or of the process of designing. For people to be a part of the nature of a design, presumably means that they are being designed. This is probably the intention of many designers and planners, who attempt explicitly to alter the actions of others through their designed products. In altering actions they may cause people to reconstrue their worlds. They are tampering with the psychological core. On the other hand, for people to be a part of the nature of design*ing* is quite a different matter. This recognises not that people should do the designing (I assume here that they cannot), but that their construct systems are an integral feature of the design process. I assume that the coining of the term "public participation" in itself suggests a rebuttal of that sense of the expression which amounts to people's lives being the object of planning. Presumably also there is no intention, at least on the part of the authorities, to have the public deny planners their role or usurp their function. One is thus left with the sense in which the public may be an integral part of the essential nature of planning at the level of being.

It is these three forms of participation—having, doing and being—which appear to be available to us today.

I have said earlier that a plan or design constitutes part of a specialist construct system. If it is to be accepted and put to use, there must be congruence between the plan and the user's constructs unless considerable strain is to result. There are various ways in which this can be achieved. The congruence may be formed at the user's despite by physical necessity or superior authority: he may be placed in a position where he must reconstrue events if he is to maintain anything like his preferred way of life. We sometimes call this "adaptation". People may come to construe a tower block dwelling as having all the essential properties of home because they have little chance of doing otherwise, unless they are to suffer hardship and disruptions in other parts of their construct system. Or the congruence may be formed insidiously. The plan may be ascribed properties which are illusory or relatively trivial in order to make it fit the public's view of the world. This is most common in the field of consumer product design.

If neither of these eminently convenient tactics is morally defensible, save perhaps in rare and exceptional circumstances, a third method of achieving congruence is for the planner to apprise himself of the public's various construct systems, and, treating them as given, to find ways of making his system maximally congruent with theirs. This is the objective of some environmental perception and evaluation studies which I discussed at the beginning of this essay.

But even this procedure falls short of what one would hope for in contact between two construct systems. The contact is one-way. There is no means by which the public can adequately inform the planner of their viewpoint, without waiting to be asked. It is rare for the public to ascertain what are the planner's constructs. Either they are not told at all, and are unable to divine them from the plan itself for lack of expertise; or they are told and are unable to understand, the constructs being sophisticated and complex and expressed in unfamiliar language.

A fourth means, then, of achieving congruence between the two viewpoints, requires that there be full two-way communication. Because one party has a set of constructs that are more complex, an expository or educative process is required in which the complexities are made fully intelligible to the public. When aid is given to an undeveloped country, it is usual to ensure that some of the population understand both the function and the long-term purposes and implications of the new financial and technical resources. In developed countries, however, very few people understand measures that are taken on their behalf and are bought from their labour.

A proper education is not a matter of learning by a particular set of conventions, but of trying on a variety of points of view to discover which gives the most convenient, interesting and progressive anticipation of events. If the viewpoint which is the subject-matter of this education is to become related to the individual's personal construct system, he needs to test it in real situations, to become personally involved with the viewpoint and committed to its implications. This cannot happen if it is merely expounded in the abstract, in relation to situations in which the learner plays no role. If the planner wishes to achieve congruence between his terms of reference and the public view, and if the public wish to understand and be understood in planning affairs, a context must be found in which they can be commitedly involved in acting for the future. The lay person is very experienced and sometimes quite good at planning other parts of his life. What is desirable in the context of participation is that he should be able to exercise that talent at some level of the more technical planning of his physical environment. I suspect that this must involve some fundamental changes in the processes employed by professionals.

Architects and Psychologists

When participation is viewed as I have tried to do, within the light of personal construct theory, one of the biggest issues becomes precisely the nature of the relation between professional and lay person, expert and non-expert. Much of the preceding discussion implies that. But participation should also be concerned with relations between people representing different professions, and with relations between non-professionals. This issue may be illuminated a little further by discussing the relation between architect and psychologist in the new field of environmental psychology which I referred to at the beginning of this essay, and their respective relations to lay people.

Whereas I have previously been looking mainly at what I would call technical and political aspects of personal construct theory, I shall now turn to moral issues.

Despite their differing subject matter, both architects and psychologists have an underlying similarity in the possibilities of their relation with non-professionals. In both fields they have frequently aborted the relation—the psychologist by insisting that, if the people he is studying realise what he is about, his investigation will be invalidated; and the architect by producing "that's it" architecture, form without people, buildings which are not to be touched, manipulated, or changed. Both attitudes come from an undue reverence for certainties and fixed states, and deny the involvement and commitment which should be central to the professional role.

Environmental psychology grew up largely in response to demands from architecture, at a time when the profession was becoming more and more anxious to draw for help on any other discipline which seemed relevant. Generally it drew on technologies and technologies tend to put a premium on certainty based on reference to the past, rather than upon reconstruction with a hazardous chance at better predictions. But when it comes to the psychologist's contribution to architecture, the interrelation of man and environment appears to be so multivariate and dynamic a network, that it is difficult to envisage how prediction of and control over relatively small, particular and circumscribed bits of behaviour can lead to the derived end-result.

An architect rarely designs small bits of isolated environment for particular, unchanging users. For most of his design briefs fragmented information is likely to be useless. A good design does not involve just slotting new elements here and there into an existing pattern, but entertaining a dynamic new look at the problem where important change is in the whole not in the parts. If the psychologist only offers the architect little additional bits of certainty, he will be encouraging him to do poor design. But he will be a great help to the architect if he offers him new ways of construing people's psychological processes in the environment. A more holistic approach will better enable the architect to formulate a theory from which he can derive the hypothesis which must inform his design activity. If he does not like the direction in which an hypothesis leads, he can happily abandon it because it should be only one of a number of alternative constructions which have been offered to him. How much more difficult is it for him to relinquish the certainties that the technological approach offers, even if he does not like their design implications.

The chances of architects and psychologists participating fruitfully in a joint endeavour will be greater if they act as constructive alternativists rather than accumulative fragmentalists. Both groups should be more interested in re-constructing the world rather than re-presenting it. The architect can represent in built form the relations he has observed between men and their environment. But it is more impressive to reconstrue the relation in such a way as to refresh man's potential experience of his environment, to help him to reconstrue his relation with it. The psychologist, by his rather different techniques and with a different focus of concern, can achieve a similar goal. Instead of offering built reconstructions, he is concerned with construct systems themselves. His skills lie in giving people insight into and understanding of their own and others' construct systems. The psychologist can be an intermediary between architects and the people who use their buildings.

An unwillingness to impose certainties on a client should come from a feeling of involvement with him. The imposition of certainties is a comfortable ploy for avoiding involvement. It is easier to be an expert than a person. It is easier to substitute so-called "knowedge" for the experience of getting to know a client; probing his construct system, and detecting where it is open to a change that can be facilitated by a particular design.

It is also a way of avoiding commitment. As soon as one takes the step as a professional from anticipating events to actually deflecting them or making changes, one is committed. One has opted to be an event oneself in the same world as one's client. The most serious responsibility of all people, professionals and non-professionals alike, is in making themselves events, producing change, and enabling others to reach a reconstruction of their world.

One of the instruments of involved and committed reconstruction is interference. We attempt to deflect anticipated events by interfering with them in the hope that thereby we may observe fresh relations which will suggest a new view of the world. We are interested in construing the replications of our own interference. We interfere because we believe that we may have missed an event or a relation.

An architect interferes with our environment to test his hypotheses and to suggest new ones to himself and others—at any rate if he can surmount the obstacles to creativity outlined above. The kind of hypothesis an architect is interested in can scarcely be tested without interfering with the environment. It is perhaps fortunate that the architect leaves his experimental instruments lying around. Their presence urges us to construe their implications. In so far as they never present the final solution to a man-in-environment problem they constantly remind us to reconstrue the problem and the relations we observe.

A psychologist's interference is less obtrusive; but only if he *does* interfere. Many of those who feel threatened by psychologists are probably disturbed by their role as uninvolved observers of behaviour. They perhaps dislike the implication that they can be understood as persons without direct approach to their construct systems. On the one hand this is an impertinence, because it is the construct system which constitutes a person as a psychological entity. On the other hand, if one admits that large parts of some construct systems can be inferred from observed behaviour, this may be taken as exploitation.

The psychologist is using the construct systems of others in order to effect reconstruction for himself in his professional role, and perhaps also personally. However, if the psychologist does interfere, to the extent of being committedly involved in the construing of others, it is far less likely that he will be seen to be interfering.

Towards a Participatory Psychology

I started this essay by describing some work that related to the field of planning. I want now to turn the discussion back to planning. Previously, I have been interpreting design and architecture in terms of psychology, and more specifically of Kelly's psychology of personal constructs. In what follows I shall use some ideas from planning theory to interpret psychology.

Three forms of legitimacy have been distinguished in the theory of planning to justify its activities—the rational, consensual and participatory. There has been in recent years a gradual development of interest away from rational, through consensual, to participatory planning. Very simply, rational planning justifies its activities as being the most efficient means to unquestioned ends. The legitimacy of consensual planning is in the endorsement of, and support for, plans by interest groups and major elites. In participatory planning there is a new regard for the customer or user: his major involvement in the planning process is its ultimate justification.

For the planner the advantage of participatory planning is that it contributes adaptivity and stability (in the systems sense; that is, not equilibrium) to the society which is being planned, through the input of new sources of relevant information. Participation facilitates the mutual adjustment of individuals, groups, communities, agencies and institutions which is necessary if planning decisions are to be accepted and effectively implemented. In the individual, participation promotes competence, by offering opportunities for effective transactions with the environment. Participation encourages the individual to be, for example, active, independent, capable of complex behaviour, of durable interests and long-range perspective, and self-aware.

Participatory processes are seen to be improvements on the two previous forms of legitimacy. The validity of the rational approach is criticised because of its assumption of certainty, predictability and of value-free facts. The introduction of values into rational planning transforms it into a consensual process. The viewpoints of different interest groups are identified, that is groups who select their factual description of the world in terms of values relevant to them. Planning then seeks an equitable and accepted distribution of resources between competing interests.

But consensus planning also has its deficiencies. In systems terminology, it favours maintenance, over adaptation, innovation and change. It assumes that one can arrive at a single expression of the public interest. It raises practical, administrative difficulties with regard to the adequacy of representation of interests, and the control of resource distribution. Do elected representatives or expert bureaucracies make the effective decisions?

A solution to these problems, some planning theorists would argue, is to broaden the consensus, so as to include all users of planning in its processes, that is all citizens. The legitimacy of participatory planning is based on policies and plans being not only endorsed and supported by the recipients, but also created by them.

I hope that the parallels that one might draw between planning and psychology are fairly obvious. Psychology is for the most part a rational discipline, in the sense in which "rational" was used above. Although most psychologists would, if challenged, admit to an inevitable lack of certainty in and predictability of human behaviour, their theories, models and techniques rarely appear consistent with that view. The knowledge of human behaviour that results from their work changes, or is intended to change, human behaviour. But the value implications of that consequence do not seem to affect their way of proceeding with their investigations. There is no procedure for decomposing the value-component of an investigation. And yet such procedures should be as commonplace as the elaborate statistical techniques which are used for decomposing quantitative data.

A consensual approach does occasionally creep into psychology, in the selection of topics to be studied. This is usually effected through the directions of funding agencies—research councils and government departments. As officers of the electorate they do exercise some control over the distribution of psychological products between competing interests. But they are a highly specialised bureaucracy with very particular viewpoints. One could scarcely claim that they were representative of the general will, or of all interest groups.

Participatory psychology in any form is virtually unknown outside the confines of certain schools of therapy. But obviously participatory psychology should not only be for a particular class, the disadvantaged, any more than participatory planning. It should aim at everyone, and involve everyone. Some of the more immediate goals of participatory planning are to collect information for planning across as wide a range as possible of those who might be affected by it—information which is more readily available to users than to planners; to invite users to evaluate alternative plans, or even to help to formulate the alternatives; and to encourage effective implementation of the plans by the prior involvement of their contributing agents. Participation in psychology would involve similar goals. Those who might be influenced by psychologists' activities might be asked to tell them what are the important questions to ask, what model of Man they have, how psychological investigations might result in action and change in our world and what actions and changes they might be. Is, for example, the concept of "attitude" of any significance to people? Do they have a mechanistic, psychoanalytic or humanistic image of Man? Does anything follow for their own behaviour from studies such as those on racial prejudice? It might be argued that an input into psychology of information of this kind is unnecessary and would be misleading, on the grounds that psychological concepts, models and techniques have their own elaborated and well-tested justification within the procedures of the discipline. Some planners produce a similar argument, and we tend not to be convinced by them!

I would not suggest that all psychology should be participatory. There might be reasons for preserving a range of approaches, from rational to participatory, and with various mixtures as well. But what might the advantages be if a more participatory element were introduced into the discipline? Apart from promoting its openness and adaptivity, participation might also bring benefits under the three heads I have used previously, the technical, the socio-political and the moral.

The first benefit, at the level of technical implementation, has to do with conducting psychological investigations, doing experiments, testing ideas. It is a continual source of surprise that, while the results of research into the social psychology of the psychological experiment seem readily accepted and sanctified by inclusion in text-books along with all the other "results", it is not appreciated how radical are the implications for traditional methodologies and for all those results. Their validity and legitimacy are in many cases so gravely impaired that a major change in direction seems inevitable. A participatory approach to psychological investigation, with the erstwhile "subject" being admitted to the decisions which have to be made, would be a means of actually exploiting or incorporating such factors as evaluation apprehension which at present are usually merely monitored or cancelled out.

At the socio-political level, participatory psychology would enable the psychologist to become more fully integrated into society in his activities. As a result of his negotiations with those whose behaviour and experience he wished to study, he would come to appreciate that his own interests must often be relinquished for co-operation to occur. Eventually, he would be sufficiently familiar with the interests of others, through purposeful and effective mutual action, as to feel no conflict between the demands of academic and naive psychology. The final integration into society would need to be mediated by integration into a variety of social institutions, of varying scope, along the way.

Finally, there would be benefits at a personal or moral level. Theories of participatory democracy hold that the more one practises it, the better one becomes at it. Accordingly, the psychologist would become educated in those very qualities which make the practice of participatory psychology possible.

Though as he improved, so what he was doing would constantly evolve and change. In this way participatory psychology can take account not only of the changes in others' behaviour which are produced, directly and indirectly, by psychological inquiry, but also of the changes in psychology itself which result. The primary quality which will be developed in participatory psychologists is their ability to relate unselfishly, actively and progressively to others. In a participatory context this would be the goal of all concerned. What would distinguish a psychologist would simply be that he has chosen to concentrate his relations around the study of human behaviour and experience.

The three-fold benefits I have proposed would apply equally to non-psychologists. Their use or implementation of a psychologist's activities would be facilitated by participatory psychology—a principle which has been rooted in the practice of action research. Participatory psychology would inevitably make its techniques and concepts instantly available to non-psychologists. Perhaps the goal of participatory psychology would be to make psychology itself, as an expertise, increasingly redundant.

Some of the technical benefits of a participatory approach were exemplified in my rep grid study of the Westow Hill Triangle redevelopment proposals, even though it fell short of enlisting members of the community as full colleagues in the enterprise. Personal construct theory, however, used imaginatively, should suggest a variety of techniques and procedures which might achieve that aim. Similarly, much of the philosophy behind personal construct theory, which Kelly expounded in his essays, is an apt basis for pursuing the moral advantages of participation, as I hope I shall have made clear in much of the discussion above of the role of architects and other experts.

On the socio-political plane, however, I am not sure that the theory is so suggestive. It evolved in a primarily individualistic culture and in a professional setting where the focus of interest appears to have been on one-to-one relations. Neither the use of a personal construct approach in group therapy nor the theory's sociality and communality corollaries can dispel this impression for me. One of the primary functions of participation is ultimately to achieve a full integration of the individual into his social and political institutions. A crucial aspect of participation is the relation-between-individuals, appreciated as such.

People working within a personal construct framework have tended to neglect an integrated view of persons actively construing one another, with some rare exceptions in the case of dyads. One of the many directions in which the theory might be developed—and the philosophy of constructive alternativism fortunately seems to demand that it be developed—would be to enable it to account for integrated and dynamic construing between persons and between persons and collectivities. The solipsistic dangers of personal constructs might be avoided by the development of the notion of public or collective constructs. It is to this point that my own meanderings through the theory in the context of participation have brought me. The next stage will be another story.

Loaded and Honest Questions: A Construct Theory View of Symptoms and Therapy*

Finn Tschudi

(In collaboration with Sigrid Sandsberg)

I. Introduction

Kelly (1955, p. 835) does not commit himself to any specific point of view on "disorders", "it represents any structure which appears to fail to accomplish its purpose". Admittedly "this is an extremely flexible definition" and Kelly is "content to let 'disorder' mean whatever is ineffectual" from *some* point of view, be it that of the client, the stuffy neighbours or even God.

This paper elaborates one point of view, derived from and compatible with Kelly's position, but hopefully somewhat more precise. "Symptom" and "symptomatic behaviour" are preferred as more specific terms than "disorder". The point of view which will be explored in this paper is: symptomatic behaviour is behaviour which obliquely gets at the issues which are important for the person.

As will be discussed in Section II, advances in Repertory methodology provide a convenient tool for exploring the personal meaning of the symptoms. A simple form of a construct network, typically consisting of three constructs (which we label *ABC*), may in many cases be helpful in locating the important issues which the symptom obliquely gets at, in the person's own terms. Several examples of such networks will be offered. The form of presentation may have the advantage of facilitating the search for therapeutic steps. This approach we label "the *ABC*-model", and it is deeply influenced by Greenwald (1973).

In earlier presentations of this model a frequent comment has been that it is too "cognitive" and does not appear to account for emotions, drama and

* This research has been supported by grant B60.01–111 from The Norwegian Research Council for Sciences and the Humanities. Sigrid Sandsberg has done the clinical work and assisted with the theoretical formulations. Of the many colleagues who have commented on previous drafts I am particularly indebted to Gudrun Eckblad and Rolv Blakar.

suffering. Part of the difficulty may be that Kelly's theory is very abstractly stated and this makes it difficult to flesh out a coherent picture of symptoms and therapy.

Section III carries the perspective further by selecting one development from learning theory which closely resembles Kelly, Goldiamond (1974). Berne's transactional analysis which both keeps some ties with psychoanalysis and also moves in directions compatible with Kelly, is also found useful.

What makes these two perspectives so compatible with Kelly is that they explicitly ask the simple but vital question: "what is the person after"? The obliqueness of symptoms may, however, call for *redefinition* of what the person is after. This may be a difficult process with several dangers on the road. To emphasise what the person is after, his project, is the essence of Kelly's paradigm *man-the-scientist*. This paradigm is extensively discussed and defended elsewhere, Tschudi (1976). There we propose that if one label should be applied to Kelly, it would not be "reluctant existentialist" as Holland (1970) would have it, but rather "existential behaviourist". Is it possible to close the apparent gap between "existentialism" and "behaviourism"? Reaching out to another person so that both simultaneously have maximal possibilities of exploring their basic constructs is perhaps an "existential" project, Bannister and Fransella (1971, p. 38) call it love. This project is seemingly far removed from what the greycoated scientist may be up to in his gadget filled laboratory, perhaps he is a "behaviourist". Kelly's unique contribution is to invite us to construe similarities between these projects. Whatever he does, Kelly insists, man asks questions, "Behaviour is a question" (1970) (see also Mair, 1976).

The bridges Kelly tries to build, the gaps he will have us close, invite us to strange and perhaps tortuous roads. This paper is only a small beginning in the necessary Kellian tightening. There are vast numbers of pitfalls and difficulties on the road. What is the place of experience? of cognition? We skirt these issues to get ahead.

Kelly almost exclusively emphasises formal aspects of constructs, and only incidentally content (*cf.* p. 935). This, we think, makes it necessary to bring other theorists to join the party.

Hopefully this will not only serve to bring Kelly closer to mainstream psychology, but more generally to further our understanding of the vital questions man poses. The aim of the present paper is to outline a way of thinking which may be a step in the direction of understanding how to help man clarify his goals and pose less oblique questions.

II. The ABC-Model

A. STATEMENT OF THE MODEL

The model to be described in this section equally draws on construct theory and Greenwald's* "Direct Decision Therapy" (1973, especially chaps. 1 and 16). The reader may want to consult the next subsection for several examples of the abstract constructs described here. The first step in Greenwald's therapy is to *define the problem.* The very first question he asks may be "what is your problem", "state your problem as clearly and completely as you can". Problems may be stated fairly specifically: "cannot have orgasm", "fear of height", "unhappy marriage" or more generally as "way of being"-problems: "inauthentic", "unhappy", "unassertive", what may be described as "lives of quiet desperation".

The symptom, or the problem, may in personal construct theory be regarded as one pole of a construct. According to the dichotomy corollary this pole will have a contrast end which may be more or less clearly defined, it may even be submerged. We use the symbol A for the construct describing the problem area. A has two valuated poles. The negative pole will be denoted a_1 and the positive pole a_2. a_1 may be "inauthentic", "unhappy", "unassertive" and a_2 may be "authentic", "happy", "assertive".† A is then a generic term for both poles. Since there generally is no separate term for the construct, it will often be described by listing its two poles: "inauthentic–authentic". The construct corresponds to a dimension and the poles to differently valued positions on this dimension. *A defines the dimension where the person wants to move.* The person finds himself at a_1 but wants to be at a_2. In a statement like "I want to stop drinking", A and its poles are immediately given: a_1: "drink too much", a_2: "stop drinking". The problem is a_1, and a_2 is the preferred alternative.

What keeps the person from moving? Greenwald assumes that the symptom always has its *payoffs*, that there are *advantages of the symptom* here and now. Depression may, for instance, imply that no one expects much of you, and give rest and possibilities for taking it easy, overload is avoided, *cf.* Example 1. Greenwald may sometimes directly ask "what's the payoff". The point of view that there always are advantages of the symptom is basic in the present paper, it will be repeated in all the examples to be discussed later.

It is not always equally straightforward to get at such advantages. We have found Kelly's Rep method, combined with Hinkle's (1965) laddering technique‡ useful in this respect, and unless otherwise stated this approach has been used in the examples reported in the next subsection. The point of departure is a Rep test where the person is always part of the triad. Having

* This presentation has also profited from several workshops and discussions with Greenwald. Some of the examples used here have only been presented orally by him.

† Would it be simpler to use a+ instead of a_2, likewise a− instead of a_1? This would imply that the valuation always was unequivocal. Quite often this will not be the case. Example 6 in the next subsection is a case in point. Associating '+' to '2' and '−' to '1' is just a first approximation.

‡ See also Bannister and Mair (1968).

elicited the constructs, the person identifies what is most valued (a_2), and what is least valued (a_1). When the person places himself at a_1 this points at a problematic area. Then the person is asked for advantages and disadvantages *both* of a_1 and a_2. Unearthing such implications gives a picture of the meaning of the construct; stated otherwise the construct network *is* the meaning. Usually it is quite easy to elicit advantages to a_2 ("what are the advantages of a_2", "why do you prefer a_2") and disadvantages to a_1 ("why do you want to avoid a_1"). The answers to these questions give a new construct which we call *B*. The implication of a_1 we call b_1, likewise b_2 is the implication of a_2. *B* gives further information on why the person wants to move in the direction *A* (it is useful, but often not necessary to have information of *B* constructs).

The crucial step, however, is to ask for evaluative reverse implications: "are there advantages of a_1", this is labelled c_2. Conversely it is asked for disadvantages of a_2, this is labelled c_1. With a relaxed, partly humorous mode of inquiry it is not too difficult to elicit such advantages, *cf.* Sandsberg (1975). The therapist may also suggest poles which the patient may then accept or refuse. This kind of inquiry may have some of the character of what Scheff (1968) has called "negotiating reality".

The general hypothesis is then that a_1, the symptom, has positive implications c_2, and likewise the "desired" alternative has negative implications, c_1. In other words, *there is a construct C which keeps the person from moving.* Hinkle calls the structure we have described an "implicative dilemma", a_1 has not only negative (b_1), but also positive (c_2) implications; likewise a_2 has not only positive (b_2), but also negative (c_1) implications. It is interesting to note that Hinkle suggests that

> "the implicative dilemmas . . . seem related to conflict and doublebind theory, and are, therefore of particular clinical interest" (Hinkle, 1965, p. 7).

Problem statement: The person wants to move along direction *A*—from a_1 to a_2. Other construct(s), *B* where a_1 implies b_1 and a_2 implies b_2, may give further reasons for the desired movement. There is, however, a construct *C* which hinders movement, *ABC* is an implicative network of a special type, an implicative dilemma. The problem statement is the first of two statements which comprise the *ABC* model.

Three is no magical number. There may be more constructs involved, for instance what Greenwald calls "life choices", *cf.* Example 4 in the next subsection. "Life choices" are related to Berne's concept "scripts", what is the theme of the drama of one's life. Furthermore one should generally consider how past decisions may be at cross purpose with the present problem. For instance, a choice which (though subtly) is at cross purpose with the problem "cannot have orgasm" is the decision "to be greatest sex partner in the world".

The basic feature of the problem statement is that the system is blocked, the person is stuck or "forced to" run in circles. If he tries to move from a_1 to a_2, he is faced with the implication c_1. Assume now that to avoid c_1 (or achieve c_2) is more important than a_2. This will then "force" the person back to a_1 where he can achieve c_2. But if C is more important than A, why does he then come for help? Does not this imply that A is more important than C? There is a contradiction here if the only way of achieving c_2 is by means of a_1. But is, e.g., the only way of avoiding overload to be depressed? A preliminary way of stating the therapeutic goal is to find other ways than a_1 to c_2, or equivalently: *find a way to combine a_2 and c_2*. Another way of stating this is that the symptom solves a problem but that the price is felt to be too high. Can one, however, "have one's own cake and eat it too?" There appears to be a gap or a cleavage separating a_2 and c_2 (since c_2 is implied by the opposite of a_2), and metaphorically the problem may be put as "how to cross the cleavage". Since there is nothing immutable by psychological implication, such changes will in most cases be possible.

But does the person want to change? An advantage of Greenwald's approach is that, having elicited payoffs, it is not facetious to ask "do you really want to change"; give up dope, homosexuality, whatever. The person gets optimal possibilities for reviewing whether a change really is a worthwhile undertaking. He, not the therapist, should be the judge of the projects he wants to pursue.

We take the concept "network" to imply that a change anywhere in the network may have repercussions for the rest of the network. But however likely this may seem in a specific case, it is a hypothesis which may be validated or invalidated. The second possibility makes it necessary to have two parts of the second statement of the *ABC* model:

Change statement:

(1) Given the network *ABC*, find a step from anywhere in the network which seems likely to lead to change and which the person decides to carry out.
(2) If this does not lead to the desired change, re-evaluate the network, and find a new step. This may call for reconstruction, forming *new* constructs.

The problem statement and the first part of the change statement, give a relatively simple model of therapy, a sequence of steps is indicated, a *linear model*. By studying the examples in the next subsection, the reader may get a feeling for the usefulness of such a linear model. The second part of the change statement, however, opens up for all kinds of complexity.

Therapy may be considered as an ongoing process where at any point the whole process may have to be repeated within any problem area. If the network has to be reconstructed so that at any step there may be a *new* network, this can no longer be described by a linear model. In technical jargon a *recursive model* is called for. The examples to come hint at the inadequacy of a linear model. In Section III we then discuss the necessity of forming new constructs and consider some of the difficulties involved. First two comments on Rep methodology.

In a therapy where all kinds of problems may turn up, the therapist (if not the client as well) may fear getting "lost", is progress being made, are relevant issues being explored? Constructs from the Rep test and laddering is one way of keeping track of the process, and new tests may be one way of assessing change. This is particularly the case for Example 5, it would be valuable with more experience on this.

Another use of Rep methodology as here discussed, is to serve as an analogue of a medical check-up. It will make a sweep through the most salient aspects of the construct system, and the person may be confronted with dilemmas where it "aches". Not all implicative dilemmas may be problematic. But sometimes the person feels "caught" in an undesirable alternative, or the symptom has an alien, "not-self" quality, as Wright (1970, p. 222) describes in a valuable paper on implicative dilemmas. These may be among the cases where the person particularly wants to change.

B. EXAMPLES OF THE MODEL

EXAMPLE 1.* DEPRESSION

a_1	being depressed	a_2	not being depressed
b_1	1. I can't do things I'd like to 2. I'm not as good a wife as I would like to be 3. I pity my husband who is so much alone	b_2	1. I could do things I like to 2. I could be a better wife 3. I could keep my husband more company
c_2	1. avoid doing unwanted things 2. avoid playing happy wife which I am not 3. avoid unwanted intercourse	c_1	1. must do unwanted things 2. must play happy wife which is disliked 3. must have intercouse when not wanting to

The crucial aspect of the depression may be the urge to avoid disliked things as stated by c_2. Perhaps at one point the patient did not have any alternatives but being depressed which implies "I can not". This would fit Haley's view of symptoms:

> "The crucial aspect of the symptom is the advantage it gives the patient in gaining control of what is to happen in a relationship with someone else" (Haley, 1963, p. 15, 19).

* This example was first described jointly by Larsen and Tschudi, see Larsen (1972).

But the price to pay is that "I can not" is made "legitimate" by the depression. This then precludes doing wanted things as well, *cf.* b_1. The tragedy of gaining control by such means is that "one cannot take either credit or blame for being the one who sets those rules". Symptoms imply that responsibility is abnegated and

> "it seems to be a law of nature that one must take the responsibility for one's behaviour in a relationship if one is ever to receive credit for the results" (Haley 1963, p. 19).

Following Ellis' (1962) Rational Emotive Psychotherapy, one may directly counteract the c_1 poles by encouraging the wife to say "no", for instance telling her husband that she does not want to have intercourse ($c_{1,3}$). This approach asks her to assume responsibility for her behaviour. If this is not an appealing choice for the patient, one can use a more indirect approach. As a response to c_1 she can be told to deliberately play being depressed. In this way she can gain the advantages of a_1, that is to achieve c_2, without the unpleasantness of *really* being depressed. This is an example of "encouraging symptomatic behaviour", a favourite procedure with the Palo Alto group, (see Watzlawick *et al.*, 1974 for a recent statement). Interestingly Kelly (1955, pp. 995–997) also discusses this procedure. He sees it as "controlled elaboration" which properly used may facilitate awareness and choice. Why, however, does she have to resort to such oblique ways as "depression" to solve marriage problems? Perhaps both spouses must be brought in. Furthermore the *C* construct is what may be called an *avoidance construct*. Such constructs do not state any positive goal, the choice is *move away from* or *suffer* a disliked state. Avoidance constructs clearly call for replacement by statements of what the person wants to *move towards*, what is she after?

EXAMPLE 2. DISHONESTY

a_1	Not particularly honest	a_2	Honest (truthful)
b_1	1. promotes painful confrontation and subsequent retreat	b_2	1. prevents painful confrontation and humiliation
	2. work may get sloppy		2. produces better work
	3. must be on guard to remember own tales		3. doesn't have to be on guard
c_2	1. allows for being more colourful	c_1	1. no nuances: things appear black/white
	2. makes for expansiveness and taking chances		2. hinders expansiveness and taking chances
	3. doesn't make difficulties for others		3. may hurt others by expecting "the ideal"
	4. makes life easy		4. makes life more bothersome

It was a relief for the person to see that the guilt-provoking behaviour, a_1, also implied acceptable wishes, c_2. It was emphasised that in order to change to a_2 it was necessary to *find other ways of taking care of the motives expressed in* c_2. Suggestions for practice, homework, were:

Accept and promote own wishes of being colourful and expansive, further to inform others of this when appropriate, *cf.* $c_{2,1}$ and $c_{2,2}$.

Protest against overdevoted, narrowly literal interpretation of "truth", and encourage reconstruction (redefinition), of *A*.

An interesting aspect of this example, is that after three years the subject no longer had any problems concerning "honesty", but remembered that it had been a very bothersome problem.

EXAMPLE 3. MONEY

a_1	can't handle money	a_2	handles money well
b_1	1. doesn't get what one wants (squanders money)	b_2	1. get things one wants
	2. too carefree, "doesn't give a damn"		2. keep control
	3. intense discomfort, stomachaches		3. avoid discomfort
c_2	avoids being boring, pedestrian, trivial dropped out of "the rat race" "the freedom of a child playing at the beach"	c_1	bourgeois, trivial, dull, sticks to the rules

The person was asked whether it might not be conceivable to gain the advantages of b_2 without necessarily losing the precious uniqueness and "freedom" implied by c_2. This was vehemently denied, but with a show of laughter. Further inquiry revealed an embarrassed attitude towards c_2. What was the worth of this goal if the *only* way of reaching it was by mishandling money? In the light of this discussion it was decided to take one concrete step towards a_2, that is every time a cheque was written out to record this carefully. Habitually neglecting this, and consequently never knowing if the cheque account might be overdrawn was a factor specifically mentioned as leading to the discomfort, $b_{1,3}$. Fairly soon, however, the person reported a failure in carrying this out when he was lining up in a queue shopping. A primary reason for the neglect turned out to be feeling embarrassed by "unnecessarily" holding up the queue. The fact that the failure was done in awareness provided a unique opportunity to revise the theory and the person has since not reported any difficulties in handling money. But we are left wondering about what may happen to "the child playing at the beach". Is he left alone when the sun is down?

Perhaps an even better strategy would have been also to promote c_2, as in the previous example.

EXAMPLE 4. READING

a_1 not read scientific literature	a_2 read scientific literature
b_1 not be proficient	b_2 be proficient in one's field
c_2 do what is fun, not strive so much	c_1 heavy duty, drudgery

After extensive discussion of the person's "time-budget", it was decided to take some small, specified steps in the direction a_2. This seemed to work for a few weeks, but then progress stopped. Further discussion revealed what was only hinted at in the first session, that what Greenwald would call a "life choice" was operating. a_2 seemed to be in the focus of convenience of "yielding" ("like a small child *having* to eat properly with the spoon"), whereas a_1 (and even more pronounced c_2) was governed by the contrast to "yielding", that is "be independent". In the light of the superordinate construct "yielding–be independent" any exhortations in the way of rational planning (move towards a_2) simply strengthened a_1 and c_2.

After conclusion of this session, however, the person "spontaneously" reported having arrived at a solution. She redefined herself in her occupational role, as previously being a *teacher* who somehow had to find time for research, but now seeing herself as a *researcher* who had to face interruptions of other chores. After this, time spent on reading scientific literature and writing research reports greatly increased.

EXAMPLE 5. BEING FORWARD. (Anne)

a_1 not-dominating	a_2 dominating
b_1 others define one, submissive, boring, reticent, repulsive	b_2 being forward, liberated, self-assertive, get one's way, stand up for one's rights
c_2 good contact with people, to be loved	c_1 dominating in condescending way, drown people in words

Anne is going to be a nurse. "Being forward" initially turned out to be a central feature of what she strived for, and a training programme to promote this as a way of being was initiated. The self-assertive tasks she had to "practice" included such things as: to be heard in class—at least once a day take the word, stick to one's position, insist on having her suggestions taken seriously. A criterion for success would be that this programme would lead to her blossoming forth and that she herself should experience active participation in defining situations and be listened to. But this did not happen. Quite the contrary, the programme led to exhaustion and depression, she felt stuck. We had worked on the "wrong" problem, *A* and *B* required redefinition. A slightly different problem (also indicated by Rep) was worked through by fixed role therapy. Then Anne once exclaimed: "it is not my nature to be 'forward', that is not *me*". Further discussion revealed that this construct was imposed on her by her boyfriend, who preferred girls to be that way. Using a new grid the *A* construct to replace the previous *B* (and *A*) was:

a_1 domineering and tendency to bully	a_2 direct, but not commanding

Here she could "be at home" under the pole a_2, and there was no longer an implicative dilemma involved.

EXAMPLE 6. LITTLE GIRL–PRIMADONNA. (Tove)

(1)	a_1 not express thoughts and feeling	a_2 express thoughts and feeling
	b_1 hold oneself in the background, alone, isolated	b_2 genuine contact
	c_2 primadonna	c_1 little girl (anxious not to measure up)
(2)	a_1 little girl (anxious not to measure up)	a_2 primadonna
	b_1 in the background, shy, boring, is exploited	b_2 centre of attention, smashing excitement and fun, is admired, keeps one's cool
	c_2 kind, generous, can see both sides of the issue	c_1 egoistic, cold, unconcerned about feelings
(3)	a_1 not sex	a_2 sex
	b_1 not contact, lack stimulation	b_2 genuine contact, feels good, gives everything
	c_2 respectable, avoid guilt and shame	c_1 casual acquaintances, exploited, thought to be rather lax, guilt (parents), shame

Tove who is 15 years old has a long story of school truancy. Therapy has proceeded for some five months, but clear progress cannot be reported since establishing a good cooperative relation has proved difficult (see comment in section on therapy). The example is intended to indicate more complex networks, and emphasise the necessity of reconstruction. *A* in (1) was the most important construct of the initial Rep. She could not clearly place herself under either pole, but the difficulties of expressing thoughts and feelings have been a repeated theme. Later in therapy "little girl–primadonna" has been elaborated,

EXAMPLE 7. TELEPHONING

a_1 fear of telephoning, travel	a_2 be able to make phonecalls, travel
c_2 hide loneliness	c_1 reveal loneliness
C transformed to *A*'	
a_1' lonesomeness	a_2' having friends
c_2' hide lack of true love	c_1' reveal superficiality of "chit-chat"
C' transformed to *A"*	
a_1'' lacks true love	a_2'' obtain true love

ABC in (2) is a summary. The "little girl" expresses thoughts through poems, but Tove also has some quite "primadonna" experience. Sex is problematic, *cf.* the summary in (3), but some progress has been made in defining simultaneously bodily and extended contact in sex. Here she is well on the way to adult goals.

Example 7 represents one of Don Bannister's favourite stories, and the network diagram will hopefully point to advantages of the present model. The story goes:

> "The young man had a phobia for telephones and travelling. About a year of systematic desensitisation treatment enabled him to travel and use the telephone. He commented on the utter pointlessness of such an achievement since he had no one to ring up and no one to travel to. He had formed no relationships with his fellows. About two years of psychotherapeutic exploration and experimentation, and the young man was going to social gatherings, visiting his newly found friends, was a member of this or that hobbies group. He then pointed out that no one could care less than he did for the kind of superficial chit-chat relationships, mainly with men, which he had now formed in great number. What he wanted was a deep passionate, intense, sexual and exclusive relationship with a woman . . . ."

Therapy seems to have started just with the *A* construct. From the account it does, however, appear reasonable to infer that there was a decided advantage of a_1, and correspondingly a threat connected with a_2. Should we say that *C* had better remain "hidden" and that after moving to a_2, the person could then face the problem implied by *C*? We think not. Perhaps one would not outright start by asking "what are the advantages of not being able to phone", but the answer to this could easily be implied by gently asking to whom the person might wish to call. *C* is a typical avoidance construct (*cf.* Example 1), and A' is a transformation of *C* to bring forth a positive goal. The same story repeats itself once more. There is a new avoidance construct C' which again can be transformed to a problem oriented construct, A''.

It seems like a waste of therapeutic effort to spend a whole year with systematic desensitisation for the first phobia. One should at least work simultaneously with *A* and A' and probably with A'' as well.* Reading the

EXAMPLE 8. EXPOSURE

a_1 indecent exposure	a_2 no indecent exposure
b_1 possibly getting caught	b_2 avoid being put in jail
c_2 express contempt	c_1 being a conformist

* A group of Norwegian clinical psychologists strongly argued for this and Bannister himself once informally suggested the "love of a good woman" as a panacea for an untold number of problems.

case story carefully, one feels quite some irritation with the step by step procedure.

Example 8 from Greenwald:

> "A radical, American university professor was committing 'indecent exposure'; occasionally he would urinate in public. Asked why he would do such a thing, the man answered that it was his way of expressing contempt for the establishment. Not without humour Greenwald pointed out that he was by no means flaunting the establishment. On the contrary he was a living proof of one of the most cherished notions of the 'silent majority', the deep conviction that 'radicals really are nothing but sex perverts'. The man stopped urinating in public and took to writing vitriolic letters to the newspapers which commented on various inequities." (These letters even caught the interest of a publisher, Greenwald reported.)

The therapeutic technique is a good illustration of redefinition, relabelling or reframing. Whereas exposure was previously seen from the point of view of expressing contempt, it was put in a quite different frame, namely supporting the prejudices of a despised group. This frame fitted the "facts" of the same concrete situation equally well, or even better, and thereby changed its entire meaning, its implications, *cf.* Watzlawick *et al.*, 1974, p. 95, on reframing. This made it easier to move away from a_1, and discover another way of achieving c_2. Perhaps then also the meaning of c_2 changed? If, however, an alternative way to c_2 had not been available, therapy would have had to proceed with caution.

EXAMPLE 9. OEDIPAL CONFLICT

a_1 fail in achieving mature sexual relations	a_2 achieve mature sexual relations
c_2 be faithful to mother	c_1 be faithless to mother, risk loosing her

This network takes its inspiration from Wachtel (1975) who very clearly describes the "running around in circles" quality of one specific type of symptomatic behaviour:

> "In looking at a man with signs of strong Oedipal conflicts and with a less than satisfying current sexual life, I could see how his conflicted ties to his mother led him to approach interactions with other women in a fearful and inhibited way, which led to frustrating and unsatisfying sexual experiences and frequent rejection by women which confirmed his anxieties about sex and his yearning for gratifying his sexual needs with the woman who had nurtured, caressed, and comforted him as a child, and hence set the stage for the repetition of the same cycle of events."

Wachtel does not discuss therapeutic steps. From his description it seems appropriate to condense the *A* and *C* constructs above and render the position of the person as "only mother is all-loving", a position which clearly calls for extensive reconstruction.

SUMMARY OF EXAMPLES

Take small step to a_2
3. MONEY
4. READING
5. BEING FORWARD
7. TELEPHONING

Controlled elaboration		Redefine
1. DEPRESSION	Change context of *A*	2. DISHONESTY
	4. READING	5. BEING FORWARD
Redefine		6. LITTLE GIRL–PRIMADONNA
8. EXPOSURE		

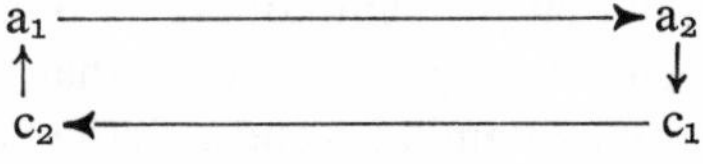

Promote independently (avoidance constructs)	Redefine *C*	Counteract
2. DISHONESTY	1. DEPRESSION	1. DEPRESSION
7. TELEPHONING	7. TELEPHONING	
8. EXPOSURE	9. OEDIPAL CONFLICT	

Arrows (starting from a_1) indicate "stuck" quality if movement attempted.

Summary of positions implied in examples, you can't have both a_2 and c_2

c_2	a_2
1. avoid intercourse	be non-depressed
2. colourful	honest
3. be a child at the beach	handle money well
4. have fun and ease	read scientific literature
5. good contact with people, to be loved	being forward
8. express contempt	no indecent exposure
6. primadonna	express thoughts and feelings
7. hide loneliness	telephone freely
9. faithful to mother	achieve mature sexual relation

(only mother is all-loving)

III. Converging Approaches to Symptoms

A. SYMPTOMS AS A SOCIALLY INADEQUATE MODEL

The position of the person is that he cannot have both a_2 and c_2. This may well have the character of a subjective law, but if we look at the six positions first listed in the summary, these can not be regarded as *intersubjectively valid positions*. The therapist cannot be expected to share the position that for instance "express contempt" *requires* "indecent exposure". Psychological implications have no absolute necessity, and this serves to make change possible. Things could be otherwise, *cf.* Kelly's philosophical point of departure: "constructive alternativism".

This point of view may contribute to give insight into the person's suffering. He cannot readily expect others to understand his predicament since he is stuck with an inadequate model from the other's point of view. This may serve to increase alienation from other people. And it may be difficult to get validational material for an inadequate model.

But is it sufficient to say that the symptom is an oblique way of going at otherwise "reasonable" ends? Or is it the case that the obliqueness also infests c_2? Perhaps means contaminate ends. Perhaps an oblique way of going at one's goals contaminate the goal. Can Anne (Example 5) really be loved by being submissive? We should remember Haley's views discussed in Example 1. Similar questions can be asked (perhaps more or less forcefully) for the rest of the six first listed positions. A combination of a_2 and c_2 may in many cases be thought of just as a *preliminary* therapeutic goal, *cf.* part 2 of the change statement.

The last three positions (Examples 6, 7 and 9) might also be said to reflect socially inadequate models, but the most salient feature of these examples is that reconstruction of c_2 is called for. This is especially obvious in the psychoanalytically inspired Example 9.

The general case may well be that c_2 needs reconstruction. This hypothesis receives further support by Berne's analysis of games (which is an important part of transactional analysis). This analysis also emphasises how the person is stuck with symptomatic behaviour.

B. SYMPTOMS AS GAMES

Example 10 is the game most extensively described by Berne (1961, pp. 99–102; 1964, pp. 45–52) cast in terms of the *ABC* model (all games may be expressed in terms of this model). It is introduced as "the most common game played between spouses". One should bear in mind that while Berne claims that all five types of advantages can be identified for all games, only two or three of them are described for most games. The trouble with transactional

EXAMPLE 10. "IF IT WEREN'T FOR YOU."

Type of advantage, "payoff"	a_1 constricted b_1 not allowed to indulge in social activities and "have fun"	a_2 explorative b_2 participate in outdoor social activities and "have fun"
External psychological advantage	c_2 1. helped to avoid phobic situations	c_1 1. social phobias exposed
Internal psychological advantage	2. receive gifts to indemnify his "severity"	2. not get gifts
Structure time	3. provide fresh spring of "if it weren't for you" resentment	3. expose void, (no common interests, no intimacy) boredom
Provide strokes	4. get social stimulation, resentment	4. withering isolation
Confirm existential position	5. prove "all men are mean and tyrannical"	5. caught short of an existential position (no unifying theory)

analysis is that the devotees may be inclined to take it too seriously. It is excellent for the heuristic inspiration it may provide, but if all the concepts do not fit a specific example, one should listen more closely to the person than to the theory.

Bearing this in mind, we turn to the game where we here have made explicit the right hand poles which Berne just implies. With this impressive list of advantages to boost, why would she relinquish the game? Similar advantages are listed for all games. Some of the titles indicate, perhaps even more clearly than in Example 10 the distress which may be involved, "Alcoholic", "Rapo", "Debtor". In transactional analysis *all* symptoms have game features.

A critical question is how we react to the type of goals implied by the c_2 poles. Are these the type of constructs we choose to elaborate (or like to see others elaborate)? You and I probably feel uncomfortable subscribing to this, so let us consider them one by one.

If we do have phobias we struggle valiantly to overcome them. If we get gifts it is certainly not by deviousness. We have "real interests" to share with our friends and fear of boredom is simply irrelevant. The social stimulation we get is, if not always heartwarming, not *that* crooked.

The preferred type of "strokes" is related to the kind of existential position to vindicate (*cf.* Berne, 1972, pp. 137–139; Steiner, 1974, pp. 109–111). In

Example 10 the position is of the type "I'm let down" where feeling of resentment is the related "stroke". Other types of "they are no good" positions may yield as strokes feelings of anger, whereas "I'm no good" provides feelings of depression. Strokes are colloquially referred to as "trading stamps". Some people amass large collections, and comparing such collections may be a favourite way of passing time ("Ain't it awful"). It is also possible to trade in a sufficiently large collection for some dramatic event (as the little green, blue or brown stamps one gets as a premium for buying groceries or gasoline may be traded in for some "free" object). Feelings of anger, "red stamps", may be traded in for a "free" physical assault or even a "free" homicide if the collection is sufficiently large.

We immediately recognise that the gamestress in the example is engaged in defeating (vicious) self-fulfilling prophecies, she gets" what she asks for". This may well be a critical feature of games. Berne suggests that the husband is probably an accomplished cogamster, shrewdly picked precisely for that reason.

You and I, on the other hand, are reluctant to spell out any "existential position", and we do not treasure such nongenuine feelings as "trading stamps". (?)

Game behaviour not only obliquely gets at the issues but the *C* constructs themselves appear "oblique", in need of reconstruction. Berne and Steiner note affinity with the theories of Erik H. Erikson, and this leads us to suggest that "basic trust" makes for "positions" which one need not continuously "defend". Game positions may tragically lose the quality of open-ended questions, and take on the quality of orders to follow: "I *must* confirm my position". Having received sufficient confirmation of one's relatedness to other people ("trust"), one can venture further, to questions beyond games.

C. SYMPTOMS AS COSTLY OPERANTS

The point of view that there are advantages of the symptom finds strong support in many approaches stemming from learning theory. Goldiamond (1974) is here considered, partly because he draws on a very broad clinical experience, but mainly because his "constructional approach" is particularly close to Kelly.

Goldiamond describes symptoms as *costly operants*. They are operants, there are maintaining or reinforcing consequences, (advantages of the symptom) and the cost is the suffering involved. Two illustrative examples:

(a) A cockroach phobia which prevents a wife from moving unaided from room to room is interpreted as

> "a highly successful instrumental behaviour which dramatically forces the husband to provide the legitimate attention which he had hitherto withheld and deprived her of." (p. 15)

(b) Uncontrollable tremors, reported as anxiety attacks, is described as a way a "competent librarian" copes with a series of urgent requests on the job (pp. 32–33).

This is not to say that Goldiamond would subscribe to any simple view of behaviour as being primarily "shaped" by external "reinforcements". His critique of token economies is here particularly relevant. He points out (p. 30) that if one knows what the patient is after, then extrinsic reinforcement in the form of tokens, points etc. is *not* necessary and in many cases even highly unethical! This fits well with Kelly's sharp critique of traditional reinforcement theories. "Avoid bribery"! is a position common to Goldiamond and Kelly.

Goldiamond's position opens for a closer comparative analysis of the partly overlapping concepts "reinforcement" and "validation" (construct theory) but that cannot be pursued at present.

Goldiamond draws a sharp distinction between an *eliminative* (pathological) and a *constructional* approach to therapy. He emphasises that the goal should be to develop less costly ways of coping and *not to eliminate the symptom.* For the cockroach phobia the (constructional) goal was to teach the husband to respond to her legitimate needs and to teach her to get these more readily across to him (and not to eliminate the phobia). Goldiamond asks how she could develop such a startling solution "to get the recognition you and I get without any effort". The high price she pays testifies to the importance of the issues and a viable alternative may not be easy to come by since she has settled for such an unconventional and costly solution.

Likewise nothing was done to eliminate the librarian's reported anxiety attacks, but steps were taken to find out how she could turn down low order requests in less costly ways.

As a final example we consider stuttering where Fransella (1972)—in the Kelly tradition—has fleshed out a constructional approach which is congruent with Goldiamond's approach to stuttering (p. 14).

(c) Fransella sees stuttering as having grown to be "a way of life" (as may also obesity, compulsive patterns and many other disorders), and a way of life is at least something to "hang on to". The alternative may be a void and that one cannot face. Fransella shows that initially "not stuttering" is practically devoid of meaning (implications) and so painstakingly meaning is built into the goal of "being a fluent speaker".

Summarising these examples:

	c_2	a_2	c_1
(a)	get attention	no cockroach phobia	no attention (?)
(b)	avoid exhaustion	no tremors	overloaded, worn out
(c)	way of life	fluent speaking	a void, no "meaning"

The concept "costly operant" will now be seen to add to our understanding of symptomatic behaviour by drawing on Kelly's structural approach to "anxiety". The most overriding advantage of the symptom is to provide structure, conversely, *the* danger of the alternative is that which cannot be faced, lack of structure, *cf.* c_1 in the three previous examples. Basically Kelly (1969, p. 264) equates "anxiety" with "loss of structure" the formal definition is:

> "the awareness that the events with which one is confronted lie mostly outside the range of convenience of his construct system" Kelly (1955, p. 485).

Kelly's position invites two different perspectives from which to view human behaviour. While admittedly difficult to separate them in concrete cases, the following outline tries to capture the essence of the two different perspectives (*cf.* p. 894 ff).

Direction of movement

↙	↘
away from anxiety	Towards optimal anticipation of events
stuck with a_1	grope towards a_2 (reconstruction)
scamper away in any direction one happens to be facing avoid chaos (failure of structure, or ultimate anxiety which is incompatible with life) avoid living in unpredictable world	probe for better ways to anticipate the future, find meaning, structure for anticipation

We venture to suggest that *symptoms may be considered from the point of view of "running away from anxiety"*, it is a way of providing structure "at any cost". In terms of the diagram above the person runs left instead of right.

We can now (perhaps somewhat broader than Goldiamond) consider the dangers of an eliminative approach and conversely the advantages of the constructional approach.

Most of the examples in Section II were preliminarily interpreted as "socially inadequate models". A therapist might be tempted to say to the client: "Look, a_1 is a stupid way to get at c_2". For the game analyst the temptation is to expose the game, "you do not really dare to have fun". Such interpretations can be seen as attempts to *eliminate* a_1 by providing "insight". This procedure is, however, strongly contra-indicated if c_2 cannot be taken care of. If the person is directly confronted with failure of structure (anxiety), the results can be disastrous. One cannot eliminate the symptom without providing an alternative. Goldiamond points to the risk that "less desirable operants may appear" so "symptom substitution" is *not* a dead issue for learning theory. Kelly's structural approach to anxiety may be valuable in dealing further with these problems.

A major advantage of the constructional approach may well be to avoid anxiety. It is noteworthy that Goldiamond has no explicit discussion of "anxiety". This may be because his approach does not leave the client "out in the cold". Notice that it would be a misinterpretation of the constructional approach to rely on "take small step to a_2", (see the summary of examples). The proper understanding of "constructional" is to take care of c_2. Provided this is done one may of course also find steps to a_2. There were no startling successes to report for "steps to a_2" for Examples 3, 4, 5 and 7 and it would not be likely to succeed in Example 9. What was more true to the constructional approach, to "promote independently" c_2 (Examples 2, 7 and 8) seemed more successful.

Berne's emphasis on structuring time readily fits with the Kellian view of anxiety. Waking hours must be filled with some activity, having no way of using time ("boredom"), clearly represents failure of structure. This again points to the dangers of the eliminative, symptom removal position. A leading learning theorist, Kanfer (1975a), did for instance point to a case where a mother spent several hours a day nagging at her small children. Removing this nagging behaviour led to psychotic behaviour in the mother. The point should now be simple. If no viable alternative is available, the symptom may be "better" than what may happen when the void is faced.

What kind of projects are most fraught with anxiety? It is significant that for most games Berne (1964) describes the psychological advantage in terms of "avoiding intimacy". Indeed, intimacy is the desirable contrast to games.

One may sometimes marvel at the beautiful choreography performed to avoid intimacy. An example from Berne (1970, p. 160) on sex:

> "He: 'How come your mother invariably knows the exact time to call? [to interrupt sex]. She: 'How come you invariably start something at the time my mother usually calls'?"

One *must*, somehow, relate to others. Fostering less oblique ways of communicating is a major challenge to a constructional therapeutic approach.

D. SYMPTOMS AND HOSTILITY. ON THE DANGERS OF INTERPRETATION

Kelly invites us to think in terms of *experience cycles*. A question is asked, a project is launched. If the answers confirm the position from which the project was launched, the person can venture further with his next project. One cycle is completed. If the answer does not confirm (validate) the position he should rephrase or ask a different question (reconstrue). This also completes a cycle.

Some projects may extend over years and the answer will then be an (irregularly) growing body of evidence. One may conceive of smaller cycles within larger cycles. (Subprojects within larger projects are well known to students of "problemsolving" who then use recursive models.)

But what if the person faces invalidating evidence in an important project and there is no structure which allows reconstruction? Lack of structure is anxiety, and this must be avoided. One way out is to "cook the book", to resort to exhortation instead of reconstruction. This is Kellian "hostility" (as with anxiety the point of view is mainly structural).

Hostility may be seen as *a blocking of the experience cycle*, the formal definition is

> "the continued effort to extort validational evidence in favour of a type of *social* prediction which has already been *recognised* as failure" (my italics).

For Kelly (1969) Procrustes is the original example of hostility. He distorts the data (his guests are either stretched or cut down to size, they are *literally* distorted) to fit his hypothesis (the excellency of his bed, that it perfectly fits everyone). This is bad science!

Both from a conceptual and an empirical point of view, however, "hostility" poses unresolved problems. To use a favourite Kellian illustration, the hostile person is like the one who tries to collect the winnings on his bet after the race is lost. Too much is at stake, he can't face losing (invalidation). But then he has not *recognised* the failure! He protests that the race "really" has been run. And his "social experiment" has not failed, he is just having one more go at validating his project. All the others are simply wrong, when they construe failure. They are not sufficiently familiar with the issues!

One approach to define "amount of hostility", however, is in terms of the amount and clarity of the invalidating experiences the person has encountered. The more there is continued exhortation in the face of massive invalidation, the more hostility. The hostile person is blocked *at the end* of the experience cycle. He clings to his project in the face of a pile of negative evidence. Too much is at stake in his project. His construct system *monolithically* (*cf.* Norris and Norris, 1973, for an attempt to operationalise this concept) centres on his project. He has no other questions to ask.

"Hostility" is here used as a descriptive concept, a blocking of experience cycles. Such blocking is a tragedy. If projects are not carried through "we may see the same experimental fragment . . . repeated over and over" (Kelly, 1969, p. 264), the person is stuck.

Hostility may be seen as characteristic of a *closed system* where the intrusion of events threatens loss of organisation. *Hostility epitomises bad science.*

Scripture and dogma reign, the evidence is distorted to fit. *Good science*, on the other hand, has characteristics of *open systems* where free interchange with the environment promotes elaboration and change. If the project involves relating to people one knows well, it is sometimes called "love". Love and intimacy is furthered by "good circles", *cf.* Benedict's concept "synergy" as picked up by O'Neill and O'Neill (1972). (*Cf.* Buckley, 1967, for further discussion of "open" and "closed" systems.)*

Do all symptoms involve hostility? Berne (1964, p. 52) suggests that children pick up games by watching their parents. This would be an example of learning from models, *cf.* Bandura (1969). Games based on modelling may be regarded as a (partial) blocking *early* in the experience cycle. If the person has not learned anything but oblique questions, he may not have received much of an answer. For the costly operants Goldiamond describes, there may, however, not have been any "models". The cockroach phobia may be seen as the result of a problem solving process where "trial and error" may have played a large part (she may have "stumbled" on the phobia as a solution). Perhaps there was a certain amount of invalidating evidence (the husband didn't give attention for "normal" behaviour), so problem solving *may* represent blocking somewhat further in the experience cycle than games.

But this is guesswork. The safest assumption seems to be that *all symptoms involve more or less hostility.*†

Amount of hostility may be important for understanding the limitations of interpretative techniques. The following hypothesis is suggested:

With little hostility involved, interpretation may be very successful. The more hostility there is, however, the more interpretation will tend to be unsuccessful or even dangerous.‡

There are a variety of interpretative techniques. Within the Palo Alto school it is called "reframing", *cf.* Example 8. Within transactional analysis, interpretation would be to point out the gamelike features. Games are dishonest, and the analyst may be tempted to expose the game. Done with humour and compassion it may be a very useful technique (and it may even be some fun as a pastime) but not if the pattern is based on hostility. Berne (1964, p. 68) describes an example where the game of a female alcoholic in group therapy was exposed. She asked for the other's "real opinion" of her, but what she was after was self derogation. When the others refused to play

* A large chunk of everyday activities seems neither "good" nor "bad" science, but rather "neutral". I brush my teeth, get dressed, read the morning newspaper and go through an "uneventful" day. This may be seen as part of a closed system in stable equilibrium.

† Further research may clarify the relation between history of learning and amount of hostility. But present learning theories seem inadequate to describe how Procrustes "learned" to fit his guests.

‡ With sufficient operationalisation the hypothesis may turn out to be circular, but it may still have heuristic value.

her game, she went on a binge, topping her previous ones, which led to prolonged hospitalisation. Unfortunately Berne does not give details which allow us to evaluate the amount of hostility in this case.

It is becoming increasingly evident that psychoanalysis, which leans heavily on interpretation ("insight") poses real dangers, *cf.* Tennov (1975). Her major theme (well documented) is that the general oppression of women is carried further by the "dynamic" therapist (colloquially "the rapist"). In the present terminology her theme may be stated that the client has massive evidence that her way of life lacks "meaning", yet she persists in her ways. Albeit in a more esoteric language this same message is repeated and repeated in therapy, the therapist adds to the pile. It may be too much.

With the present terminology, it is perhaps a shade easier than from other positions to see that hostility is not something reserved for the client. Though he may exude "warmth", the therapist is hostile if he keeps on repeating a message which has not produced any change. We may put the game of "archeology" (Berne, 1966, p. 324) in terms compatible with this view: "If you don't hit the ore with your shovel, you must dig more and dig deeper." If the patient is hooked to this game, we may have a galloping spiral as in other cases of hostility. There is no end to the analysis. Deeper and deeper it goes!

Hostility poses difficult problems in therapy. For Procrustes to change would imply losing face since everyone else has seen how inadequate his behaviour is. It is very hard to admit that one has been basically wrong! It might be profitable to explore other areas not directly related to the core topics (e.g., the bed and the guests for Procrustes). The Rep test may be profitable in this approach. (The therapy with Anne provides an example which will be discussed elsewhere.)

E. SYMPTOMS AS LOADED QUESTIONS

How do the perspectives discussed so far add up? The following features characterise (in varying degrees in specific cases) symptomatic behaviour, a_1.

(a) *Obliqueness* in getting at the issues, c_2, the person has a socially invalid position.
(b) *Cost*: There is *suffering* involved, this speaks to the *importance* of the issues involved, c_2 (Goldiamond).
(c) *Infested goals*: Obliqueness prevents healthy feedback, *cf.* the cockroach phobia, and Haley's views of symptoms. (Example 1.)
(d) *Contaminated goals*: Regardless of the characteristics of a_1, the analysis of games tells us that the goal may in itself be *contaminated*. There may be two sides to c_2, a viable one, and a "nongenuine" one, (*cf.* e.g., the feelings discussed as "trading stamps"). The less dominant the viable side, the more reconstruction of c_2 is called for.

(e) *Negative direction of movement*: Symptoms may be seen as "running away from anxiety" (provide structure at *any* cost, avoid the challenge of intimacy by settling for games). Related to this is *amount of hostility* which *prevents* movement or blows up in vicious circles.* Hostility is further characterised by defeating self-fulfilling prophecies, these are also involved in vindication of existential positions in games.
(f) *Lack "steering" of behaviour*: Kelly emphasises the direction of movement as primarily towards anticipation. Symptoms may, however, primarily involve avoidance of anxiety, see (e).

But another possibility is that there is no unequivocal "steering" in the positive direction, and that oscillating courses are likely.

All these features are clearly revealed in the Oedipal conflict in Example 9.

Games are dishonest, Berne says. Those familiar with double bind theory will wisely nod their head and note the subtly incompatible communications in Example 10, "give me freedom", "protect me".

One way to suggest an integrated perspective is to introduce the concept "*loaded questions*". Kelly is there since "behaviour is a question". Tribute is paid to Berne since "loaded" (dice) carries the connotation "dishonest". The dishonesty is revealed by (a) obliqueness, furthermore by the fact that (d) the goals may be fake, (c) perhaps also *made* fake by the way of going about them.† The sadness of the drama is that there may be high costs to pay, *cf.* (b). The points (e) and (f) capture the "stuck" quality (negative or oscillating direction) of loaded questions, the person is "not getting anywhere".

Loaded questions are poorly posed, they carry all the marks of *bad science*. *Good science* is characterised by questions that go directly at vital issues. Good scientists pose *honest questions*.

We now formulate two hypotheses which will be of value in describing interpersonal relations.

I. Loaded questions invite evasive answers.

Any scientific adviser is stomped by questions which he cannot see pointing anywhere in particular, and if not in an unusually helpful mood, will tend to be evasive. This, we hypothesise, will be the general rule in interpersonal relationships.

It is now straightforward to formulate the goal of therapy: *Help the person*

* Kelly (1955, p. 881) mentions that the importunity, the exhortation of the hostile person *may* pay off, the other may try appeasement. ("Yes, you really are right dear") But if the inflection is not satisfactory (and how could it be?) a heated quarrel may be on ("you don't sound as if you mean it") and the reader may fill out the conversation. Anyone happening to eavesdrop will see that real issues are avoided, but plenty of strokes provided. For the next round odds are worse for straight questions to get across: "How can I believe you, when you got so mad at me last time"?

† "The Balcony" by J. Genet, provides a chilling illustration of fake goals.

to replace loaded by honest questions. Train him to be a good scientist. An interchange of honest questions and honest answers, is here called "dialogue". The goal of therapy can also be said to be to "decontaminate" or "purify"* questions and answers.

In the field of close (intimate) interpersonal relations, the present sense to "dialogue" is one approach to "love". It is interesting that Steiner refers of "uncontaminated", "pure" strokes as "love" (1974, p. 26).

The second hypothesis is the converse of I:

II. Honest questions will facilitate, but cannot guarantee honest answers. This is intentionally chosen as a guarded formulation, there may be little reason to overvalue the extent of "good circles". Hypothesis II will be useful in raising the problem of the possibilities and limits of therapy. The present perspective will be illustrated by discussing rape.

IV. Loaded Questions and Evasive Answers: On Rape. Dialogue: A Goal Beyond Therapy?

Suppose we regard rape as a set of loaded questions. This perspective first points to the tragedy of the victim who is "forced" to give evasive answers. But the rapist is also stuck and prevented from intimacy. The analysis illustrates Hypothesis I and indicates that supplementary variables which deal with power, should be added.

The present analysis is inspired by Brögger (1973) and draws on her descriptions. She starts by stating that the really vicious violence is not just the genital assault but equally the structure of the preliminary communications which lead up to the rape. She considers cases where the jury (in the relatively few cases of this sort which lead to trial) is led to conclude that "she was a willing accomplice, she must have known what would happen". Consider the lonely woman who tentatively accepts the first "invitation". She wants some company and is willing to settle for a cup of coffee near by, but not more. The second step is that the man "gallantly" opens the car door without in any other way asking whether she is willing to take the step of entering his car. Brögger points out that this is principally the same sort of violence as in rape. The apparent innocence of the invitation is highly deceptive. The woman is stuck. If she refuses, it implies that she can take no risks and thus is not "free", further it turns down a seemingly "accepted" social contract (for company). Refusal would further invite nasty comments on prickliness and prudishness. If she accepts, this is taken as a sexual invitation, or rather it allows for later reinterpretation as a sexual invitation. So this is a highly

* From a Kellian point of view it is not irrelevant to point out that "purification" is a concept which may be used to great advantage in very different fields (Tschudi, 1972, 1975).

pertinent example of a loaded or dishonest question.* It is a sexual invitation, cleverly veiled as gallantry. We might add that even her experience of the sexual overtones may be stifled, so that she can later say: "I never knew this would happen". The viciousness of the question is that it prevents an honest answer. So she turns to an evasive answer or as Brögger puts it: "she is forced to use a *code*"—"it is more convenient to walk". Breaking the code reveals that "she is afraid of being eaten", but that sounds a bit stupid to say, and she is stuck with the really peripheral issue of car versus walk. She is then bullied into entering the car. Ten successive steps with the same structure as in the second step lead up to the genital penetration, a 30 seconds ugly rape.

A pattern of evasive answers may lead to compromising one's autonomy and dignity and is part of a pattern where even the contact with one's own body is undermined.† In construct theory: the evasiveness prevents elaboration of one's own questions, she has no honest reply.

Honest questions would have made it possible for her to refuse when she would go no further. They facilitate honest answers and the sombre "inevitability" of the vicious snowballing Brögger describes is avoided.

But why could she not openly have refused? She might even have exposed the game *in his own terms.* Brögger suggests as an example: "You little whoring prick, and for free! it can't be much good." But (fortunately) she does not find this kind of talk at all appealing. This would presuppose the woman in a one-up position (she might also be in need of advanced judo self-defence techniques), and she is in the one-down (oppressed) position. This makes it impossible for her to expose the game both for psychological and physical reasons.

This kind of analysis may hopefully be useful in other areas of interpersonal behaviour. It calls attention to the lack of power of the victim—the oppressed —and the further degrading by being treated in oblique ways. But the oppressor is also stuck! There is for instance no reason to imply, as does Brögger, that rapists "enjoy" their conquest and show of power (as there is no reason to impute "sadistic glee" to Procrustes). The rapist cannot risk rejection and turns to exhortation. This bars him from intimacy, and he is stuck with the pathetic position that "she really enjoyed it". Perhaps other oppressors also believe they are doing "what is best" for their victims?

This analysis does not deny that the victims are "worse off" than the oppressors. It does, however, question an "eliminative" approach towards the oppressors. Severe sentences for rapists would exemplify this type of solution.

* Flirt is not a loaded question. It is not *dishonest* but based on the mutual contract "this is play", *cf.* Tschudi (1972, p. 230). If, however, there is reason to doubt whether there were two "*consenting* adults" to the contract, flirt may shade into seduction which may shade into rape.

† The most complete study of rape today, aptly titled "Against Our Will" (Brownmiller, 1975), supports the generality of Brögger's perspective.

But the problem is to find a *constructional* approach to intimacy, to foster *dialogue* between woman and man.

How far can therapy take us towards this goal? Consider Tove (Example 6). We hope that she will not be stuck in a one-down position. Building up self-confidence may provide a viable self-assertion to protect her from being caught in vicious circles when she is faced with loaded questions.

But this does not necessarily lead to dialogue, *cf.* Hypothesis II. The men she will meet must on their own start the long road to posing honest questions. The burgeoning literature on men's liberation (*cf.* Steiner, 1974, chap. 27 for an example) may be a small, but promising sign that at least some men are willing to reconsider their ways.

The general point is that *therapy can not guarantee a dialogue*. The client(s) —the couple, the family whatever group the therapist is working with—is communicating with people outside. The therapist has no possibility of teaching *them* honest questions. It would be highly desirable to have good descriptions of the "laboratory" out there, but that is a further task. As a therapist you can do nothing but teach your client(s) honest questions, help them to decipher the communications (more or less loaded) they receive, and not compromise with themselves when faced with loaded questions.

A further issue is to probe into the origin of loaded questions. Berne (1970, chap. 5) draws on ethological material and suggests that "this is the way things are", nature pulls her mean tricks on all of us.

Steiner (1974), however, is not content with this answer and puts the blame on the "artificial" lack of pure strokes to go round (the "stroke economy") so that the alternative is to exhort contaminated strokes. Drawing on, among others, Marcuse and Reich, he suggests that the stroke economy is related to the modes of production. Clarifying these issues may bear on such problems as: How much of the responsibility for the surroundings can the client possibly face? Can a "successful therapy" isolate the client from other people? To what extent can psychology be useful in fighting oppression?

V. The Therapeutic Situation

What can be added to Kelly's (1969, pp. 60–61) observations of the similarities between the roles of psychotherapist and the role of scientific adviser? The key issues centre around providing control, that is to reduce the number and complexity of variables as much as possible, and above all to be pertinent. This is elaborated in the three points below:

(1) *Keep to the issues.* Projects with which clients want therapeutic help are likely to involve intimacy in interpersonal relations. To help with such projects the therapist must be able to provide purified strokes, and to teach the client to recognise such feedback. If the person is emotionally starved or deeply stuck, a handshake, pat on the shoulder etc. *may* be appropriate.

Tune in to the client. This is related to but perhaps not identical to Rogers' (1957) "unconditional positive regard" and "emphatic understanding". We warn against any exhortations in terms of generally exuding "warmth". Honest children teach us this. If we marvel at the ten year old: "what a nice plane you have built, sonny", he stoops to point out that the point of gravity is somewhat displaced, and asks how long ago (if ever) *you* built a plane. Tuning in was a real problem in the therapy with Tove. Praising her poems was not of any help. "Warmth" was of no use to her. She had not asked for this, did not know what to do with it, but it was not for her to tell the therapist how *she* goofed. This problem may not be special to teen-agers who have no great reason to trust "helpful" adults. So here is a caveat: Do not prematurely try to reduce the distance between you and your client. Should you persist in this you are showing hostility and may end up with a large category of patients "unsuited for therapy".

(2) *Recognise a gap to be crossed—offer an "as if" stance.* Carving out new pathways may leave the client stranded in the wilderness with no roads to point out directions (anxiety). Faced with this possibility, he is likely to require guarantees that he does not have to give up his ways. The therapeutic problems this raise may be expressed by drawing on Berne (1972, pp. 37, 349, 351) who takes inspiration from fairy tales of princes and princesses disguised as frogs. The client, the frog, is steeped in his ways and does not believe that he will ever become a prince. He wants to be a braver frog. The neurotic (so say "dynamic" clinicians) doesn't come to get well, but to learn to be a better neurotic. "I want to live more comfortably while holding on to the sides of a tunnel", the client might say. In the present terminology: Frogs ask loaded questions and settle for evasive answers. They are unwilling to become princes and change to honest questions. But the therapist will believe that frogs ("losers") *can* be turned into princesses or princes ("winners"). There is thus a gap between the therapist and the client. Metaphorically the therapist has the prince point of view, and the client is at the other side of the river, unwilling to jump the gap. But the therapist should have a precise understanding of the client's position. She should understand the point of view from which the client's solution is not "stupid" or "irrational". (One of the steps in Greenwald's therapy is to understand the context of the decision.) There exists a point of view from which even the cockroach phobia is good science, but the client can do better$_1$

Perhaps the most important help the therapist can provide to make the jump safer is to reduce the gap by drawing on the "as if" character of hypotheses (constructs). New constructs may be "tried on for size" and not necessarily be "for keeps". This is the essence of fixed role therapy, a technique which should be subject to much further innovation. Knowing that "there is a way back" may be a decisive factor in making the client jump some of the gaps facing him. The complexity of the situation is reduced to manageable size.

Advising students poses similar problems. There will be a point where the student has to be on his own, the adviser may then reduce anxiety by carefully pointing out that everything is not at stake.

(3) *Provide a laboratory with more than one tool.* The Kellian view of therapy requires "orchestration of techniques" (1969, p. 223). There is no reason why the Kellian therapist should not use systematic desensitisation, Gestalt techniques, biofeedback, self-management techniques etc. etc. A basic point to bear in mind, however, is that the results of *any* technique should be evaluated jointly with the client. He is the principal investigator!

The classic psychoanalytic situation is a curiously restricted laboratory with basically only one tool ("follow the first rule, and I will at proper times interpret"). While there may be cases where this is sufficient, the present stance is: "it is no shame to have more than one tool!"

VI. Concluding Comments

Loaded questions may not be all there is to disorders of construction. There may be cases where there simply are no advantages of the problematic behaviour. Kanfer (1975b) draws a distinction between behaviour with "conflicting consequences" and "behaviour deficits". In the latter case there are no advantages. A case in point may be provided by Argyle *et al.* (1974) who devised a social skills training programme for young adults equipped with psychiatric labels like "schizoid", "immature" etc. The report does not disclose any advantages of patterns of behaviours described as "vicious circles of rejection and withdrawal" (p. 64) and the successful training programme followed the formula "take small steps to a_2".

Further analysis may reveal that many therapies involve *both* behaviour with conflicting consequences, loaded questions, *and* behaviour deficits. It may even be the case that a specific behaviour can be regarded as belonging to the one *or* the other class depending on the abstractness of the point of view from which it is seen.

A second example where there may be no advantages to the complaint is psychosomatic disorders. Wear and tear, stress and sleeplessness is not always in order to achieve something, though there are those who "seek suffering, like polysurgery addicts" (Berne, 1964, p. 98). Criteria for differentiating these types may have therapeutic implications.

Finally, there are ambiguities in the present use of "question" and "answer". It has implicitly been taken for granted that behaviour can be seen as either a question or an answer (to the behaviour of others). But this is just a rough first approximation. As Bateson (1975, especially pp. 250–280) makes clear, any behaviour can be seen as both a response to preceeding behaviour (an answer) and also a stimulus for further behaviour (a question). This is evident for the remarks in conversations.

It may be worth while to explore the suggestion that in concrete analyses any given behaviour may be placed on a continuum where the end points are "pure" question and "pure" answer respectively.

A somewhat different issue is whether the behaviour of some persons is at all appropriately described as "questions". Perhaps it seems to lack the autonomous qualities suggested by "question", and is better described as "forced". Some people seem to "follow orders" in all their undertakings. This is partly captured by "loaded question". The more the behaviour is "loaded" (or "evasive"), the more there may be a compulsive aspect to it, and conversely honesty will imply "autonomy".

But fostering honesty may, as previously discussed, take us beyond psychology, and there may be aspects of behaviour falling outside the range of convenience of the present conceptual framework.

References

Argyle, M., Trower, P. and Bryant, B. (1974). Explorations in the treatment of personality disorders and neuroses by social skills training, *British Journal of Medical Psychology*, **47**, 63–72.

Bandura, A. (1969). "Principles of Behaviour Modification", Holt, New York.

Bannister, D. and Fransella, F. (1971). "Inquiring Man", Penguin, London.

Bannister, D. and Mair, J. M. M. (1968). "The Evaluation of Personal Constructs", Academic Press, London.

Bateson, G. (1973). "Steps to an Ecology of Mind", Paladin, Suffolk.

Berne, E. (1961). "Transactional Analysis in Psychotherapy", Grove Press, New York. (Ballantine, New York, 1975.)

Berne, E. (1964). "Games People Play", Grove, New York. (Penguin, London, 1975.)

Berne, E. (1966). "Principles of Group Treatment", Grove Press, New York.

Berne, E. (1970). "Sex in Human Loving", Simon and Schuster, New York. (Pocket books, New York, 1971.)

Berne, E. (1972). "What do You say after You say Hello?" Grove Press, New York. (Bantam, New York, 1973.)

Brownmiller, S. (1975). "Against Our Will", Simon and Schuster, New York.

Brögger, S. (1973). "Fri oss fra kärligheten", Rhodos, Köbenhavn.

Buckley, W. (1967). "Sociology and Modern Systems Theory", Prentice Hall, New Jersey.

Ellis, A. (1962). "Reason and Emotion in Psychotherapy", Lyle Stuart, New York.

Fransella, F. (1972). "Personal Change and Reconstruction. Research on a Treatment of Stuttering", Academic Press, London.

Goldiamond, I. (1974). Toward a constructional approach to social problems, *Behaviourism* **2,** No. 1, 1–84.

Greenwald, H. (1973). "Decision Therapy", Wyden, New York.

Haley, J. (1963). "Strategies of Psychotherapy", Grune & Stratton, New York.

Hinkle, D. N. (1965). The change of personal constructs from the viewpoint of a theory of construct implications. *Unpublished Ph.D. Thesis*, The Ohio State University.

Holland, R. (1970). George Kelly: Constructive innocent and reluctant existentialist, *In* "Perspectives in Personal Construct Theory", (Bannister, D., ed.), pp. 111–132. Academic Press, London.

Kanfer, F. H. (1975a). Personal communication.

Kanfer, F. H. (1975b). Self-management methods, *In* "Helping People Change", (Kanfer, F. H. and Goldstein, A. P., eds.), pp. 309–356. Pergamon, New York.

Kelly, G. A. (1955). "The Psychology of Personal Constructs", Vols. 1, 2, Norton, New York.

Kelly, G. A. (1969). "Clinical Psychology and Personality: The Selected Papers of George Kelly", (Maher, B., ed.), Wiley, New York.

Kelly, G. A. (1970). Behaviour is an experiment, *In* "Perspectives in Personal Construct Theory", (Bannister, D., ed.), pp. 255–270. Academic Press, London.

Larsen, E. (1972). Valget—Et strategisk og terapeutisk virkemiddel i psykoterapi. *Unpublished M.A. Thesis*, University of Oslo.

Mair, J. M. M. (1976). Metaphors for living. To be published in "Nebraska Symposium on Motivation", Vol. 24, (Cole, J., ed.), The University of Nebraska Press.

Norris, F. M. and Norris, H. (1973). The obsessive compulsive syndrome as a neurotic device for the reduction of selfuncertainty, *British Journal of Psychiatry* **122,** 277–288.

O'Neill, N. and O'Neill, G. (1972). "Open Marriage", Avon, New York.

Rogers, C. R. (1957). The necessary and sufficient conditions of therapeutic personality change, *Journal of Consulting Psychology* **21,** 95–109.

Sandsberg, S. (1975). Valg og forandring i lys av Kelly's Personal construct theory. *Unpublished M.A. Thesis*, University of Oslo.

Scheff, T. J. (1968). Negotiating reality, *Social Problems* **16,** 3–17.

Steiner, C. H. (1974). "Scripts People Live", Grove Press, New York (Bantam, New York, 1975.)

Tennov, D. (1975). "Psychotherapy: The Hazardous Cure", Abelard-Schumann, New York.

Tschudi, F. (1972). The latent, the manifest and the reconstructed in multivariate data reduction models, *Unpublished Ph.D. Thesis*, University of Oslo.

Tschudi, F. (1975). Evaluation of multidimensional scaling. A general approach to testing models with application to multidimensional scaling (MDS), *Göteborg Psychological Reports* **5,** No. 29, 35–40.

Tschudi, F. (1976). Constructs *are* hypothesis. Prediction *is* the goal. A belated response to T. Mischel's critique of Kelly. *Manuscript*, University of Oslo.

Wachtel, P. L. (1975). Plus ça change . . . Self-perpetuating interaction cycles as a unit for psychotherapists and personality researchers. Paper presented at: *Symposium on Interactional Psychology*. Stockholm, June 22–27.

Watzlawick, P., Weakland, J. and Fisch, R. (1974). "Change. Principles of Problem Formulation and Problem Resolution", Norton, New York.

Wright, K. J. T. (1970). Exploring the uniqueness of common complaints, *British Journal of Medical Psychology* **43,** 221–232.

Author Index

Subject Index